D1228614

Pollution Ecology of Freshwater Invertebrates

WATER POLLUTION

A Series of Monographs

D. T. O'Laoghaire and D. M. Himmelblau. *Optimal Expansion of a Water Resources System*. 1974

C. W. Hart, Jr., and Samuel L. H. Fuller (eds.). *Pollution Ecology of Freshwater Invertebrates*. 1974

Pollution Ecology of Freshwater Invertebrates

EDITED BY

C. W. Hart, Jr., and Samuel L. H. Fuller

Department of Limnology
The Academy of Natural Sciences of Philadelphia
Philadelphia, Pennsylvania

ACADEMIC PRESS *New York and London* *1974*

A Subsidiary of Harcourt Brace Jovanovich, Publishers

ACADEMIC PRESS, INC.
111 Fifth Avenue, New York, New York 10003

United Kingdom Edition published by
ACADEMIC PRESS, INC. (LONDON) LTD.
24/28 Oval Road, London NW1

Library of Congress Cataloging in Publication Data
Main entry under title:

Pollution ecology of freshwater invertebrates.

Includes bibliographies.
1. Freshwater invertebrates. 2. Fresh-water ecology. 3. Pollution. I. Hart, C. W., ed. II. Fuller, Samuel L. H., ed.
QL141.P63 592'.05'2632 73-18939
ISBN 0-12-328450-3

PRINTED IN THE UNITED STATES OF AMERICA

To Ruth Patrick

whose foresight, imagination, and
courage have led to new ways of
looking at environmental problems

Contents

Chapter 4 **Leeches (Annelida: Hirudinea)**

Roy T. Sawyer

Chapter 5 **Aquatic Earthworms (Annelida: Oligochaeta)**

R. O. Brinkhurst and D. G. Cook

Chapter 6 **Bryozoans (Ectoprocta)**

John H. Bushnell

Chapter 7 **Crayfishes (Decapoda: Astacidae)**

Horton H. Hobbs, Jr. and Edward T. Hall, Jr.

List of Contributors

Numbers in parentheses indicate the pages on which the authors' contributions begin.

R. O. BRINKHURST* (143), Department of Zoology, University of Toronto, Toronto, Ontario

JOHN H. BUSHNELL (157), Department of Environmental, Population, and Organismic Biology, Division of Environmental Biology, University of Colorado, Boulder, Colorado

JOHN CAIRNS, JR. (1), Biology Department and Center for Environmental Studies, Virginia Polytechnic Institute and State University, Blacksburg, Virginia

D. G. COOK† (143), National Museum of Natural Sciences, Ottawa, Ontario, Canada

SAMUEL L. H. FULLER (215), Department of Limnology, The Academy of Natural Sciences of Philadelphia, Philadelphia, Pennsylvania

EDWARD T. HALL, JR. (195), Georgia Department of Natural Resources, Environmental Protection Division, Atlanta, Georgia

WILLIAM N. HARMAN (275), Biology Department, New York State University College at Oneonta, Oneonta, New York

FREDERICK W. HARRISON‡ (29), Department of Biology, Presbyterian College, Clinton, South Carolina

*Present address: Biological Station, St. Andrew's, New Brunswick, Canada.

†Present address: Great Lakes Biolimnology Laboratory, Canada Center for Inland Water, Bulington, Ontario, Canada.

‡Present address: Department of Anatomy, Albany Medical College of Union University, Albany, New York.

HORTON H. HOBBS, JR. (195), Department of Invertebrate Zoology, National Museum of Natural History, Smithsonian Institution, Washington, D. C.

ROMAN KENK (67), Department of Invertebrate Zoology, National Museum of Natural History, Smithsonian Institution, Washington, D. C.

SELWYN S. ROBACK (313), Department of Limnology, The Academy of Natural Sciences of Philadelphia, Philadelphia, Pennsylvania

ROY T. SAWYER (81), Biology Department, College of Charleston, Charleston, South Carolina

Preface

Ideas about pollution indicator species are almost as numerous and diverse as are the people concerned with them. They range from the simplistic (yet tantalizing) quest for an all-purpose aquatic "canary" that will warn of pollution to complex mathematical models of community interrelationships.

None of these ideas is to be scorned, for each, in its own way, contributes to the ultimate goal of measuring amounts and kinds of pollution. But there is no aquatic canary. Neither is there any one index or model that can be depended upon to describe a community completely. Rather, a balance of many techniques—including studies of species compositions, population sizes, and the physical-chemical environments to which they are exposed—must be employed.

In studying water pollution, one finds that the lack of data on just the normal ecology of most organisms—let alone their pollution ecology and the extent to which individuals or species communities indicate degrees of pollution—is disheartening. While considerable work has been done, concern has been with relatively few organisms. Some of the information is not readily available, some of it is not easily useable because of nomenclatural problems, and broad syntheses are scarce. And it is foolish to try to understand abnormal conditions without first knowing what is normal.

The purpose of this volume is therefore (1) to present in concise form at least a basic outline of the normal ecology of certain invertebrate groups; (2) to bring into focus the most recent systematic interpretations; (3) to discuss the present status of knowledge concerning the pollution ecology of the groups; and (4) to point the way to ancillary background material by means of extensive, and sometimes annotated, references. The most important aim of this volume is to discuss, in as much detail as possible, the current status of knowledge concerning pollution ecology. This encompasses some discussions of the indicator species concepts—ranging from the straight forward presentation of Roback to the more theoretical discussion by

Harman. Summaries of the pertinent literature are included, together with the personal field observations of the contributors. This latter approach is stressed because people who work closely with organisms over a considerable period of time, in addition to accumulating quantifiable data, often gain certain insights into their behavior which they may not be able to quantify.

It may be pointed out that not all major invertebrate groups are represented here. This is true, and is due to a combination of factors involving a scarcity of people who wished to commit themselves, taxonomic chaos at the alpha level, and certain space constraints initially placed upon the size of this volume.

This is not, therefore, intended to be a final word. The chapters that follow neither define the canary nor point to a specific index. They are, however, an attempt to pull together old data, introduce new ideas and information, and synthesize these wherever possible with the latest systematic interpretations—a sadly neglected facet, and one which we believe to be indispensible.

C. W. HART, JR. *

*Present address: Office of International and Environmental Programs, Smithsonian Institution, Washington, D.C.

Pollution Ecology of Freshwater Invertebrates

CHAPTER 1
Protozoans (Protozoa)

JOHN CAIRNS, Jr.

I. Current Uses of Protozoans in Pollution Assessment

There are three primary ways in which protozoans are used to assess the impact of waste discharges: (1) In surveys of rivers, streams, lakes, and other receiving systems, which are carried out by a field team consisting of a variety of specialists working with organisms ranging from bacteria through fish; (2) in laboratory bioassays in which protozoans are exposed to various concentrations of a particular waste or toxicant to determine the relationship between the concentration and the response; and (3) in laboratory microecosystems, artificial streams, and the like which are designed to fill the gap between the single species laboratory tests and the analysis of the complex systems found in nature.

Because these microecosystems are somewhat akin to the scale models used by engineers—and may have many of the same weaknesses, such as a scale effect—this chapter is primarily concerned with aspects of protozoan ecology related to stress surveys. None of these uses are very common or widespread at present, although limnological survey teams, such as those sent out by the Academy of Natural Sciences of Philadelphia, customarily include a protozoologist. In some cases, such as the well-known study by Lackey (1938), protozoans were the primary group studied. Nevertheless, compared to the number of studies involving fish and macroinvertebrates, studies involving protozoans are hardly noticeable. One need only to glance through "Water Quality Criteria" (McKee and Wolf, 1963) to see that protozoans have not been commonly used in pollution studies.

Despite substantial evidence to the contrary, many biologists persist in the belief that microbial communities are erratic in structure and function and thus are unreliable indicators of any environmental condition. Possibly this prejudice results from their experiences as beginning students of biology, when a container of species collected from a natural environment was placed on a laboratory bench and left for a day or two, after which the students found a completely different assemblage of species present than that which the jar originally contained. Often this secondary assemblage would be dominated by one or two species such as *Paramecium*, *Halteria*, *Euplotes*, and the like. As a result, protozoan populations are thought to be extremely unstable, although one would not expect an assemblage of higher plants or animals to persist in its original state when ripped from its natural habitat and placed in new and artificial surroundings.

In addition, the rate at which microbial species are replaced in natural systems is so many orders of magnitude more rapid than the replacement rate of higher organisms that the idea of an interdependent, interlocking system with multiple cause-effect pathways seemed almost incongruous.

However, attitudes are changing and many investigators recognize the distinct advantages of protozoans and other microbial species for assessing pollution as well as for environmental and ecological studies. A few of the more important advantages are (1) They are small, easily handled, and require relatively small containers in contrast to those needed for fish and other larger organisms; (2) They multiply rapidly so that one can test effects of potential pollutants upon reproduction, growth, metabolism, etc., of several generations without waiting months or years for the results; (3) They can be grown on synthetic media so that the conditions of a bioassay are completely reproducible—a condition that is difficult to achieve with the higher organisms; (4) Since they have both sexual and asexual reproduction, it is possible to get a large number of individuals which have the same genetic makeup and which are descended from a single pair of cells. Such uniformity of test organisms is presently difficult to even approach using higher organisms; (5) Since protozoans and other microbial organisms are unicellular, they are in very close contact with the environment and are thus exposed to unfavorable stresses of many kinds much more intimately and with less of a time lag than are many of the higher organisms; (6) Since most free-living microorganisms are thought to have a cosmopolitan distribution and are thus likely to be found wherever ecological conditions are appropriate (they are carried by wind and water current and by other organisms), one can use the same species on different continents and thereby eliminate questions such as how much of the difference in results between an investigator in Europe and an investigator in the United States is due to the nature of the methodology used and how much is due to inherent differences in the test organisms; (7) Since protozoans and other microorganisms can be kept in constant culture at a slow rate of growth without frequent reinoculation of the culture system, it is possible to keep a "library" of different species found in different ecological systems available much more easily than one can when using fish or other higher organisms. This is particularly useful during the winter months when ponds are frozen and other organisms are difficult to obtain. It is also useful when one wants to test organisms under conditions which exist only at some other time period of the year than the one available to run the tests; (8) The difference in tolerance to various waste materials between fish, invertebrates, and microbial species is not as great as most people suppose. Patrick *et al.* (1968) have shown that diatoms are sometimes more sensitive than fish, sometimes less, and sometimes quite comparable. The same statement holds when comparing diatoms to invertebrates or invertebrates to fish. Therefore, fish tolerance is no more or less representative of the general aquatic community than is a microbial species or an invertebrate species; (9) Protozoans and other microbial species make up the largest portion of the total biomass of any aquatic system. Therefore, in terms of weight per unit of volume or area they are a much greater constituent of natural systems than are fish and other higher organisms. Despite this they get comparatively little legislative attention or research project funding; and (10) In many states a collecting permit is not required.

For a field survey, protozoans and other microbial species are relatively easily collected without specialized equipment. Transportation back to the laboratory is inexpensive, as are the containers required. Obtaining collections requires a relatively short period of time compared to the time required for collections of higher organisms. In addition, protozoan community structure is as likely to be changed by waste discharges, sediment from land runoff, etc., as are communities of fish and higher organisms.

However, there are some serious problems when dealing with protozoan collections. For one thing, they are difficult or virtually impossible to preserve without destroying or so distorting the more fragile and even some of the "tougher" species that identification becomes nearly impossible. In addition, it is difficult to separate protozoans from various types of debris—organic and inorganic—and the ability to detect them (particularly in low densities) when mixed in with all this extraneous material is seriously hampered when they are no longer alive and moving. Furthermore, the location and number of various organelles (such as the contractile vacuole and locomotor organelles) as well as other taxonomic characteristics may be eliminated completely or rendered virtually undetectable by preservatives. Even protozoans maintained in the environmental conditions in which they were collected are likely to die or reproduce very quickly after collection, thus distorting the community structure.

Although microecosystem studies involving protozoans have not been particularly common (nor for that matter have microecosystem studies of almost any kind), these are likely to increase dramatically in the near future as a result of the interest in the fate of microbial species entrained in the coolant water of steam electric power generating plants. Federal and state laws regarding environmental impact assessment are beginning to require some indication of power plant operation effects upon microbial communities, and these studies will undoubtedly have a profound effect upon the amount of information, the development of new methodology and the number of specialists working in this particular field. Since this is a new and rapidly developing market for professionals specializing in protozoology and related fields, it is quite likely that the current production of persons with graduate degrees will not be adequate to meet the future demand. Unfortunately, it is also true that most of the graduate programs of the few students being trained in protozoology are not designed to meet this new demand.

Stream Surveys

One of the most comprehensive stream surveys involving protozoans was carried out by Lackey (1938). Protozoans were also used in the landmark river basin study of Patrick (1949) and have been used in a variety of other ways. Nevertheless, as is the case for bioassays, the literature on the subject is not extensive, nor is there a standard method or standard series of methods that are generally accepted and widely used. Marine and freshwater studies fall into the major categories: (1) Those involving planktonic species, and (2) those involving species associated with substrates. The latter group may contain predominantly motile species which are not planktonic in the strict sense of the term, but which may be swept free of the substrate and thus appear in plankton samples of flowing waters. Because planktonic species are moved about primarily by forces beyond their control, because these forces may take them in one direction during one period of a day or season and in an entirely different direction in another, and furthermore because of the speed at which the plankton is carried may vary substantially, I do not feel that planktonic species are as useful in assessing pollution from a point source discharge (i.e., wastes that enter streams from a pipe or a relatively narrow discharge area) as are those associated with substrates. Planktonic species collected near discharges may have only very recently arrived there and may be in admirable condition, while specimens several miles away may have spent a considerable amount of time in the discharge area and may just be responding to the stress to which they were previously exposed, even though they are some distance from the outfall. In contrast, species associated with a substrate are likely to have been in that general position essentially permanently (although they might move about within it) and thus furnish more reliable evidence about the environmental conditions of that particular area. In rivers, there seems to be little justification for studying planktonic species since most of the species collected in plankton nets were swept from various substrates and are not truly planktonic. In lakes, oceans, and other large bodies of "standing waters" the biomass of planktonic species may be a substantial percentage of the total microbial biomass, and thus effects of waste discharges upon plankton should be a part of every environmental impact study despite the difficulties of associating a particular effect with a particular waste discharge. Even in these situations, the use of artificial substrates (substrate here is used in a very general sense as a surface on which protozoans and other microbial species may associate, rather than in a nutritive sense) may enable one to estimate the biological quality of the water in a particular area through time by comparing the response of a community associated with an artificial substrate anchored in or near a waste discharge mixing zone to that of a reference area well outside of any possible effects from the waste discharge. Thus, substrate-associated species may serve a useful purpose in water quality monitoring in lakes and oceans because they are associated with a fixed location relative to the waste discharge rather than moving rather erratically throughout the system. In other respects, plankton samples may be treated from an analytical viewpoint very much like the collections obtained from various artificial and natural substrates in that the analysis may be either qualitative, quantitative, or a combination of the two.

Although protozoans colonize artificial substrates quite well and may be used in monitoring systems for waste surveillance, they are not as easily preserved as diatoms nor are they stored and shipped as readily. For artificial substrate monitoring units, I recommend diatoms if one wants a microbial species (see Patrick *et al.*, 1954, for an illustration of this use). If macroinvertebrates are desirable, the discussion by Beak *et al.* (1973) should be consulted.

The evaluation of protozoan communities in streams and lakes is complicated by several factors not troublesome with most other aquatic groups. Probably the most important of these is the difficulty of preserving protozoans without eliminating the more fragile species or destroying and distorting important taxonomic characteristics in the more durable species (particularly when low in density and associated with detritus and other materials). Anyone who has examined a collection of protozoans taken from a substrate will recall that movement is the quickest and most reliable way to distinguish protozoans from the other material collected. In the absence of movement, a quick scanning of the collection to estimate the numbers and kinds of species present would be virtually useless. Thus all of the major problems almost unique to collections of protozoans are related to their perishable nature and difficulty of preserving them.

In order to get a reliable estimation of the protozoan species in a survey area, one should collect from each of the major habitats seen within the station or collecting area. These will generally include mud surfaces, surfaces of various types of vegetation (both living and dead), foams, scum, and the like. It is possible, of course, to composite these samples, but this would further reduce the density of the low density species and make their identification more difficult. Getting a good view of key characters is difficult when dealing with moving specimens, and reducing their density reduces the possibility of speedily finding one in the proper position. In addition, compositing does not permit an analysis of patchiness and other characteristics of species distribution, so that you do not know whether all species are found everywhere in the system, are only associated with particular microhabitats, are of low density throughout the system, or in high density in one place and scarce in others. However, keeping these samples at least partially separate means that one will end up taking between 10 and 20 half-pint screw-top jars back to the laboratory. This large number will almost impossibly expand the work involved.

Identifying protozoans in a freshwater community is a race to get as many species as possible identified before the collection becomes distorted by death, encystment, reproduction, or predation. I generally try to keep the number of sample jars to be examined between eight and ten, collecting the samples primarily with a large rubber bulb and placing them in half-pint, wide-mouth, screw-top jars with a half-inch or more of air space and transport them as rapidly as possible to the field laboratory in a wet cloth or insulated container. Freezing may cause rapid changes in the population while in transit, and these, in my experience, have generally been more severe than moving the collections while wrapped in damp cloths or insulated containers. Immediately upon return to the laboratory, the screw tops should be loosened and the collection placed in a fixed position where the specimens may respond to a single light source such as a window which will thus concentrate them and make identification easier. Passing them through filters is not advisable because, among other difficulties, it also concentrates the debris and wastes which, if brought into more intimate contact with protozoans, may injure them. Usually it is a good idea to allow the jar to remain standing for a half hour so that the protozoans can concentrate.

The initial examination should consist of a quick scan of material composited from the miniscus and bottom of each sample jar so that one has a general idea of: (1) Which samples contain the highest densities of individuals and the highest numbers of species; (2) what kinds of species are present; and (3) which jars contain species likely to be perishable, to reproduce, or to rapidly change their abundance relative to those of the other species present. Once these determinations have been made, an informal priority system can be established regarding which samples should be examined first, which samples might be eliminated because they furnish redundant information, and which species should be identified as rapidly as possible. It is not advisable to utilize collections more than 48 hours old, and thus you must apportion your time as well as possible to accomplish as much work within that time span and to be certain to identify the well-established species.

Although you may employ several foot-lockers of taxonomic keys and other assorted information, it is inevitable that one will not have all the literature necessary. In such cases, or where time does not permit identification of the specimens, quick sketches and notes can be made that will make later identification possible. Photomicrography may be helpful also. When you are visiting a stream or lake for the first time, it sometimes is advisable to spend a few days getting acquainted with the species present before beginning the critical collecting. Even though communities change, many of the species found in the preliminary collections will also be found in subsequent collections. Having carried out previous identifications for some species will markedly improve the identification time of the investigator.

In addition, a voice or foot-activated tape recorder should always be available so that density observations, comments on structural characteristics of undetermined species, and other important information can be recorded without the investigator removing his eyes from the microscope or losing any time from the main task. Since protozoans have a cosmopolitan distribution, being found throughout the world whenever the ecologically suitable conditions occur, a protozoologist will find many "old friends" wherever he goes. If one works regularly at this task for a number of years, it is startling how rapid the determinations will be and how reproducible the results.

During the summer of 1948, Ruth Patrick had each of the two field teams sample the same area during the same day but not at the same time. Mary Gojdics sampled a station with one team, and I sampled the same station a few hours later with another. We carried out our identifications separately and did not help each other or exchange information in the process. We had about a 95% overlap in species identified and, predictably, the differences were due to our individual interests since Mary found the most Euglenoidina and I found the most ciliates. This was done at the end of the summer, when both teams had been working on the project for several months, and during which Mary and I had worked at adjacent laboratory benches, continually exchanging information and helping each other with identifications. Thus, we had much the same views on what characteristics should be used in identification and which taxa should be called which names, etc.

During the following year I spent nearly all of my working hours identifying protozoans, since the survey continued with a single team. For a number of years after that I spent about half my time identifying protozoans. During those years, I developed an ability to recognize many species at sight, and for many others could go to almost the right place in the key immediately. Later, as my interests in other areas developed, work with protozoans diminished and so did my ability to identify them as rapidly as I did in the early days of my career. Like any other skill, this one must be constantly used in order to be at maximal effectiveness. Therefore, I would advise that field teams involving protozoologists have this as a specific assignment, with the protozoologists having little or no responsibility other than the identification of protozoans.

You can somewhat reduce wasted effort by determining, with preliminary collections, how many samples should be evaluated. Thus you can tell whether two different algal mats will have the same or different species of protozoans. Any simplification of effort along these lines is highly desirable. The number of perishable species can also be estimated by carrying a small microscope to the station and examining a few of the collections before they are taken back to the laboratory. Such efforts take little time, and may substantially reduce the amount of work. Also, working closely with an algologist can be extremely helpful, since many algal collections are often excellent sources of protozoans.

Although it is difficult to determine precisely how many protozoans are in a sample or on a given slide because they are constantly moving and frequently obscured by debris, it is possible to estimate relative abundances by placing them in estimated ranges of density. I have usually used a modification of the system originally suggested by Sramek-Husek (1958). Information of this kind can be quite useful (Cairns *et al.*, 1969). Just because the precision possible with diatoms and other species that are comparatively easy to preserve and count cannot be achieved does not mean that all hope of obtaining useful quantitative information should be abandoned.

The previously described protozoan sampling is usually carried out as a part of a team operation which covers all of the major groups of aquatic organisms from bacteria through fish. The rate at which the team functions partly determines the time available for protozoan determinations. This coincides quite well with the time of sample deterioration. These surveys are usually carried out at one or more reference stations above the waste discharge and usually at two, three, or more stations below the waste discharge. A more extended discussion of this can be found in Cairns (1965). However, this is much like the complete annual physical given to humans, in that it is both time consuming, expensive, and not likely to be repeated on a routine weekly or monthly basis. As a supplement to this information, one should consider the installation of an instream biological monitoring system, usually using artificial substrates which can be sampled more frequently to ascertain biological response to water quality in the receiving system (Patrick *et al.*, 1954; Cairns *et al.*, 1973). This might be considered analogous to taking one's temperature to see if something has gone wrong—it will tell you whether something has or not but not precisely what has gone wrong or the extent of the damage. Since no single group of organisms can be used to represent the response of the entire aquatic community, incomplete information will be obtained no matter which taxonomic group is used.

On the other hand, since the entire aquatic community is an interlocking, interdependent system with multiple cause-effect pathways, most pollutants will have some effect on almost all aquatic organisms, even though this may vary in magnitude. It is almost impossible to select a group of organisms to use in biological monitoring that will not provide some indications of major effects. Therefore, one might reasonably select the group of organisms for convenience in sampling, preservation, ease of counting, and collectability. Generally, fish are immediately eliminated by these considerations because they are difficult to collect, particularly in large numbers. Many fish species move about frequently, thus making it uncertain precisely what set of conditions their presence or absence reflects. Probably, as mentioned previously, the two best groups to use are diatoms and aquatic insect larvae. Of these two, the insect larvae are probably the easiest to use if the person doing the counting has an analytical mind but no formal taxonomic training. Under these circumstances one can use the sequential comparison index (Cairns *et al.*, 1968; Cairns and Dickson, 1971) and get a good approximation of the *relative condition* of the various areas being studied. If you are going to employ a professional taxonomist I would recommend more sophisticated methods be used.

In certain circumstances, however, you may wish to use protozoans in conjunction with artificial substrates as part of the monitoring system, possibly because a protozoologist has been employed for other purposes and has no specific assignment during certain portions of the year. In such circumstances, I would recommend that polyurethane foam substrates as described by Cairns *et al.* (1969) be used for lakes or the same substrates, anchored to styrofoam floats or attached to stakes inserted in the stream bed, be used for flowing systems. These should be sufficiently large to get a substantial number of species (Cairns and Ruthven, 1970).

The polyurethane foam substrate can be sampled merely by removing it from the water, holding it over a sample jar, and squeezing the contents into the jar. The substrate may then be returned to the water and used over and over again (Cairns *et al.*, 1971). Analysis of the community can then be carried out in the manner described for all other samples. It should be noted that the first harvesting of protozoans from the polyurethane substrate is followed by a slight increase in the number of species present during the second harvesting—probably because the first squeezing brings organic material into the interior that was not previously present (Cairns *et al.*, 1971). However, following the first harvesting this effect was not noticeable. Although the sequential comparison index can be used for protozoans or any other group it is more difficult to use when moving organisms are involved. However, if you wish to do so, use a gridded slide or cover slip and proceed to examine in a linear series, one square after another, comparing the organism found in the next square containing one with the organism present in the last square to contain one. (For further details, see either Cairns *et al.*, 1968 or Cairns and Dickson, 1971.) Essentially what you are doing is assuming that the movement is completely random and treating the organisms found in the squares sequentially as if they were not moving. Certain studies of predator-prey relationships where both the predator and prey are moving (e.g., Salt, 1967) show that one can effectively assume that the prey are randomly distributed and are not moving and come out with a workable equation describing the relationship.

If, as I expect will be the case in the future, a series of commercial laboratories are set up throughout the country in major drainage basins to do routine chemical, physical, and biological analyses (transporting the materials from the collecting site to the laboratory quickly and having the counting done by a resident full-time staff), the disadvantages of protozoans will be minimized and the advantages of using them in biological monitoring systems enhanced. In this case I would expect to see them much more commonly used than they are presently.

II. Temperature Effects

All living cells, including algae and protozoans, are exposed to some normal thermal variation in aquatic ecosystems. Optimum temperatures for one cellular function may not be optimum for another. In addition, a temperature optimum for a specific function during brief exposure of a cell may not be identical to that for longer exposure times. Living cells usually are active metabolically within a narrow range of temperatures ($\simeq 10°$–45°C) referred to as the biokinetic range. This generalization of course excludes cryophilic and thermophilic organisms which tolerate temperatures at somewhat greater extremes.

The exposure of algal and protozoan communities to various forms of thermal pollution can be anticipated when aquatic systems are used for cooling. Temperature alterations which exceed the tolerance ranges of individual populations will induce changes, directly and indirectly, on the structure of aquatic communities.

The majority of protozoans are capable of living within specific ranges of temperature variation, and certain species posses the ability to extend their survival capacity through the process of encystment. In addition, any one species may possess various temperature tolerances under different environmental conditions. This fact is clearly demonstrated by *Ochromonas malhamensis*, which was successfully grown above its "maximum" tolerable temperature of 35°C in minimal medium by addition of extra vitamin B_{12}, thiamine, metals such as Fe, Mg, Mn, and Zn, and several amino acids (Hutner *et al.*, 1957). Thus, certain protozoans may be ecologically restricted, in terms of limits of temperature tolerance, due to a lack of available nutrients required at elevated temperatures. Also, a phospholipid requirement, along with morphological abnormalities (i.e., extreme variation in size and shape, fractures and displaced kineties, and abnormalities of karyokinesis) could be induced by growth of *Tetrahymena pyriformis* at supraoptimal temperatures (Rosenbaum *et al.*, 1966). These studies show that at supraoptimal temperatures, nutritional requirements of protozoa may be altered along with a variety of other metabolic responses and critical organism functions (i.e., reproductive rate, growth and development, regeneration, encystment, etc.).

Dallinger (1887) subjected cultures of *Tetramitus rostratus*, *Dallingeria drysdali*, and *Monas dallingeri* to a gradual increase in temperature. In this experiment the temperature in a water bath surrounding the cultures was elevated from about 22° to 70°C over a period of more than five years. Less gradual temperature increases may allow for the natural selection for the tolerance of increased temperature, and this possibility can be considered as one mechanism for survival (Noland and Gojdics, 1967). An example in support of this is a thermal strain of *Tetrahymena*, which was found in a New Mexico hot spring environment and successfully cultured at 41.2°C while other strains succumbed at temperatures above 40°C (Phelps, 1961).

Kasturi Bai *et al.* (1969) studied the effect of temperature increases upon the activity, structure, and reproductive rate of the ciliate, *Blepharisma intermedium.* Initially, increases in temperature resulted in an increase in the organisms's reproductive rate, i.e., at 26° C one division per 24 hours occurred, at 27°C two divisions occurred, at 28°C the maximum of three divisions in 24 hours was attained. Temperature increases beyond 29°C induced a decline in the fission. Between 31° and 36°C a leveling effect occurred, and at 37°C, fission again declined, halting completely at 38°C. *Blepharisma* also acclimated to its maximum reproductive temperature level of 28°C. Cytochemical studies revealed conditions optimal for cell division (i.e., glycogen, basic proteins, unsaturated lipids, enzymes, and alkaline and acid phosphates present in required quantities). At 38°C, alterations occurred in metabolism with a decrease in glycogen, protein, lipids, and activities of some enzymes; also changes in the free amino acid pattern and inhibition of cell division occurred. When returned to room temperature with the addition of fresh culture medium, *Blepharisma* resumed normal metabolism and reproduction by fission after a three-day lag phase. Studies by Thormar (1959) indicated that division in protozoa may be synchronized through heat-induced cellular alterations and that such synchronization in *Tetrahymena pyriformis* appears related to thermal sensitivity which is dependent on cell age. Thus, delays in the division of *Tetrahymena* result from heat shocks and the length of the delay increases with increasing cell age.

Gross and Jahn (1962) studied *Euglena* and *Chlamydomonas* with reference to their cellular response to interacting thermal and photo stresses. Both organisms were cultured in an organic medium in the dark and at several light intensities and temperatures ranging from 20° to 35°C. At temperatures below 32.5°C, *Euglena* growth became inversely light dependent; that is, growth decreased with increasing illumination. Giant, multinucleated *Euglena* cells occurred at 35°C with higher percentages of abnormal cells at the higher light intensities. *Chlamydomonas* did not develop such abnormalities. It was postulated that at elevated temperatures a dark-formed thermosensitive protein, essential for normal cellular division, is denatured in *Euglena*. Furthermore, light could enhance the harmful effect of heat on chlorophyll and the chloroplast.

Gross (1962) studied three apochlorotic *Euglena* substrains in light and dark cultures at 25°C and 33°–35°C. Permanently chlorophyll-less with different carotenoid distributions, the strains were derived originally from a single photosynthetic stock culture. Also employed were *Astasia longa* (a naturally pigmentless euglenoid) and the ciliate *Tetrahymena*. At 25°C, none of the organisms were light dependent, while at higher temperatures, only the three *Euglena* substrains demonstrated an effect of light on growth. One substrain (PBZ-G4) was inhibited as was the photosynthetic parent strain, while the other two (SM-L1, HB-G) were stimulated. At elevated temperatures, abnormal forms developed in all three substrains; two became multinucleate (SM-L1, HB-G), while the other (PBZ-G4) developed an enlarged nucleus. Multinucleate forms occurred in a small percentage of the *Astasia* cells, while almost all of the *Tetrahymena* cells became giants in high temperature culture. Neither *Astasia* nor *Tetrahymena* were affected by light in contrast to the unexplained light—temperature interactions of the nonphotosynthetic substrains of *Euglena*. While the physiological explanation for these phenomena is lacking, it is clear that the major factor causing the abnormalities was thermal stress. Temperature-induced developmental abnormalities even occur in the microtubular organelle of the ciliate *Nassula*, in the flagellum of *Naegleria gruberi* (Tucker, 1967; Dingle, 1967), and in other instances. No doubt such temperature-induced abnormalities accompany functional changes.

The relationship of temperature range and cell size in protozoa has indicated that such adaptation to environmental change has considerable implications. The growth of *Tetrahymena pyriformis* at three different incubation temperatures (10°, 20°, and 30°C) resulted in three classes of cell size (James and Read, 1957). That cell size increases at low growth temperatures, as well as the highest temperature utilized in later research, was shown by Thomar (1962).

Another temperature effect of ecological significance is the change in the regenerative rate of organisms. For example, Giese and McCaw (1963) studied the effects of temperature and other factors on the ability of *Blepharisma americanum*, *B. undulans*, and *B. japonicum* to regenerate removed hypostomes. These protozoa were cultured at various temperatures under continuous yellow light. At 30°C cells accelerated regeneration, while at 35°C they showed retarded regeneration. Brief exposures to a temperature of 40°C also resulted in a retardation of regeneration. Inhibited or retarded regeneration could be caused by enzyme denaturation at increased temperature. Temperatures below 25°C retarded regeneration down to 10°C where *Blepharisma* lost pigment in a "shock" reaction (many dying at 5°C). In all cases, nutrient-starved cells had a slower regeneration rate. No diurnal rhythms of the rate of regeneration were noted. High and low temperature limits for division were like those for regeneration. On the basis of preliminary evidence, Giese and McCaw noted *Blepharisma* adapted to some degree to temperatures higher and lower than those usually tolerated.

One protozoan response to adverse conditions is the mechanism of encystment, which temperature also influences. For example, *Oxytricha fallax* showed an accelerated rate of change into precystic forms on exposure to temperature increases (Hashimoto, 1962). To what degree the encystment process has survival value to protozoan communities subjected to pollution, including thermal types, remains to be determined; knowledge in this area would prove extremely valuable in future community interaction analysis. Numerous algae also produce resistant cellular stages in response to environmental stimuli. These include cysts, hypnospores, akinetes, auxospores, and zygospores.

Cairns (1969) studied the rate of species diversity restoration in freshwater protozoan communities following stress in the form of pH and temperature shocks. Freshwater protozoan communities inhabiting plastic troughs through which unfiltered Douglas Lake (Michigan) water flowed were investigated to ascertain both temperature effects and the time required to restore the community (by recolonization) to its original species diversity (i.e., number of species). Generally, the recovery rate was comparatively rapid. For example, a shock from 20° to 50°C for 7–8 minutes caused a decrease from 26 to 7 species, recovering to 18 species in 24 hours and to the original diversity of 26 species in 72 to 144 hours. In contrast, the thermal shock from 18° to 40°C for 7–8 minutes reduced the species from 31 to 22, with recovery to approximately control diversity in 72 hours. A thermal regime of slightly over 30°C for about 24 hours reduced the diversity from 34 to 21 species in the initial 4-hour period, and then while this thermal regime was sustained, there was an increase of three species at 12 hours. This latter variation is sufficiently small to constitute sampling error. However, the rapid restoration of initial species diversity after return to normal temperatures suggests that some temperature-tolerant species began to invade the sampling site while still under thermal stress. Two major conclusions of this preliminary study are that: (1) The magnitude or intensity of the shock is more important in reducing protozoan species numbers than its duration; and (2) restoration of protozoan species diversity takes only a few hours following mild thermal shock, but as much as five or six days following severe shocks. The latter statement is probably only true of temperature, pH, and other shocks where there is no test material or toxicant residual left in either substrate or organisms.

III. Toxic Substances

Toxic substances can occur as heavy metals, halogens, pesticides, surface-active agents, oils, inorganic reducing agents (sulfides, sulfites), and a variety of other materials. The introduction of toxic material to aquatic ecosystems produces a variety of complex responses governed by several basic factors: (1) nature of the toxicant; (2) concentration; (3) exposure time; (4) environmental characteristics of the receiving system; (5) age, condition, etc., of exposed organisms; and (6) the presence of other toxicants. (Antagonistic additives or synergistic interactions with other substances, changes in dissolved oxygen concentration, redox potential, pH, nutrient, and salt balance, etc., alter the response of aquatic organisms to toxic substances.)

A. Heavy Metals

Excess heavy metals are often introduced into aquatic ecosystems as by-products of industrial processes and acid mine drainage residues. Recently, the significance of airborne nonferrous metals and their accumulations in plants and soils have received increasing attention (Goodman and Roberts, 1971).

Bringmann and Kuhn (1959a,b) evaluated threshold effects of zinc added to the water containing algae and protozoa during a two- to four-day exposure. *Scenedesmus* had a median threshold response at 1.0–1.4 mg/liter zinc while the protozoan, *Microregma*, was more sensitive with a median threshold response at 0.33 mg/liter.

No available evidence explains protozoan differences in sensitivity to copper at the biochemical level. Such information would aid prediction and possible diminution of the effects of copper toxicity. Because copper is a constituent of certain oxidizing-reducing enzymes (e.g., tyrosinase and ascorbic acid oxidase), an enzyme activator to catalyze transfer of some amino acids, and a component of metalloflavoproteins (Dixon and Webb, 1958), these and other cellular processes may involve copper toxicity mechanisms in the different protozoans.

The effect of lithium on morphogenesis of protozoa has been investigated. *Naegleria* amoeba stages displayed characteristic "tadpole" forms when subjected to lithium (Willmer, 1956), while abnormal, delayed regeneration occurred in *Condylostoma* (Suhama, 1961). Various effects, including a broadening of cells with a great increase in the number of lateral stripes and kineties leading to conversion into doublets in *Stentor* also have been attributed to lithium exposure (Tartar, 1957).

B. Halogens

While extremely useful in controlling certain human health hazards in water, chlorine has been implicated in situations where the biota has been degraded, both when present in thermal effluents and alone.

Kott and Kott (1967) found that exposure to 8 mg/liter of chlorine at 10°C for one hour killed cysts of *Endamoeba histolytica*. While the destruction of this human parasite is desirable, the concurrent destruction of free-living forms representing constituent parts of aquatic ecosystems also must be carefully evaluated.

C. Surfactants

Paramecium aurelia lost its ability to respond to a difference in electrical potential when treated with anionics. Treatment with cationics and non-ionics did not affect the protozoan's normal movement toward the cathode (Butzel *et al.*, 1960).

D. Oils

Oil can form films on the surface and thus interfere with gaseous exchange, photosynthesis, and a multitude of other phenomena. Settleable oily substances may coat bottom surfaces, substrate surfaces, or protozoans themselves and thus lead to indirect or direct death. Certain oils also contain toxic substances (i.e., naphthenic acids, phenol, etc.) which intensify the danger of these types of pollutants to aquatic ecosystems.

E. Suspended Solids

Suspended solids often pollute aquatic ecosystems even though the range of variation of suspended solids concentrations in natural systems is enormous. For example, McCarthy and Keighton (1964) reported sediment concentrations for the Delaware River at Trenton, New Jersey, ranging from 1 ppm to 4100 ppm. The daily loads of suspended sediment in the Delaware River at Trenton ranged from half a ton per day to as much as one million tons per day (on August 20, 1955). In the six years from October, 1949 to September, 1955, a total of 6.7 million tons of suspended sediments was carried by the Delaware past Trenton; 26% of this accompanied the flood caused by Hurricane Diane.

Cairns (1968b) listed such ecological effects of suspended solids as: (1) mechanical or abrasive action (i.e., clogging or irritational); (2) blanketing action or sedimentation; (3) loss of light penetration; (4) availability as a surface for growth of microorganisms; (5) adsorption and/or absorption of various chemicals; and (6) change in temperature fluctuations. While offsetting the balance of suspended solids in the system can produce any or all of these ecological effects, certain of them are more likely to endanger the integrity of the community structure of algae and protozoa. Any reduction in the penetration of visible radiation into aquatic ecosystems due to increases in suspended solids would restrict or prohibit the growth of photosynthetic organisms. Predator–prey relationships (i.e., zooplankton grazing on phytoplankton) might change, leading to abnormal increases or decreases of individuals and thus offsetting the system's population balance and stability.

F. Reducing Agents

Inorganic reducing agents such as sulfides and sulfites are often components of various industrial and sometimes sewage wastes.

Bick (1958) demonstrated that in systems with anaerobic bottom conditions and with H_2S concentrations and limited oxygen conditions above, protozoans capable of tolerating low oxygen but intolerant to H_2S occurred only in the limited oxygen layer. Examples of this type are *Spirostomum ambiguum*, *Halteria grandinella*, and *Paramecium caudatum*. In another study, Rylov (1923) noted that both H_2S and oxygen influenced the movement of *Loxodes rostrum* in a pond. The protozoan moved out of the H_2S-containing bottom layer, also avoiding the zone of high oxygen content, and as a result of such restriction occurred in a narrow stratum in sufficient numbers to be visible as a white layer with the naked eye. In general, with the exception of sapropelic species, H_2S is more toxic to protozoa than other gases such as H_2, CO_2, or CH_4 (Nikitinsky and Mudrezowa-Wyss, 1930).

G. Radioactive Wastes

Radioactive materials have received increasing attention with the growth in plans to utilize atomic power plants as energy sources and with increased use of isotopes in hospitals, industry, and research. The following generalizations were made by Krumholz and Foster (1957):

1. Radioactive materials are taken into the body of an organism either through physiological processes and incorporated directly into the tissues, or they are adsorbed on the surfaces of the organisms. In general, adsorption and absorption are governing mechanisms for the lower forms of life, while ingestion is the principal route for predators.
2. The direct concentration of certain radioelements reaches a higher level in the lower plant and animal forms, such as bacteria, protozoa, and phytoplankton, than in higher forms such as vertebrates. In such instances there is an inverse correlation between the complexity of body structure and the direct concentration of the radioelement in question. (This generalization, of course, does not apply to accumulation of radioactivity through the food chain, as with ^{14}C.)

TABLE I

LETHAL AND TOLERANT CONCENTRATIONS OF 12 TOXIC COMPOUNDS TESTED ON 13 SPECIES OF PROTOZOA[a]

Test organism	Toxicant				
	Cr^{6+} (as $K_2Cr_2O_7$)	Phenol	Cu^{2+} (as $CuSO_4 \cdot 5H_2O$)	Pb_2+ [as $Pb(NO_3)_2$]	Mn^{7+} (as $KMnO_4$)
Chilomonas paramecium	1000 (> 18)	1500 (560)	0.056 (0.024)	320 (5.6)	3.2; 5.6; 7.5 (1.0; (1.0); (1.0)
Peranema trichophorum	160 (100)	2500 (1000)	> 100 (1.8)	(1000)	> 32 (3.2)
Tetrahymena pyriformis	1000; > 1000 (180); (750)	3200 (1000)	10 (0.32)	> 100 (24)	3.2 (0.75)
Paramecium caudatum	5000 (1000)	10 (1.35) (< 1.35)			
Paramecium multimicronucleatum	> 1000 (320)	(1000)	0.1; > 1.0 (0.032); (0.24)	56 (24)	10 (0.65)
Stentor coeruleus			1.0		
Euglena gracilis	> 1000 (180)	> 1000 (750)	> 100; 500 (0.1); (5.6)	(1000)	10; 100 (3.2); (3.2)
Chlamydomonas sp.			56 (0.1)		
Chlamydomonas reinhardi			56 (18)		
Blepharisma	1000 (32)	(1000)	3.2; 3.2; 1.8 (0.1); (0.18); (0.32)	100 (42)	18; 5.6 (1.0); (0.32)
Amoeba proteus		(1000)	10		18
Euglena acus			(1.0) 1.0 (0.18)		(< 1.0)
Chaos carolinensis			2.4 (0.1)		

TABLE I (*continued*)

Test organism	Toxicant						
	Zn^{2+} (as $ZnSO_4 \cdot 7H_2O$)	Co^{2+} (as $CoCl_2 \cdot 6H_2O$)	Nitric acid	Acetic acid	Al^{3+} (as $AlCl_3 \cdot 6H_2O$)	Sn^{2+} (as $SnCl_2 \cdot 2H_2O$)	HCl
Chilomonas paramecium	> 10; 3.2 (18); (5.6)	> 2500 (1000)	10 (7.5)	100; 32 (180); (100)	2.4 (1.0)	10 (7.5)	
Peranema trichophorum	(1000)	> 5000 (2500)	56 (5.6)	1000 (75)	> 1000 (560)		
Tetrahymena pyriformis	5.6 (1.0)		18 (10)	320 (180)	3.2 (1.0)	32 (7.5)	3.2 (2.4)
Paramecium caudatum	32 (15.5)						
Paramecium multimicronucleatum	10 (0.56)		13.5 (10)				
Stentor coeruleus	42						
Euglena gracilis	1000 (5000)		(1000)	1000 (560)	(1000)	(100)	
Chlamydomonas sp.	1.8						
Chlamydomonas reinhardi	(1.0) (100)						
Blepharisma	100; 32; 56 (10); (5.6); (5.6)		32 (54)				
Amoeba proteus	(1000)						
Euglena acus							
Chaos carolinensis	> 1000 (320)						

3. Certain plants and animals have a predilection for concentrating specific radionuclides in certain tissues or organs. Iodine, for example, is concentrated in the thyroid, silicon in the frustules of diatoms, calcium in the shells of mussels, calcium and phosphorus in the bony skeletons of vertebrates, strontium in bones, and cesium in soft tissues.
4. Although certain radioelements may occur in acceptable concentrations for drinking water, many freshwater organisms can concentrate them to levels that might be harmful.

IV. Comparative Response to Toxic Materials

Most of the experiments previously cited were carried out under considerably different conditions and one could hardly call the array of methods and procedures used "standard." Of course, in some instances, more than one organism was tested, but rarely was a series of species exposed to a series of toxicants under reasonably comparable conditions. However, Ruthven and Cairns (1973) have provided such information, and this is summarized in Table I. Note that there are considerable differences from one species to another in response to toxic materials, and also a considerable difference

TABLE II

Average Percent Survival of *Vorticella campanula* Exposed to Zn^{2+} and Cu^{2+}

	No. of tests	Average percent survival	Range, percent survival
Zn^{2+} (mg/liter)			
Control	5	99.4	97–100
1.0	5	93.0	76.3–100
1.35	6	87.4	66.7–100
1.8	7	61.5	22.0–100
2.4	5	22.8	11.4–32.9
3.2	4	16.6	13.5–17.9
5.6	3	4.2	0–12.5
10	2	0	
Cu^{2+} (mg/liter)			
Control	4	100	
1.0	7	79.7	56–100
1.8	8	65.9	30.2–100.
2.4	6	51.1	27.2–73.3
3.2	8	41.4	9.1–70.6
5.6	10	32.0	5.8–64.3
7.5	2	0	
10	2	0	

within a species in the response to various toxicants. It is evident even from this limited amount of information that protozoans are sensitive to some toxicants and relatively tolerant of others, and some species are more sensitive than others to a particular toxicant. This is, of course, what one would expect for higher organisms. This similarity would not be worth emphasizing except that many persons involved in evaluation of waste discharges believe no special consideration need be given to microbial communities and that concentrations sufficiently low to protect higher organisms will also protect lower organisms. However, it is conceivable, given the range of variability even in the small group of species covered in Table I, that a much larger range variability exists for the entire protozoan community. One might reasonably assume that there are many extremely sensitive species which could be eliminated by concentrations of toxicants which would not kill fish or other higher organisms. Thus a community imbalance might be created within the microbial community at concentrations presumably safe for higher organisms. The higher organisms might then be indirectly affected by these presumably safe concentrations, and the protection would be less than assumed. These are questions for which

TABLE III

Percent Survival of *Vorticella campanula* Based on Total Number of Individuals from Tests with Zn^{2+} and Cu^{2+}

	No. of tests	Total no. of individuals		Percent survival
		0 hours	3 hours	
Zn^{2+} (mg/liter)				
Control	5	414	560	100
1.0	5	450	479	100
1.35	6	293	257	87.7
1.8	7	969	562	58
2.4	5	357	91	25.5
3.2	4	577	96	16.6
5.6	3	288	19	6.6
10	2	180	0	0
Cu^{2+} (mg/liter)				
Control	4	316	393	100
1.0	7	237	118	79.3
1.8	8	791	401	50.7
2.4	6	715	295	41.3
3.2	8	475	151	31.8
5.6	10	693	108	15.6
7.5	2	428	0	0
10	2	118	0	0

answers based on sound data should be available so that enrivonmental management decisions will be scientifically justifiable.

Table II (from Ruthven and Cairns, 1973) shows the graded response of *Vorticella campanula* exposed to copper and zinc. These graded responses are comparable to the graded responses for higher aquatic organisms as well as diatoms and confirmed the hypothesis that the responses of protozoans to pollutional stresses are similar to the responses of higher organisms, despite their taxonomic differences. Table III depicts the graded response in a somewhat different manner.

Figures 1–10 illustrate dose-response curves for various species of protozoans. The reason for including all these curves is to indicate the variability of response.

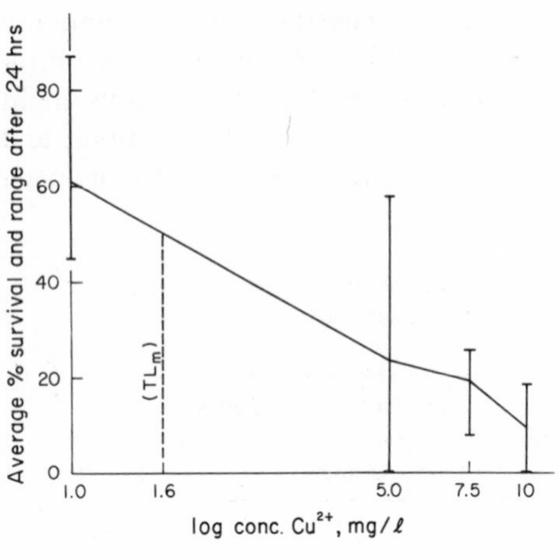

Fig. 1. Average percent survival of protozoan species, range of percent survival, and calculated TL_m values based on 24-hour tests with Cu^{2+}. Samples taken at end of the 24-hour exposure period.

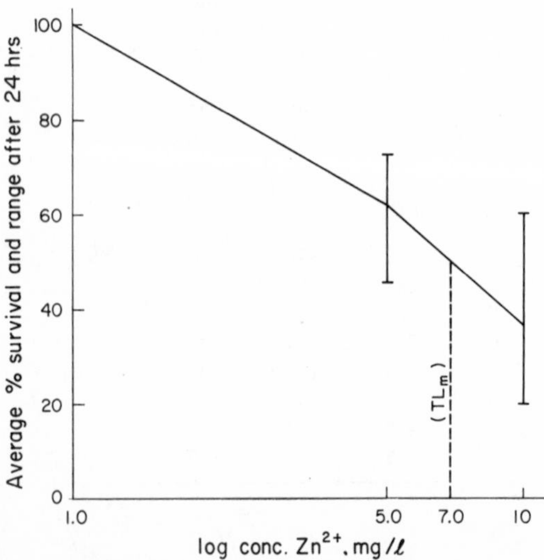

Fig. 2. Average percent survival of protozoan species, range of percent survival, and calculated TL_m values based on 24-hour tests with Zn^{2+}. Samples taken at end of the 24-hour exposure period.

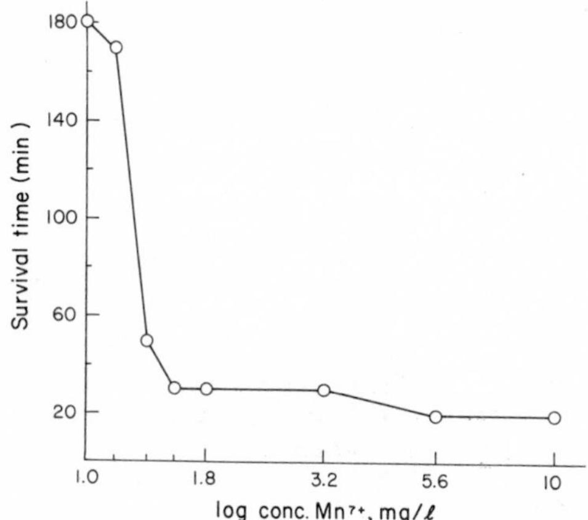

Fig. 3. Concentration of Mn^{7+} vs. survival time for *Blepharisma* sp.

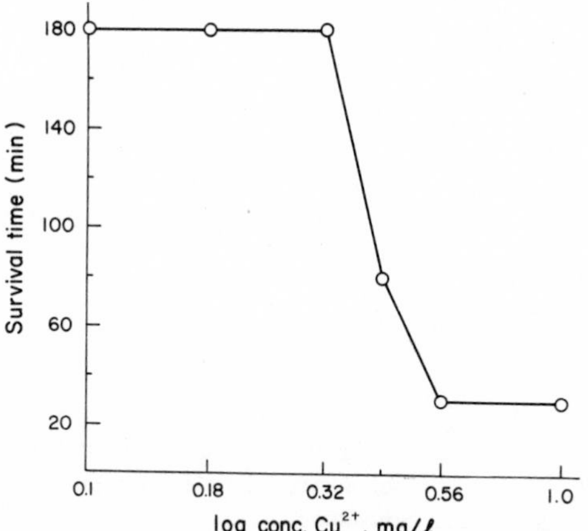

Fig. 4. Concentration of Cu^{2+} vs. survival time for *Blepharisma* sp.

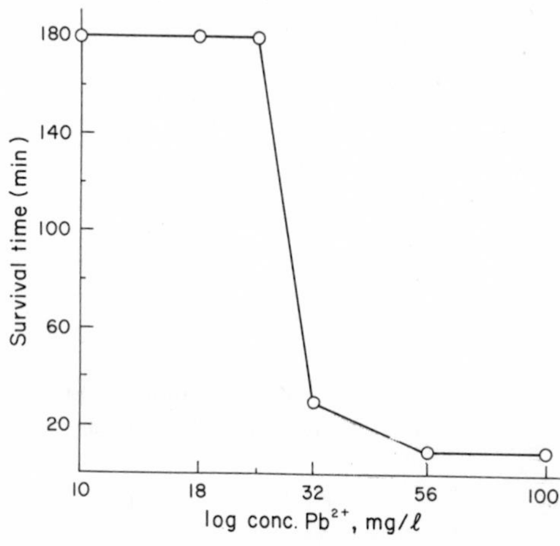

Fig. 5. Concentration of Pb^{2+} vs. survival time for *Paramecium multimicronucleatum.*

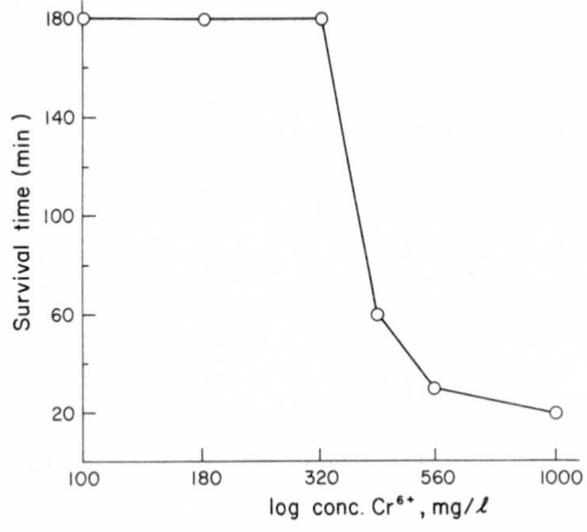

Fig. 6. Concentration of Cr^{6+} vs. survival time for *Paramecium multimicronucleatum.*

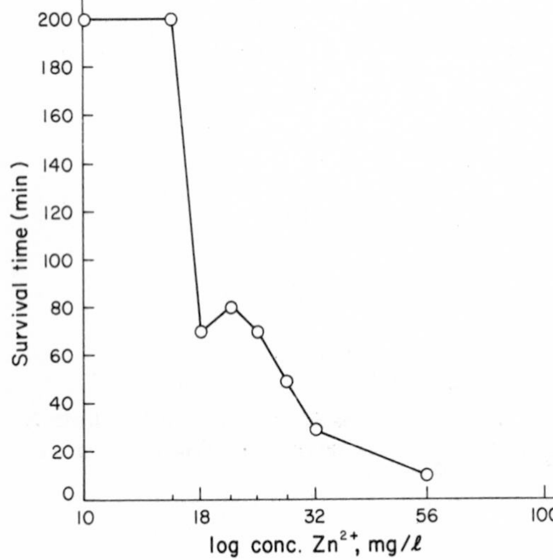

Fig. 7. Concentration of Zn^{2+} vs. survival time for *Paramecium caudatum.*

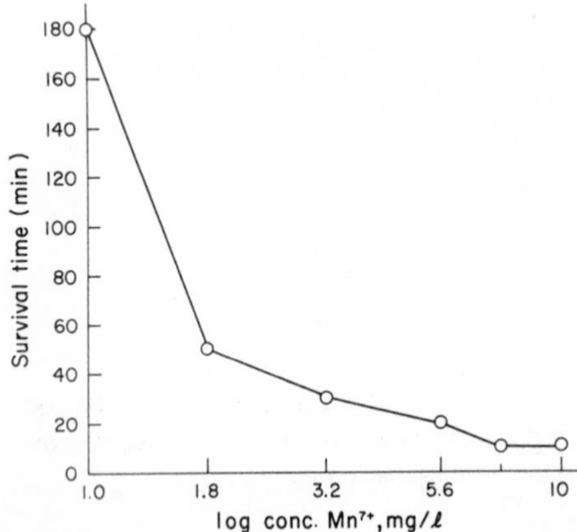

Fig. 8. Concentration of Mn^{7+} vs. survival time for *Chilomonas paramecium.*

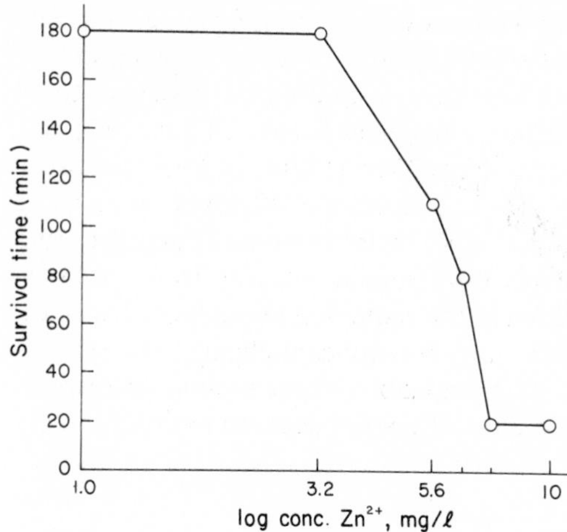

Fig. 9. Concentration of Zn^{2+} vs. survival time for *Chilomonas paramecium*.

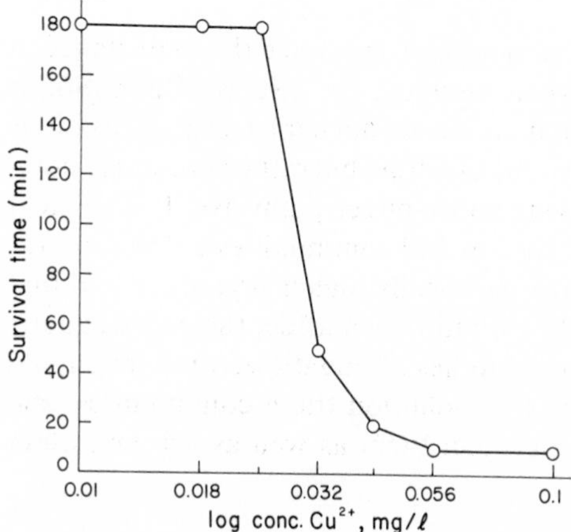

Fig. 10. Concentration of Cu^{2+} vs. survival time for *Chilomonas paramecium*.

V. Recovery Rate

Cairns and Dickson (1970) exposed protozoan communities, established in plastic troughs through which unfiltered Douglas Lake (Michigan) water was pumped, for 24 hours to either 24 ppm of Zn^{2+} or 24 ppm of Cu^{2+}. The number of species and approximate density of protozoans in each trough were determined before the zinc or copper was introduced, at the end of the 24-hour exposure period, and at each 24-hour period after that for either 120 or 144 hours. These results were then compared to those obtained for controls receiving identical treatment except for the omission of heavy metal. A subsidiary test was carried out by exposing a protozoan population in a plastic trough to 24 ppm of zinc with an exposure period of one hour. The results for the 24 ppm Cu^{2+} test solution show a reduction from 46 to 7 species at the end of the 24-hour exposure, with a partial recovery to 14 species 144 hours after the test began. The results for the 24 ppm Zn^{2+} test solution show a reduction in species from 35 to 11 after 24 hours exposure with a comparatively rapid recovery to 20 species at 48 hours, 30 at 72, 32 at 96, and 34 at 120. Control species number varied between 32 and 36 during the zinc tests. Copper test controls were similar, varying from 37 to 48. Density changes were not as striking as the reduction in number of species, although more detailed studies may reveal significant changes. The effects of the one-hour exposure to zinc were negligible. These preliminary experiments indicate that the residual effects of copper seem to be considerably greater than the residual effects of zinc.

VI. Conclusions

The past and present tendency to ignore or minimize the importance of aquatic microbial communities when assessing the effects of pollution is clearly short-sighted. Microbial communities are not haphazard aggregations of species thrown together by the whims of nature but rather structured communities with numerous interlocking cause-effect pathways. It is evident that the requirements of microbial species and communities are as complex or nearly as complex as those of taxonomically higher organisms and that disruption of these communities by pollution can affect the entire aquatic food web. All environmental impact studies of aquatic ecosystems should include an evaluation of the effects of pollution these communities, and standards should be developed to protect them as well as fish and other organisms.

References

Beak, T. W., Friffing, T. C., and Appleby A. (1973). The use of artificial substrate samples to assess water pollution. *In* "*Biological Methods for the Assessment of Water Quality.*" (J. Cairns, Jr. and K. L. Dickson, eds.), Amer. Soc. for Testing and Materials, Spec. Tech. Publ. 528, 227–241.

Bick, H. (1958). Okolgische Untersurchungen an ciliaten fallaubreicher kleingewasser. *Arch. Hydrobiol.* **54**, 506–542.

Bringmann, G., and Kuhn, R. (1959a). The toxic effects of waste water on aquatic bacteria, algae, and small crustaceans. *Gesundh. Ing.* **80**, 115.

Bringmann, G. and Kuhn, R. (1959b). Water toxicology studies with protozoans as test organisms. *Gesundh. Ing.* **80**, 239.

Butzel, H. M., Jr., Brown, L. H., and Martin, W. B. Jr. (1960). Effects of detergents upon electromigration of *Paramecium aurelia. Physiolog. Zool.* **33**, 39.

Cairns, J., Jr. (1965). The protozoa of the Conestoga Basin. Notulae Naturae, Acad. of Natur. Sci. of Philadelphia, No. 375.

Cairns, J., Jr. (1968a). The effects of dieldrin on diatoms. *Mosquito News* **28(2)**, 177–179.

Cairns, J., Jr. (1968b). Suspended solids standards for the protection of aquatic organisms. Purdue Univ. Eng. Bull. **129(1)**, 16–27.

Cairns, J., Jr., M. L. Dahlberg, K. L. Dickson, Nancy Smith, and W. A. Waller (1969). The relationship of fresh-water protozoan communities to the Mac Arthur-Wilson equilibrium model. *Amer. Naturalist*, **103**(933), 439–454.

Cairns, J., Jr. (1969). Rate of species diversity restoration following stress in freshwater protozoan communities. *Univ. Kansas Sci. Bull.* **48(6)**, 209–224.

Cairns, J., and Dickson, K. L. (1970). Reduction and restoration of the number of fresh-water protozoan species following acute exposure to copper and zinc. *Trans. Kansas Acad. Sci.* **73(1)**, 1–10.

Cairns, J., Jr., and Dickson, K. L. (1971). A simple method for the biological assessment of the effects of waste discharges on aquatic bottom dwelling organisms. *J. Water Pollut. Contr. Fed.* **43(5)**, 755–772.

Cairns, J., Jr., and Kaesler, R. L. (1969). Cluster analysis of Potomac River survey stations based on protozoan presence-absence data. *Hydrobiologia* **34**, No. 3–4, 414–432.

Cairns, J., Jr., and Ruthven, J. (1970). Artificial microhabitat size and the number of colonizing species. *Trans. Amer. Microsc. Soc.* **89(1)**, 100–109.

Cairns, J., Jr., Albaugh, D. W., Busey, F., and Chaney, M. D. (1968). The sequential comparison index—a simplified method for non-biologists to estimate relative differences in biological diversity in stream pollution studies. *J. Water Pollut. Contr. Fed.* **40(9)**, 1607–1613.

Cairns, J., Jr., Dickson, K. L., and Yongue, W. H. (1971). The consequences of non-selective periodic removal of portions of freshwater protozoan communities. *Trans. Amer. Microsc. Soc.* **90(1)**, 71–80.

Cairns, J., Jr., Sparks, R. E. and Lanza, G. R. (1973). Developing a biological information system for water quality management. *Water Resources Bull.* **9(1)**, 81–99.

Dallinger, W. H. (1887). "The President's address." *J. Roy. Microsc. Soc.* Ser. 2, **7**, 185–199.

Dingle, A. D. (1967). Evidence for a temperature-sensitive control over flagellum development in transforming cells of *Naegleria gruberi. J. Protozool. Suppl.* **14**, 12.

Dixon, M., and Webb, E. C. (1958). "Enzymes". Longman, Green, London.

Giese, A. C. and McCaw, B. (1963). Regeneration rate of *Blepharisma* with special reference to the effect of temperature. *J. Protozool.* **10(2)**, 173–182.

Goodman, G. T., and Roberts, T. M. (1971). Plants and soils as indicators of metals in the air. *Nature (London)* **231** (5301), 287–292.

Gross, J. A. (1962). Cellular responses to thermal and photo stress. II. Chlorotic Euglenoids and *Tetrahymena. J. Protozool.* **9(4)**, 415–418.

Cross, J. A., and Jahn, T. L. (1962). Cellular responses to thermal and photo stress. *Euglena and Chlamydomonas. J. Protozool.* **9(3)**, 340–346.

Hashimoto, K. (1962). Relationships between feeding organelles and encystment in *Oxytricha fallax* Stein. *J. Protozool.* **9(2)**, 161–169.

Hutner, S. H., Baker, S., Aaronson, S., Natuan, H. A., Rodriguez, E., Lockwood, S., Sanders, M., and Peterson, R. A. (1957). Growing *Ochromonas malhamensis* above 35°C. *J. Protozool.* **4**, 259–269.

James, T. W., and Read, C. P. (1957). The effect of incubation temperature on the cell size of *Tetrahymena pyriformis. Exp. Cell. Res.* **13(3)**, 510–516.

Kasturi Bai, A. R., Srihari, R., Shadaksharas-wamy, M., and Jyothy, P. S. (1969). The effects of temperature on *Blepharisma intermedium. J. Protozool.* **16(4)**, 738–743.

Kolflat, T. (1968). Thermal pollution—1968. Hearings before the subcommittee on air and water pollution of the committee on public works, United States Senate, Ninetieth Congress, Second Session, February, 1968. Washington, D. C. U.S. Government Printing Office, 63.

Kott, H., and Kott, Y. (1967). The fate of *Entamoeba histolytica* cysts in chlorinated sewage. *J. Protozool. Suppl.* **14**, abstr. 179. 44.

Krumholz, L. A., and Foster, R. F. (1957). Accumulation and retention of radio-activity from fission products and other radiomaterials by fresh-water organisms. *Nat. Acad. Sci. Nat. Res. Council Publ.* **551**, 88.

Lackey, J. B. (1938). Protozoan plankton as indicators of pollution in a flowing stream. *Public Health Reports* **53**, 2037.

McCarthy, L. T., Jr., and Keighton, W. B. (1964). Quality of Delaware River water at Trenton, New Jersery. Geological Survey Water-Supply Paper 1779-X, pp. 36–37. U.S. Gov. Printing Office, Washington, D. C.

McKee, J. E., and Wolf, H. W. (1963). Potential pollutants. *In* "Water quality criteria," (J. E. McKee and H. W. Wolf, eds.), pp. 123–298. The Resources Agency of California, State Water Quality Control Board, Publ. 3–A.

Nikitinsky, J., and Mudrezowa-Wyss, F. K. (1930). Under die wirkung der kohlensaure, des schwefelwasserstoffs, des methans und der abwesenheit des sauerstoffs auf wassert-organismen. *Centralbl. Bakteriol.* Abt. 2, **82**, 167–198.

Noland, L. E., and Gojdics, M. (1967). Ecology of freeliving protozoa. *In* "Research in Protozoology" (T. Chen, ed.), pp. 215–266. Permagon, Oxford.

Patrick, R. (1949). A proposed biological measure of stream conditions, based on a survey of the Conestoga Basin, Lancaster County, Pennsylvania. *Proc. Acad. Natur. Sci. Philadelphia.* **101**, 277–341.

Patrick, R., Hohn, M. H., and Wallace, J. H. (1954). A new method for determining the pattern of the diatom flora. *Notulae Natur. Acad. Natur. Sci. of Philadelphia* No. 259.

Patrick, R., Cairns, J., and Scheier, A. (1968). The relative sensitivity of diatoms, snails and fish to twenty common constituents of industrial wastes. *Prog. Fish Cult.* **30**(**3**), 137–140.

Phelps, A. (1961). Studies on factors influencing heat survival of a ciliate, a mite, and an ostracod obtained from a thermal stream. *Amer. Zool.* **1**, 467.

Rosenbaum, N., Erwin, J., Beach, D., and Holz, G. G. Jr. (1966). The induction of a phospholipid requirement and morphological abnormalities in *Tetrahymena pyriformis* by growth at supraoptimal temperatures. *J. Protozool.* **13**(**4**), 535–546.

Ruthven, J. A., and Cairns, J. Jr. (1973). The response of freshwater protozoan communities to concentrations of various toxicants, particularly the heavy metals, zinc and copper. *J. Protozool.* **20**(**1**), 127–135.

Rylov, V. W. (1923). Uber den Einfluss des im wasser gelosten sauerstoffs und schwefelwasserstoffs auf den lebenszyklus und die vertkkale verteilung des infusors Loxodes rostrum. *Int. Rev. Ges. Hydrobiol.* **11**, 179–192.

Salt, G. W. (1967). Predation in an experimental protozoan population (*Woodruffia–Paramecium*). *Ecol. Monogr.* **37**, No. 2, 113–144.

Sramek-Husek, R. (1958). Die rolle der ciliaten-analyse bei der biologischen kontrolle von flussverunreinigungen. *Verh. Int. Ver. Limnol.* **13**, 636–645.

Suhama, M. (1961). Experimental studies on the morphogenesis of *Condylostoma spatiosum* Ozaki and Yagiu. *J. Sci. Hiroshima Univ.* (b) **20**, 33–91.

Tartar, V. (1956). Reactions of *Stentor coreuleus* to certain substances added to the medium. *Exp. Cell Res.* **13**, 317–332.

Tucker, J. B. (1967). Abnormal development of a microtubular organelle induced by heat treatment of the ciliate *Nassula*. (Abstract). *J. Protozool. Suppl.* **14**, 28.

Willmer, E. N. (1956). Factors which influence the acquisition of flagella by the amoeba, *Naegleri gruberi*. *J. Exp. Biol.* **33**, 583–603.

CHAPTER 2

Sponges (Porifera: Spongillidae)

FREDERICK W. HARRISON

I. Introduction

Shortly after beginning this study it was found that little was known about pollution ecology of the sponges. Also, the "normal" ecology of all the North American freshwater sponges had never been presented in one discussion. Since an understanding of the normal ecology of sponges is an absolute prerequisite to any study of sponges and pollution, the normal ecology of the freshwater sponges of North America has been compiled. In the majority of cases, credit must be given to previous workers whose pioneering studies of various spongillid species are scattered throughout the literature.

The examination of sponge "pollution ecology" involved, in the absence of adequate literature, personal correspondence with the spongologists of the world. Without their overwhelming interest and cooperation this section could never have been written.

Hopefully, the following material will provide a base from which to utilize the freshwater sponges in studies of environmental pollution. In view of their intimate relationship with the aquatic environment, the sponges should prove to be delicate indicators of both population alterations and pollution-induced cytopathology.

II. Systematics

The known freshwater sponges of North America are all members of the family Spongillidae, including those species "occurring normally in fresh or occasionally in brackish water" (de Laubenfels, 1936). The insufficiency of this definition, as pointed out by Penney and Racek (1968), lies largely in the fact that the freshwater sponges are a conglomerate group, likely derived from several evolutionary lines. The affinities of several species of freshwater or brackish water sponges that form atypical gemmules or, apparently, do not form gemmules, are not clearly understood. These forms, found primarily in the deep freshwater lakes of ancient origin are currently being examined to determine the possibility of a polyphyletic origin of the freshwater sponges.

The North American species all form "typical" gemmules, are closely related, and are probably descended from a common ancestor. The recent report (Ott and Volkheimer, 1972) of a gemmule-bearing fossil spongillid from the mesozoecium of Argentina indicates the antiquity of this evolutionary line.

The systematics of the freshwater sponges have been, until recently, chaotic. The pioneering revisionary work of Penney and Racek (1968) now allows workers of various disciplines to utilize this important but neglected group of organisms at varying levels of study without fear of entrapment in a taxonomic morass.

The freshwater sponges of North America belong to nine genera and consist of 33 species (Table I) at this time. The studies of Poirrier (1969) indicate that some of these species may be ecomorphic forms exhibiting phenotypic variation in response to habitat conditions. While recognizing the importance of Poirrier's conclusions, for reasons of taxonomic priority all terminology herein is based on Penney and Racek (1968). These authors also provide a synonomy covering the majority of the freshwater sponges of the world. A synonomy for those few species not contained in the Penney collection may be found in Penney (1960).

TABLE I

FRESHWATER SPONGES OF NORTH AMERICA

Genus *Anheteromeyenia* Schröder, 1927	Genus *Heteromeyenia* Potts, 1881
Anheteromeyenia argyrosperma (Potts, 1880)	*Heteromeyenia baileyi* (Bowerbank, 1863)
A. biceps (Lindenschmidt, 1950)	*H. latitenta* (Potts, 1881)
A. pictovensis (Potts, 1885)	*H. longistylis* (Mills, 1884)
A. ryderi (Potts, 1882)	*H. tentasperma* (Potts, 1880)
Genus *Corvomeyenia* Weltner, 1913	*H. tubisperma* (Potts, 1881)
Corvomeyenia carolinensis Harrison, 1971	Genus *Radiospongilla* Penney and Racek, 1968
C. everetti (Mills, 1884)	*Radiospongilla cinerea* (Carter, 1849)
	R. cerebellata (Bowerbank, 1863)
	R. crateriformis (Potts, 1882)
Genus *Dosilia* Gray, 1867	Genus *Spongilla* Lamarck, 1816
Dosilia palmeri (Potts, 1885)	*Spongilla aspinosa* Potts, 1880
D. radiospiculata (Mills, 1888)	*S. cenota* Penney and Racek, 1968
Genus *Ephydatia* Lamouroux, 1816	*S. heterosclerifa* Smith, 1918
Ephydatia fluviatilis (Linneaus, 1758)	*S. johanseni* Smith, 1930
E. millsii (Potts, 1887)	*S. lacustris* (Linneaus, 1758)
E. muelleri (Lieberkühn, 1885)	*S. sponginosa* Penney, 1957
E. robusta (Potts, 1887)	*S. wagneri* Potts, 1889
E. subtilis (Weltner, 1895)	Genus *Trochospongilla* Vejdovsky, 1883
Genus *Eunapius* Gray, 1867	*Trochospongilla horrida* Weltner, 1893
Eunapius fragilis (Leidy, 1851)	*T. leidyi* (Bowerbank, 1863)
E. mackayi (Carter, 1885)	*T. pennsylvanica* (Potts, 1882)

III. The "Normal" Ecology of Freshwater Sponges

A. Life Cycle

In most cases North American spongillids exhibit an annual life history involving a cycle of gemmule production, degeneration or death of adult tissues, gemmule germination, and renewed growth. Gemmule production may occur in late spring, early fall, or throughout the year in perennial forms. In species normally following a seasonal pattern of asexual reproduction, gemmulation may be induced by factors coinciding with seasonal flooding and drying (M. A. Poirrier, personal communication). Cheatum and Harris (1953) found gemmulation was stimulated throughout the year by warm days following several days of low temperatures.

Many freshwater sponge species may enter into a period of reduction (Penney, 1933) during their life cycle as a response to adverse environmental conditions. The reduction process, involving a withdrawing and rounding-up of the sponge tissues, normally entails the loss of several cell types and the cytological degeneration of others. At the ultrastructural level, the process is characterized by the degeneration of the rough endoplasmic reticulum and formation of "myelin figures" typical of cytopathologies (F. W. Harrison and N. Watabe, unpublished results). Following the restoration of favorable environmental conditions, the sponge may be restored to its normal morphological form through somatic embryogenesis (Korotkova, 1970). Similar phenomena have been observed in marine sponges. A drop in water temperature to and below 10°C was sufficient to initiate a regressive overwintering phase in specimens of *Microciona prolifera* (Ellis and Solander, 1786) from the Connecticut coast. A rise in temperature above 10°C was necessary to initiate reorganization and growth (Simpson, 1968).

Although little is known of environmental effects upon the sexual reproductive phase of spongillid life cycles, gametogenesis and larval production in marine sponges have been correlated with water temperatures. Simpson (1968) found temperatures of 10°–18°C stimulated egg production in *M. prolifera* while temperatures of 15°C were necessary for production of spermatozoa. Larval production was, consequently, dependent upon temperatures of 15°C or above while maximum production and release of larvae occurred at temperatures of 20°–25°C. Storr (1964) found larval production in the wool sponge, *Hippiospongia lachne* de Laubenfels, 1936, of the Cedar Keys, Florida area to begin with mean monthly surface water temperatures of 23°C, reaching a maximum at 29°C. At temperatures 2–3 degrees above this level the number of sponges producing larvae fell drastically.

It seems probable that either sexual or asexual phases of reproduction may assume dominance in the life history of freshwater sponge species. Sexual reproduction may dominate in species typically inhabiting more stable habitats. This is evidenced by the absence of gemmules or either development of atypical gemmules in sponges of the deep ancient freshwater lakes. Conversely, the production of larvae by asexual means and gemmule production in intertidal marine species (Hopkins, 1956a; Wells *et al.*, 1960; Bergquist *et al.*, 1970) reflects the tendency toward asexual reproductive patterns in ecologically variable habitats.

B. Habitats

Freshwater sponges will attach and grow upon almost any type substrate. They may be found on vegetation, submerged logs, iron waterpipes (Old, 1932b), concrete, or glass. In the laboratory they have grown well on plastic dishes and on Epon.

1. Current

Although Potts (1887) found the finest specimens of sponges where they were subjected to the most rapid currents, there are definite species preferences for lentic or lotic environments. Lotic environments, although adversely affecting sponges by buffeting actions of excessive turbulence and by smothering them with suspended silt accompanying flooding, compensate by bringing food and oxygen, removing waste, and washing off sediments (de Laubenfels, 1947; Racek, 1969).

The degree of current has considerable influence on sponge morphology. *Radiospongilla sceptroides* (Haswell, 1882), one of the few Australian spongillids able to withstand the direct force of current without positional shelter, is highly modified (Racek, 1969). The oscula are surrounded by an intricate star-shaped canal system, likely an adaptation to the extensive water pressures on each osculum. Additionally, the number of oscules and the development of chimneys may vary with current conditions (Wells *et al.*, 1960). In quieter waters the numerous oscules are usually elevated on chimneys. In surf or strong currents the oscules are fewer and less conspicuous. These specializations are apparently related to efficient dispersal of waste products away from the incurrent pores.

2. Light

As might be expected, there is no hard and fast rule defining the relative tolerance of various sponge species to light. The effects of light intensity are greatly modified by two other environmental factors, water color and transparency. In less transparent waters a sponge may grow in a relatively exposed position while in less highly colored or more transparent habitats the same species may be found at greater depths or in a sheltered location.

The relationship of light to larval attachment is poorly understood but has been studied in some marine sponges. Bergquist *et al.* (1970) found larval behavior was closely correlated with the location of adult colonies. Those species with markedly negatively phototactic larvae, i.e., *Mycale macilenta* Bowerbank, 1866, grew in deeply shaded locations or under stones. *Microciona coccinae* Bergquist, 1961 and *Ophlitaspongia seriata* Grant, 1826, species growing in full light to deep shade, had larvae that exhibited no light response. Various studies (Wilson, 1935; Ali, 1956; Simpson, 1968) indicate similar relationships in other species. Additionally, Rasmont (1970) has found constant illumination to inhibit gemmule formation in the spongillid *Ephydatia fluviatilis* (Linneaus, 1758).

3. Temperature

In addition to the relationships of temperature to sponge life cycles (Section III, A), temperature may also influence morphological features or may actually be a limiting factor.

The development of thick gemmular pneumatic layers armed with numerous gemmoscleres is temperature related (Penney and Racek, 1968). *Spongilla lacustris* (Linneaus, 1758), a Northern Hemisphere species with subarctic and cold-temperate distribution shows this clearly. The pneumatic layer of subarctic specimens is either absent or poorly developed, a feature also seen in *Spongilla arctica* Annandale, 1915. Specimens from southern or central Europe display the well-developed gemmular form.

The degree of development of the gemmule pneumatic layer may not entirely be a response to climatic conditions. Poirrier (1969) induced ecomorphic changes in pneumatic layer dimensions in some species by experimentally altering habitat water mineralization.

The temperature range for sponges is uncertain at this time. de Laubenfels (1932) observed that typical marine sponges succumb to water temperatures as low as 12°–18°C. Simpson (1968) found temperatures of 10°C initiated degenerative overwintering in specimens of *M. prolifera* but also stressed the physiological adaptation of sponges to particular temperature ranges. *Microciona prolifera* colonies from Beaufort, North Carolina waters grow best during winter, with water temperatures of 3°–12°C (McDougall, 1943).

Higher temperatures (30°C+) may affect freshwater sponges, particularly in the initiation of gemmulation in seasonal forms (Poirrier, 1969), but sponges may be collected regularly at temperatures up to 37°C. However, the combination of high temperatures (above 30°C) and declining water levels is frequently associated with sponge degeneration and death (Moore, 1953 and personal communication).

4. *Climatic Factors*

Freshwater sponge species surviving under harsh climatic conditions show, in several cases, adaptations that allow them to be characterized as "ephemeral" forms. Specimens of *Dosilia palmeri* (Potts, 1885) were collected in the branches of "screw-bean" trees, *Strombocarpus pubescens*, growing along the borders of the Colorado River, near Lerdo, Sonora, in northwestern Mexico. The sponges, appearing like dark "wasp-nests," were present by the thousands suspended two to three feet above ground (Potts, 1887). This particular area of the river border is submerged only during the annual floods during May or June, with water coverage for approximately six weeks at a time. The sponges are, therefore, suspended for nine to ten months of the year over some of the "hottest, driest, and most barren" (Potts, 1887) land in North America.

This sponge has a precocious development, rapidly producing gemmules that have a thick pneumatic layer and are heavily armed with gemmoscleres. Its congener, *Dosilia plumosa* (Carter, 1849), shows similar features as an adaptation to the Indian dry season (Potts, 1887).

Racek (1969) found that the harsh climatic conditions of the Australian interior had profound effects upon not only the life history of spongillids but also upon their morphology. *Eunapius sinensis* (Annandale, 1910), in addition to exhibiting the characteristics of an ephemeral, e.g., precocious development, is very rigidly constructed, almost stony. An additional adaptation to the aridity of the Australian interior is the extended viability of gemmules. Gemmules of interior sponges, on the shelves of the Australian Museum for 25 years, have germinated after being placed into water (A. A. Racek, personal communication).

5. *Chemical Factors*

The distribution of freshwater sponge species in North America can be correlated with the chemical properties of the habitat and the individual tolerances of the species to those factors. Jewell (1939) considered the calcium content of the water to be the most important single factor in determining distributional patterns in species she studied. In other cases (Racek, 1969), the hydrogen ion concentration is apparently the most important determining factor. The silicon concentration is important in determining the degree of skeletal development, particularly spicule thickness (Jorgensen, 1944), in spongillids. It is, however, not usually a limiting factor except in the case of some strongly skeletoned forms (Jewell, 1935).

Because habitat color mainly results from dissolved organic matter (Hutchinson, 1957), and thus is indicative of nutriment available to sponges, it is a factor in sponge distribution and abundance. The inverse relationship between habitat color and light penetration also is a factor (Jewell, 1935). Although water color is primarily related to the abundance of a species in an otherwise suitable habitat (Jewell, 1935), low water color (25 or less) may limit sponge colonization of an area (Cheatum and Harris, 1953).

Conductivity, an index of dissolved inorganics of a habitat, may be dependent primarily upon salts of calcium or magnesium as in lakes studied by Jewell (1935) or may reflect sodium chloride levels in brackish water habitats. In either case, there are instances of species limitations at either end of the scale. The problem of species distribution in the brackish water habitat will be treated as a separate entity.

6. *The Brackish Water Habitat*

Redeke (1932) placed the division between "seawater" and "brackish water" at 30 ppt salinity. Although the brackish water zone of an estuary contains primarily marine sponges, there are also some sponge species present as invaders from the freshwater habitat. These are limited in number, of course, and occur only in the low salinity areas of the estuary. *Ephydatia fluviatilis* is apparently the most euryhaline spongillid species. Potts (1889) found the species at the margin of the Everglades in a region flooded by salt water during southwest gales. It was found most often in brackish water areas in Louisiana (Poirrier, 1969), occurring in waters with salinities up to

4.7 ppt. *Ephydatia fluviatilis* has also been collected in brackish water areas in Denmark from waters of up to 5 ppt salinity (Tendal, 1967).

Members of the *Spongilla alba* evolutionary complex (*S. alba, S. wagneri, S. cenota*) all show high levels of tolerance to either brackish or highly mineralized habitats. *Spongilla alba* has been reported (Annandale, 1915) from brackish waters of Lake Chilka, India with a specific gravity of 1.0065 and as being extremely abundant in brackish water in the Gangetic delta. In Australia it is restricted to coastal waters (the tidal Brisbane River) of low to fairly high salinities (Racek, 1969). Racek suggests that this habitat possibly resulted from gemmular transport from Asian countries by intercontinental ships. The optimal habitat of *S. wagneri* is slightly to strongly brackish water (Eshleman, 1950; Penney and Racek, 1968), while *S. cenota* is found in the limestone cenotes of Yucatan in waters of relatively high carbonate content (Hall, 1936; Penney and Racek, 1968).

Other spongillids reported from brackish water regions of estuaries are *Ephydatia robusta* (Moore, 1953), *Spongilla lacustris* (Tendal, 1967), *Trochospongilla leidyi*, *Heteromeyenia baileyi*, and *Trochospongilla horrida* (Poirrier, 1969). Additionally, *Heterorotula capewelli* (Bowerbank, 1863), a species apparently restricted in its distribution to the salt-pan areas of the Australian interior, inhabits slightly brackish athallassic waters (Racek, 1969).

The euryhaline marine sponges, although penetrating into brackish water regions, are seldom able to survive salinities below 15 ppt for any length of time. Some members of the Clionidae, particularly *Cliona truitti* Old, 1941, seem to be tolerant of low salinities. Hopkins (1956a) found *C. truitti* was the dominant clionidid in Louisiana estuarine waters exhibiting salinity below 15 ppt at least 10% of the time. It occurred sparsely in areas where salinities were less than 15 ppt 75% of the time or less than 10 ppt nearly 50% of the time. He was able to characterize Louisiana estuarine waters into zones on the basis of the presence or absence of *C. truitti* (low salinity) or *C. celata* Grant (high salinity). Old (1941) found similar distributional patterns in Chesapeake Bay water with *C. celata* absent from salinities below 15 ppt and *C. truitti* most abundant in lower salinities of 11–13 ppt. However, neither Hopkins (1956b) nor Wells (1959, 1961) were able to detect this salinity-related distributional pattern among clionids of Carolina waters.

Annandale (1915) reported the following nonspongillids from brackish water areas of Lake Chilka, India: *C. vastifica* Hancock (specific gravity of water 1.000–1.0265); *Suberites sericeus* Thiele, 1898 (sp. gr. 1.000–1.0145); *Laxosuberites aquae-dulcioris* (Annandale, 1914) (sp. gr. 1.000–1.0150); and *Tetilla dactyloidea* var. *lingua* (Annandale, 1915) (sp. gr. 1.000). Sarà (1960) described *Tedania anhelans* (Lieberkuhn) and *Hymeniacidon sanguinea* (Grant) from brackish water environments.

It is of interest that *C. truitti* and *C. vastifica*, both clionids penetrating into zones of low salinity, reproduce asexually by gemmule formation (Annandale, 1915; Hopkins, 1956a; Wells *et al.*, 1960). Similarly, Annandale (1915) reported gemmules from *L. aquae-dulcioris*, *L. lacustris*, *S. sericeus*, and *T. dactyloides* var. *lingua*—all being species found in brackish or fresh waters during parts of the year. Gemmule formation may allow these species to colonize low salinity ranges inimical to other forms (Hopkins, 1956a).

IV. Species Ecology of North American Spongillidae

1. *Anheteromeyenia argyrosperma* (Potts, 1880)

This species is usually found in open areas, exposed to some sunlight (Old, 1932a) and growing on upper or lower surfaces of stems, timber, or vegetation (Old, 1932b) in waters of varying degrees of current. While it has been found in running water habitats (Jewell, 1935; Neidhoefer, 1940), it may also occur in standing roadside waters, ponds, or sinkholes (Old, 1932a; Eshleman, 1950; Poirrier, 1969). It prefers slightly acidic waters of moderate to low alkalinity that are high in conductivity (Old, 1932a; Jewell, 1939; Penney, 1954; Poirrier, 1969).

Summarizing the data presented by Old (1932a, b), Jewell (1939), Wurtz (1950), Penney (1954), and Poirrier (1969), *A. argyrosperma* is reported from waters with the range of conditions as tabulated below. Unless otherwise indicated, data are given in parts per million.

Alkalinity (methyl orange)	12.7–85.0
Calcium	1.2–22.0
Color	60–80
Conductivity	80–750 micromhos/cm
Free CO_2	0.5–22.0
Hardness (as $CaCO_3$)	40.0–80.0
pH	4.2–7.5
SiO_2	3.8–7.8
Temperature	24°–31°C

2. *Anheteromeyenia biceps* (Lindenschmidt, 1950)

This species, reported only from the type locality, the inlet and outlet of Douglas Lake, Michigan, shows certain spicular malformations suggesting hybridization or adverse environmental factors (Penney and Racek, 1968). Further collecting is needed to determine the ecology of this species if, indeed, it warrants species status.

3. Anheteromeyenia pictovensis (Potts, 1885)

Potts (1887) collected *A. pictovensis* in summer and during the last week of December (on sticks pulled up through breaks in the ice) from Nova Scotia. Although Potts reported the presence of a few gemmules he gave no indication as to period of gemmulation. As Nova Scotia temperatures may range from 35.5°C in summer to −31°C in winter with an annual average of 6.5°C (Potts, 1887), gemmulation could occur in response to rising summer temperatures or lowering autumnal temperatures.

The species is apparently most often found in lakes and shuns light and calcareous rocks (Stephens, 1912).

4. Anheteromeyenia ryderi (Potts, 1882)

Anheteromeyenia ryderi is usually found in lightly shaded locations in swamps, ponds, or slowly moving streams, never in rapids (Old 1932a,b; Neidhoefer, 1940; Hoff, 1943; Eshleman, 1950; Poirrier, 1969). M. A. Poirrier (personal communication) noted that the species grows rapidly in the fastest moving Louisiana streams during fall and winter. It is more common in acid areas of relatively low bicarbonate content (Neidhoefer, 1940; Old, 1932a; Poirrier, 1969). It is somewhat limited by higher temperatures, forming gemmules with the approach of 30°C water temperatures (Poirrier, 1969) but thrives during winter. It was plentiful throughout winter in lakes and ponds near Cold Spring Harbor, New York (Potts, 1887). Local variations are apparently present, however. Old (1932a) found the species limited by both high and low temperatures, with a range of 19°–26.5°C. Jewell (1939) found it to tolerate the entire range of calcium in Wisconsin waters (1.2–22 ppm), but Stephens (1912) never found it on limestone rocks. It is always absent from polluted waters (Old, 1932a) and brackish water (Porrier, 1969).

Summarizing the data presented by Old (1932a,b), Jewell (1935, 1939), Wurtz (1950), Moore (1953), Penney (1954), Author Anonymous (1961), and Poirrier (1969), *A. ryderi* is reported from waters with the range of conditions tabulated on page 39. Unless otherwise indicated, data are given in parts per million.

5. Corvomeyenia carolinensis (Harrison, 1971)

Known only from the type locality, Adams Pond, Columbia, South Carolina, this species is to an extent perennial. Although forming gemmules in late spring in response to rising temperatures, it survives through the summer months and may be collected throughout the year. It is most abundant in fall and spring, although it has been collected while air temperatures

Alkalinity	
Methyl orange	4.0–128.0
Phenolphthalein	0.00
Aluminum	0.06–0.27
Bacteriology	
Total Count[a]	285
Coliform Count[b]	300
BOD	2.17
Calcium	1.2–22.0
Total hardness (as $CaCO_3$)	28.8–80.0
Calcium hardness (as $CaCO_3$)	24.5
Chloride	1.0–5.3
Chromium	0.004–0.057
Color	20–160
Conductivity	37–170 micromhos/cm
Copper	0.09–0.45
Dissolved oxygen	7.10
Free CO_2	3.15–12.0
Iron	< 0.002–0.033
Magnesium	1.05
Magnesium hardness (as $MgCO_3$)	4.30
Nickel	0.002–0.005
Ammonia nitrogen	< 0.001
Nitrate nitrogen	0.192
Nitrite nitrogen	0.001
pH	4.2–8.5
Phosphate	0.028
SiO_2	3.05–13.0
Sulfate	1.5–5.06
Total solids	164
Fixed residue	68
Volatile solids	96
Temperature	19°–32°C
Turbidity	12.42

[a] Colonies per milliliter, 20 hours incubation at 35°C, membrane filter method (Author Anonymous, 1961).

[b] Colonies per milliliter, 20 hours incubation at 35°C membrane filter method (Author Anonymous, 1961).

were below freezing. It is found in quiet waters in the littoral, growing on submerged portions of the emergent vegetation. Intracellular zoochlorellae aid in survival during the hot summers (air temperatures 38°C or more) and are a definite factor contributing to the ease with which this species is maintained in the laboratory.

Water of the type locality exhibits the following properties (Harrison, 1971). Except for pH, values are given in parts per million.

Calcium	12.0
pH	6.7
SiO_2	3.9

6. *Corvomeyenia everetti* (Mills, 1884)

Corvomeyenia everetti is a light-positive form, normally absent from less transparent waters (Jewell, 1935). In the one collection from a lake of high color (optimum habitat, color 6–12), Deadwood Lake, color 65, Jewell (1939) found the species at depths of 0.25 to 1 m. In transparent waters the species was found at depths of 1.5–3.5 m. The 3.5 m depth, the greatest depth from which Jewell obtained any sponge, was in Crystal Lake (color 0, disk 12.7) and still within the range of light penetration.

This delicate sponge covers submerged vegetation with meandering threads passing from leaf to stem (Potts, 1887). It is restricted in nature to still waters of low mineral and inorganic content (Jewell, 1935). *Corvomeyenia everetti* is found only in waters containing less than 4.0 ppm calcium and less than 3.0 ppm magnesium (Jewell, 1939). Juday and Birge (1933) showed that, for the Wisconsin lakes investigated by Jewell, conductivity depends primarily upon the salts of calcium and magnesium present in soluble form in the water. Ordinary distilled water (Juday and Birge, 1933) has a conductivity of 3–9 micromhos/cm. *Corvomeyenia everetti* is restricted to waters of conductivity below 19 micrómhos/cm and has an optimum for growth in waters of conductivity below 19 micromhos/cm, approaching the level of distilled water.

The results of experiments in which Jewell (1939) controlled the mineral content of water containing specimens of *C. everetti* were totally unexpected in view of its known distribution in the field. The species grew profusely in water with a bound CO_2 content (25 ppm) five times that of any water from which is had been collected, in water with a silicon content (20 ppm) over twenty times the species optimum, and a conductivity (150 micromhos/cm) twelve times that of any Wisconsin habitat known to contain the species. Apparently, *C. everetti* has inherent capabilities of growth beyond those indicated by its restricted distribution in the field.

Summarizing the data presented by Jewell (1935, 1939), *C. everetti* is reported from waters with the range of conditions tabulated below. Unless noted, all data are given in parts per million.

CO_2	
Bound	0.5–4.5
Free	0.5–9.0
Color	0–65
Conductivity	7.5–19 micromhos/cm
pH	5–6.6
Residue	11.0–29.0
SiO_2	0.0–0.8

7. Dosilia palmeri (Potts, 1885)

Habitat data and the peculiar mode of existence (Potts, 1887) of this species (see above, Section III,B,4), suggest that it is an ephemeral form, adapted to withstand arid environments.

8. Dosilia radiospiculata (Mills, 1888)

Dosilia radiospiculata is apparently limited to alkaline waters of high bicarbonate alkalinity and conductivity (Poirrier, 1969).

In the Dallas, Texas area, *D. radiospiculata* is a perennial form, although growing best in the colder months. Gemmules are produced throughout the year with maximum production in late fall and early spring. Gemmulation is apparently stimulated by increasing temperatures following several days of cold due to "northers" (Cheatum and Harris, 1953).

Summarizing the data presented by Cheatum and Harris (1953) and Poirrier (1969), *D. radiospiculata* is reported from waters with the ranges of conditions tabulated below. Unless noted, all data are given in parts per million.

Alkalinity (methyl orange)	62.0–135.0
CO_2 (free)	3.0–9.0
Conductivity	118–293 micromhos/cm
pH	7.3–7.9
SiO_2	5–24.9
Temperature	3°–31°C
Turbidity	20.0–150.0

9. Ephydatia fluviatilis (Linneaus, 1758)

This cosmopolitan species (Penney and Racek, 1968), although generally displaying a wide tolerance of environmental conditions, may be limited to waters of high conductivity as in Louisiana waters high in carbonates or sodium chloride (Poirrier, 1969). This species is often found in brackish water (see above, Section III,B,6). *Ephydatia fluviatilis* may be found in both lentic and lotic habitats, in situations varying from swamp pools (Eshleman, 1950) to "boiling, rushing water" (Potts, 1887). It may colonize either upper or lower sides of a substrate but prefers partially shaded locations (Old, 1932b).

Ephydatia fluviatilis has an annual life cycle (at least in the southeastern United States) with gemmulation in late spring followed by degeneration and death of the colony (Poirrier, 1969). Water temperatures approaching 30°C initiate gemmulation. With gemmule germination in August throughout early autumn, active growth commences. Larval development and release occurred from November through May in colonies maintained in the laboratory (Poirrier, 1969).

In its ability to withstand periods of high siltation, i.e., Secchi disk 0.5 feet (Poirrier, 1969), it is similar to congeners of other regions of the world (Racek, 1969). It is of particular interest that Old (1932a), investigating the spongillids of Michigan, most frequently found *E. fluviatilis* in polluted waters.

Summarizing the data presented by Old (1932a), Moore (1953), Wurtz (1950), and Poirrier (1969), *E. fluviatilis** is reported from waters with the ranges of conditions tabulated below. Unless otherwise noted, all data are given in parts per million.

Alkalinity (methyl orange)	19.0–230.0
Chlorides	0.016–2.6 ppt
CO_2 (free)	6.5–9.0
Color	40–120
Conductivity	350–2, 950 micromhos/cm
Dissolved oxygen	1.0–4.3
Hardness	
Temporary	0.0–30.0
Total	140.0–180.0
pH	5.9–8.3
Temperature	15°–33°C

10. Ephydatia millsii (Potts, 1887)

This rather insufficiently known species is reported as being fairly common in cypress, pine, and gum swamps of Florida (Eshleman, 1950).

*See Poirrier (1969) for a discussion of possible synonomy with *E. subdivisa* and *E. robusta*.

11. Ephydatia muelleri (Lieberkühn, 1885)

Ephydatia muelleri is most frequently found in smoothly flowing streams (Old, 1932a,b; Jewell, 1935; Neidhoefer, 1940) of relatively high mineral and organic content. It is limited by both upper and lower levels of calcium, is restricted to high silica concentrations, and to higher conductivities (Jewell, 1935). Imlay (personal communication) found dissolved oxygen concentrations of 2.5 ± 0.5 ppm were required for normal growth over a two month period in the laboratory. Colonies exhibited a slower growth rate in waters containing less than 2.5 ppm oxygen. Its growth is favored by partly shaded habitats and, as such, water transparency affects the relative positions of colonies. In waters of low transparency (Secchi disk 0.75–1.1 m), *E. muelleri* is abundant growing at the surface or at a depth of a few centimeters in unshaded habitats. In more transparent waters (Secchi disk 2.3–7.3 m) the species may be collected near the surface in partly shaded habitats or at depths of 1–1.5 m in less shaded areas (Jewell, 1935). Simon (personal communication) noticed that *E. muelleri* occurs more frequently in polluted shore waters or in relatively more eutrophic waters of Lake Constance, Switzerland.

Summarizing the data presented by Old (1932a), Jewell (1935, 1939), and Wurtz (1950), *E. muelleri* is reported from waters with the range of conditions tabulated below. Unless otherwise noted, all data are given in parts per million.

CO_2	
Bound	6.5–28.0
Free	0.0–13.95
Color	6–118
Conductivity	31–100 micromhos/cm
Hardness	
Temporary	0.0–40.0
Total	60.0–160.0
pH	6.1–8.5
Residue	42–75
SiO_2	0.7–11.6
Temperature	16°–24°C

12. Ephydatia robusta (Potts, 1887)

Ephydatia robusta, an insufficiently known species, has been collected in still water growing on submerged logs or twigs or on the underwater stems of aquatic plants. It has an annual life cycle, being limited by rising water temperatures coinciding with falling water levels (Moore, 1953 and personal communication, see above Section III,B,3). The species was present during periods of high, but decreasing, organic content, color, and salinity. Dissolved oxygen concentrations as low as 1.00 ppm apparently are not limiting to this species. Moore (1953) suggested that zoochlorellae might create a microenvironment but also observed that the color of the sponge was gray or brown.

*Ephydatia robusta** is reported (Moore, 1953) from waters with the range of conditions tabulated below. Unless otherwise noted, all data are given in parts per million.

Alkalinity (methyl orange)	43.0–118.0
Chlorides	0.515 ppt
CO_2 (free)	16.0–26.0
Color	50–120
Dissolved oxygen	1.00–4.30
pH	5.9–6.8
SiO_2	8.0–17.0
Temperature	15°–30°C
Total dissolved solids	180.0–900.0
Volatile solids	102.0–408.0

13. Ephydatia subtilis (Weltner, 1895)

This virtually unknown species has been reported once from the type locality, Lake Kissimmee, Florida, by Weltner (1895). It is extremely desirable that further collecting be done in the area to verify the continued presence of this species.

14. Eunapius fragilis (Leidy, 1851)

This truly cosmopolitan species is found in all continents and climates, in caves, and at high altitudes (Penney and Racek, 1968). It exhibits, as would be expected, a rather wide range of tolerance to most environmental factors. It is virtually insensitive to hydrogen ion concentration (Jewell, 1935; Racek, 1969), tolerates a wide range of water flow rates, and is little affected by siltation. It grows equally well in rapidly flowing streams, currentless lakes, or even in nearly stagnant pools (Potts, 1887; Old, 1932a; Jewell, 1935; Hoff,

*See Poirrier (1969) for a discussion of possible synonomy with *E. fluviatilis*.

1943). Its preference for growing on the underside of objects (Potts, 1887; Old, 1932b; Jewell, 1935) no doubt is of value in surviving high levels of siltation inimical to most sponges. As Potts (1887) noted, natural selection favors those sponges growing on the lower surfaces of a support, thus protected from the intrusion of particulate material into canal systems. *Eunapius fragilis'* tolerance of high siltation levels is not entirely a matter of substrate preference. Cheatum and Harris (1953) found colonies growing on the upper surfaces of objects, submerged in mud. On washing, the sponges appeared as healthy as those from more usual habitats. Racek (1969), also, has noted that in Australia the species withstands considerable flooding accompanied by high siltation without perceptible structural change or damage.

The species is favored by waters of high mineral and organic content, is limited by low silica levels, and attains its finest development in waters of relatively high color (Jewell, 1935). Jewell (1935) described one habitat, Harvey Lake, as follows:

> "A small lake with relatively large inlet and outlet, dark water and bog margin. Several wagon loads of potatoes, recently dumped, float along the margin in various stages of decomposition. *S. fragilis* (syn. = *E. fragilis*) exceedingly abundant coating sticks, dead wood, and even stones. Twigs of marginal *Chamaedaphne* completely coated where submerged, and, even above the water level, encrusted with dry sponge and gemmules."

The species is favored by high conductivities, usually in waters of moderate alkalinity (Jewell, 1935) but is never found in brackish water (Poirrier, 1969). Its rate of growth decreases in waters containing less than 2.5 ppm dissolved oxygen (Imlay, personal communication). It will, however, survive in highly polluted waters. It has been collected from waters with a coliform count of 24,500 colonies/ml (Author Anonymous, 1961).

Eunapius fragilis is a perennial species, at least in habitats not subject to seasonal drying. Cheatum and Harris (1953) observed gemmule production throughout the year with maximum production in late fall and early spring. In seasonal habitats (Poirrier, 1969) it produced gemmules in late spring with autumnal germination and winter maximum growth. In streams, however, gemmules germinated in the spring with maximum growth throughout the summer and gemmulation in the fall.

Summarizing the data presented by Old (1932a), Jewell (1935, 1939), Wurtz (1950), Cheatum and Harris (1953), Penney (1954), Author Anonymous (1961), Patrick *et al.* (1966), and Poirrier (1969), *E. fragilis* is reported from waters with the range of conditions tabulated below. Unless otherwise noted, all data are given in parts per million.

Alkalinity	
Methyl orange	14.7–230.0
Phenolphthalein	0.00
Aluminum	0.06–0.27
Bacteriology	
Total count[a]	285–2,956
Coliform count[b]	200–24,500
BOD	0.26–2.17
Calcium	1.6–45.6
Total hardness (as $CaCO_3$)	8.0–200.0
Calcium hardness (as $CaCO_3$)	0.0–30.0
Chlorides	0.6–7.0
Chromium	0.004–0.057
Color	0–202
Conductivity	16–760 micromhos/cm
Copper	0.09–0.45
Dissolved oxygen	2.5–10.74
Free CO_2	0.0–22.0
Iron(Fe^{3+})	$<$ 0.002–0.6
Magnesium	1.05–3.0
Magnesium hardness (as $MgCO_3$)	4.3–10.5
Nickel	0.002–0.005
Ammonia nitrogen	$<$ 0.001–0.169
Nitrate nitrogen	0.001–0.385
Nitrite nitrogen	0.001–0.232
pH	4.2–9.2
Phosphate	$>$ 0.01–0.1
SiO_2	0.3–24.9
Sulfate	1.0–10.0
Total solids	86–164
Volatile solids	38–96
Fixed residue	48–68
Temperature	11°–34°C
Turbidity	$>$ 18–500

[a] Colonies per milliliter, 20 hours incubation at 35°C, membrane filter method (Author Anonymous, 1961).

[b] Colonies per milliliter, 20 hours incubation at 35°C, membrane filter method (Author Anonymous, 1961).

15. *Eunapius mackayi* (Carter, 1885)

Eunapius mackayi is restricted to soft acid waters in shaded locations (Jewell, 1935, 1939; Wurtz, 1950; Moore, 1953; Poirrier, 1969). It is favored by standing or sullenly creeping waters of high color and organic content (Potts, 1887; Jewell, 1935). It is apparently limited by strong light (Moore, 1953). This limitation is borne out by its habit of colonizing the undersides of supports in transparent waters, even in partially shaded locations (Old,

1932b; Jewell, 1935). Water transparency affects its relative position. In less transparent lakes (Secchi disk 0.8–1.0 m) it may also be found on upper surfaces of submerged logs, for example, even a few centimeters below the surface (Jewell, 1935).

Summarizing the data presented by Jewell (1935, 1939), Wurtz (1950), Moore (1953) and Poirrier (1969), *E. mackayi** is reported from waters with the range of conditions tabulated below. Unless otherwise noted, all data are given in parts per million.

Alkalinity (methyl orange)	5.0–9.9
Calcium	0.0–3.16
CO_2	
Bound	0.5–4.5
Free	1.0–26.0
Color	6–320
Conductivity	7–20 micromhos/cm
pH	5.0–6.2
Residue	12.5–71.0
SiO_2	0.0–4.4
Temperature	12°–23°C

16. Heteromeyenia baileyi (Bowerbank, 1863)

This species, exhibiting a widely scattered distribution throughout eastern North America (Penney and Racek, 1968), tolerates a wide range of environmental conditions. It is a quiet water form, not occurring in rapids (Old, 1932a). In temporary habitats, gemmulation occurs in late spring (Poirrier, 1969). It is adversely affected by the combination of rising temperatures and falling water levels (W. G. Moore, personal communication). In permanent habitats it may be found throughout the year. Its toleration of extremely low dissolved oxygen levels (0.80 ppm) and high free CO_2 levels (26.5 ppm) is very likely related to the presence of zoochlorellae in this bright green (Moore, 1953; Penney and Racek, 1968) species. Old (1932a) noted that the optimum habitat of *H. baileyi* is pollution free. It has been collected once in slightly brackish water, conductivity 3000 micromhos/cm chloride 1.1 ppt (Poirrier, 1969).

Summarizing the data presented by Old (1932a), Jewell (1935, 1939), Wurtz (1950), Moore (1953), Penney (1954), and Poirrier (1969), *H. baileyi* is reported from waters with the range of conditions tabulated below. Unless otherwise noted, all data are given in parts per million.

**Eunapius mackayi* = *E. igloviformis*, see Poirrier (1969).

Alkalinity (methyl orange)	18.0–150.0
Aluminum	0.06–0.27
Calcium	1.5–72.2
Chlorides	0.001–1.1 ppt
Chromium	0.004–0.057
CO_2 (free)	2.1–26.5
Color	20–100
Conductivity	14.5–3000.0 micromhos/cm
Copper	0.09–0.45
Hardness	
Temporary	0.0–10.0
Total	40.0–216.0
Iron	0.004–0.033
Nickel	0.002–0.005
pH	4.2–8.4
SiO_2	0.25–13.0
Total solids	239.0–360.0
Volatile solids	109.0–140.0
Sulfate	1.5–3.7
Temperature	21°–31°C

17. Heteromeyenia latitenta (Potts, 1881)

This species, distributed in the northeastern United States, has been collected on stones in rapidly running water (Potts, 1887). It is recorded as being green in life (Penney and Racek, 1968).

18. Heteromeyenia longistylis (Mills, 1884)

This species, type locality, Lehigh Valley, Pennsylvania (Penney, 1960), is insufficiently known (Penney and Racek, 1968).

19. Heteromeyenia tentasperma (Potts, 1880)

Heteromeyenia tentasperma apparently prefers alkaline waters rich in silicates and carbonates (Neidhoefer, 1940). Potts (1887) collected it from a "softly murmuring" stream. Neidhoefer (1940) collected it from a lake. It is pale green (Neidhoefer, 1940), and may colonize upper or lower surfaces of a support (Potts, 1887; Neidhoefer, 1940).

Heteromeyenia tentasperma is reported (Wurtz, 1950) from waters of the conditions tabulated below. Unless otherwise noted, all data is given in parts per million.

Calcium	14.6
CO_2 (bound)	20.66
Conductivity	114.0 micromhos/cm
Hardness	47.0
pH	7.3
SiO_2	12.05

20. *Heteromeyenia tubisperma* (Potts, 1881)

This species prefers smooth to rapidly flowing streams (Old, 1932a,b; Jewell, 1935; Neidhoefer, 1940). It is relatively light-positive, usually colonizing upper surfaces (Potts, 1887; Old 1932b) in alkaline waters of high silicate and carbonate content (Old, 1932a; Neidhoefer, 1940). It is indifferent to pollution (Old, 1932a).

Summarizing the data presented by Old (1932a) and Wurtz (1950), *H. tubisperma* has been collected from waters with the range of conditions tabulated below. Unless otherwise noted, all data are given in parts per million.

Alkalinity (methyl orange)	110.0–170.0
Calcium	8.0–15.0
CO_2 (free)	0.0–12.0
Color	40–80
Hardness	
Temporary	0.0–20.0
Total	80.0–200.0
pH	6.6–8.5
Temperature	13.5°–26.5°C
Turbidity	$<$ 20.0

21. *Radiospongilla cerebellata* (Bowerbank, 1863)

This species, known from the tropical and subtropical regions of Africa, the Indo-Pakistani subcontinent, Indonesia, the Philippines, and New Guinea, through China to the U.S.S.R. (Penney and Racek, 1968), has been collected only from Texas in the Western Hemisphere (Poirrier, 1972).

22. *Radiospongilla cinerea* (Carter, 1849)

The North American record of this species is from Ohio (Landacre, 1902), reported only as "ashen gray in color . . . found on floating timber." In view of the known distribution of the species, i.e., the vicinity of Bombay and the Himalayas (Penney and Racek, 1968), the North American record is likely based on a misidentification.

23. *Radiospongilla crateriformis* (Potts, 1882)

Radiospongilla crateriformis, a species preferring alkaline waters high in bicarbonates and conductivity (Wurtz, 1950; Poirrier, 1969), is unusual in its ability to tolerate very stagnant or very turbid waters (Smith, 1921; Eshleman, 1950; Cheatum and Harris, 1953; Poirrier, 1969). Its tolerance of siltation is related to some extent to its preference for colonizing undersurfaces of supports (Hoff, 1943; Cheatum and Harris, 1953). However, (see *E.*

fragilis, Section IV,14) in view of collections of healthy colonies buried in mud (Cheatum and Harris, 1953), other factors must also be involved.

Radiospongilla crateriformis is seasonal, at least in some areas, with gemmulation usually occurring in fall and active colonies present May through September (Poirrier, 1969). Considering the widely discontinuous distribution of this species (Penney and Racek, 1968), life histories may vary in distant populations.

24. Spongilla aspinosa (Potts, 1880)

Spongilla aspinosa prefers clear standing waters (Potts, 1887; Eshleman, 1950). It apparently is favored by acidic conditions, having been reported from cypress swamps (Eshleman, 1950) and an acid lake of pH 5.0 (Jewell and Brown, 1929).

The sponge, bright green due to zoochlorellae (Penney and Racek, 1968), is a perennial form in some localities. Potts (1887) found it thriving in the heat of summer and in February under a thick sheet of ice. He considered the scarcity of gemmules throughout the year to be related to its ability to withstand wide temperature ranges.

25. Spongilla cenota (Penney and Racek, 1968)

This species is apparently restricted in distribution to the vicinity of the type locality, the cenotes of Yucatan (Penney and Racek, 1968). The living sponge varies in color from green to light gray (Old, 1936). The cenotes have no currents and have water of relatively high carbonate content (Hall, 1936).

Spongilla cenota is reported (Old, 1936; Penney and Racek, 1968) from waters of the conditions tabulated below (Hall, 1936).

$CaCO_3$	220–300 ppm
Dissolved oxygen	3.26–4.68 ppm
NaCl	0.09–0.21 ppt
pH	7.0–8.6
Temperature	24.7°–27.1°C
Transparency	3.0–28.0 m

26. Spongilla heteroschlerifa (Smith, 1918)

This species has been reported from the type locality, Oneida Lake, New York (Penney, 1960).

27. Spongilla johanseni (Smith, 1930)

This species has been reported from the type locality, a tundra lake on a bog four miles west of Shippigan, New Brunswick, Canada (Penney, 1960).

28. *Spongilla lacustris* (Linnaeus, 1758)

Spongilla lacustris, a moderately light-positive form (Potts, 1887; Old, 1932a), tolerates a wide range of environmental conditions. It is relatively independent of the calcium content of the habitat (Jewell, 1935, 1939). It exists throughout a wide pH range with a slight preference for alkaline waters (Old, 1932a). The species may be common, however, in acid areas low in bicarbonate (Poirrier, 1969). Its optimum habitat is relatively high in color,

Alkalinity	
Methyl orange	90.0–170.0
Phenolphthalein	0.00
Bacteriology	
Total count[a]	285
Coliform count[b]	300
BOD	2.17
Calcium	0.16–178.0[c]
Total hardness (as $CaCO_3$)	28.80–250.0[c]
Calcium hardness (as $CaCO_3$)	0.0–24.50
Chlorides	5.3–9.8[c]
Color	0–202
Conductivity	9.4–470.0 micromhos/cm[c]
Dissolved oxygen	2.4–15.2[c]
Free CO_2	0.0–13.95
Bound CO_2	0.85–74.8
Iron	< 0.002–1.2[c]
Magnesium	1.05–9.4[c]
Magnesium hardness (as $MgCO_3$)	4.30
Manganese	0.1–1.2[c]
Ammonia nitrogen	< 0.001
Nitrate nitrogen	0.192–9.9[c]
Nitrite nitrogen	0.001–0.05[c]
pH	5.3–9.0
Phosphate	0.028–4.34[c]
Potassium	1.3–3.7[c]
SiO_2	0.0–20.5
Sulfate	5.06–37.0
Total solids	20–164
Volatile solids	96
Fixed residue	68
Temperature	12°–37°C
Transparency (Secchi disk)	0.9–7.3 m
Turbidity	2.0–200.0[c]

[a] Colonies per milliliter, 20 hours incubation at 35°C, membrane filter method (Author Anonymous, 1961).

[b] Colonies per milliliter, 20 hours incubation at 35°C, membrane filter method (Author Anonymous, 1961).

[c] Unpublished data, Water Resources Laboratory, University of Louisville.

silicon, and conductivity (Jewell, 1935). *Spongilla lacustris* has been collected from brackish water areas (Tendal, 1967).

Silicon levels are important, not only in determining the level of skeletal development, but also in the attainment of sponge consistency. In higher concentrations specimens of *S. lacustris* are rather brittle while at low levels (0.3 ppm) the sponge is soft, slimy, collapsing when removed from the water (Jewell, 1935).

Although generally considering skeletal development to be reduced in specimens of any species in waters of high altitudes, Potts (1887) observed that one of the most robust forms of *S. lacustris* was collected at 7000 feet in the Sierra Nevada mountains. It has been collected from higher altitudes than any other North American sponge, 10,800 feet in West Forest Lake near Tolland, Colorado (Smith, 1921).

Spongilla lacustris may have a seasonal cycle with autumnal gemmulation or may be a perennial form (Poirrier, 1969).

Summarizing the data presented by Old (1932a), Jewell (1935, 1939), Wurtz (1950), Penney (1954), Author Anonymous (1961), Poirrier (1969), and Water Resources Laboratory, University of Louisville (unpublished), *S. lacustris* is reported from waters with the range of conditions tabulated below. Unless otherwise noted, all data are given in parts per million.

29. Spongilla sponginosa (Penney, 1957)

The species is known only from the type locality, a cypress pond (Weeks Pond, Manchester State Forest, Sumter County, South Carolina). It was collected on sticks and twigs at the bottom of two feet of water during low water, normal depth 6–7 feet (Penney, 1957).

30. Spongilla wagneri (Potts, 1889)

The optimal habitat of *S. wagneri* is slightly to strongly brackish water (Eshleman, 1950; Penney and Racek, 1968). Potts (1889) observed many specimens coated with barnacles and calcareous tubes of *Serpulae* sp. Gemmules were hidden within barnacles or among *Serpulae* sp. tubes.

Sponges in the type locality were adhering to barnacles on the bottom of a rapidly flowing creek 4 or 5 feet deep (Potts, 1889).

31. Trochospongilla horrida (Weltner, 1893)

Trochospongilla horrida is favored by habitats exhibiting high levels of color, pH, alkalinity, and mineral content (Wurtz, 1950; Cheatum and Harris, 1953; Poirrier, 1969). It will tolerate slightly brackish water (Poirrier, 1969) and thrives in exceptionally high turbidities, 5300 ppm (Cheatum and Harris, 1953). As the species normally grows on lower surfaces (Cheatum and Harris, 1953), it is somewhat sheltered from the effects of siltation. However, like *E. fragilis* and *R. crateriformis*, it has been buried in mud and remained healthy (see above, Sections IV,14 and 23).

Trochospongilla horrida is a perennial species, at least in some areas, producing gemmules throughout the year with maximum gemmulation in late fall and early spring (Cheatum and Harris, 1953). It will withstand waters polluted by organic wastes, having been collected from habitats with a coliform count of 7750 colonies/ml (Author Anonymous, 1961).

Summarizing the data presented by Wurtz (1950), Cheatum and Harris (1953), Penney (1954), Author Anonymous (1961), Patrick *et al.* (1966), and Poirrier (1969), *T. horrida** is reported from waters with the range of conditions tabulated below. Unless otherwise noted, all data are given in parts per million.

Alkalinity	
Methyl orange	7.0–185.0
Phenolphthalein	0.00
Bacteriology	
Total count[a]	1965–2820
Coliform count[b]	5600–7750
BOD	0.5–1.0
Calcium	4.1–45.6
Total hardness (as $CaCO_3$)	10.0–156.0
Calcium hardness (as $CaCO_3$)	10.9–11.0
Chlorides	4.0– < 8.0
Conductivity	0.068–322.0 micromhos/cm
Dissolved oxygen	6.2–8.0
Free CO_2	0.0–15.0
$Iron^{3+}$	0.02–0.3
Magnesium	1.31–3.0
Magnesium hardness (as $MgCO_3$)	5.4–7.1
Ammonia nitrogen	0.005–0.047
Nitrate nitrogen	0.07–0.2
Nitrite nitrogen	0.001– < 0.007
pH	5.5–8.7
Phosphate	0.024–0.1
SiO_2	5.0–24.9
Sulfate	2.28–11.02
Total solids	78.0–190.0
Volatile solids	20.0–60.0
Fixed residue	58.0–130.0
Temperature	18.5°–34°C
Turbidity	20.0–5300.0

[a] Colonies per milliliter, 20 hours incubation at 35°C, membrane filter method (Author Anonymous, 1961).

[b] Colonies per milliliter, 20 hours incubation at 35°C, membrane filter method (Author Anonymous, 1961).

*See Poirrier (1969) for a discussion of possible synonomy with *T. pennsylvanica*.

32. Trochospongilla leidyi (Bowerbank, 1863)

Trochospongilla leidyi is most often found in alkaline waters high in bicarbonate alkalinity, relatively high in conductivity due to sodium chloride, and, often, very turbid due to siltation (Poirrier, 1969). It has been found growing in the interior of iron water pipes in total darkness (Potts, 1887).

The species, at least in the southeastern United States, is seasonal. Gemmules, produced in the fall, germinate in the spring to form colonies that thrive throughout the summer (Poirrier, 1969).

Summarizing the data presented by Patrick *et al.* (1966) and Poirrier (1969), *T. leidyi* has been collected from waters with the range of conditions tabulated below. Unless otherwise noted all data are given in parts per million.

Alkalinity	
Methyl orange	14.7–185.0
Phenolphthalein	15.0
BOD	> 0.1–0.5
Calcium	3.0– < 10.0
Chlorides	0.575–1.1 ppt
CO_2 (free)	0.0–18.0
Conductivity	80–3000 micromhos/cm
Dissolved oxygen	9.0– < 11.0
Hardness	10.0– < 50.0
$Iron^{3+}$	< 0.01–0.01
Magnesium	< 3.0–3.0
Ammonia nitrogen	0.009–0.03
Nitrate nitrogen	0.2–0.7
Nitrite nitrogen	< 0.001–0.001
pH	6.5–8.7
Phosphate	> 0.01–0.05
SiO_2	> 6.0–12.0
Sulfate	> 1.0–10.0
Temperature	23°–34°C
Turbidity	> 10–25

33. Trochospongilla pennsylvanica (Potts, 1882)

This species is found in standing or relatively quiet waters high in color, organic content, and carbon dioxide. It is limited by waters high in alkalinity, conductivity, and pH (Old, 1932a,b; Jewell, 1935, 1939; Poirrier, 1969). In waters with calcium above 6.0 ppm it is usually found in areas of rapid organic decomposition, a condition likely reducing the calcium bicarbonate content of the water in the immediate vicinity of the sponge (Jewell, 1939). *Trochospongilla pennsylvanica* is favored by shaded locations or areas of low light intensity. In waters of low transparency, or in shaded locations, it may be found on the upper surface of a support. In more transparent waters it is always found on lower surfaces unless the habitat is deeply shaded.

It is indifferent to silica content except in the degree of skeletal development (Jewell, 1935). Potts (1887) observed variations in skeletal form in specimens of *T. pennsylvanica* from different altitudes. Sponges from higher altitudes exhibited poorly developed rotules, etc., while those near sea level were robust. He considered that waters from higher altitudes would be deficient in silica while violent contact with rocks and stream beds would increase silica levels.

Trochospongilla pennsylvanica tolerates both silting (Jewell, 1935; Poirrier, 1969) and seasonal drying of habitats. Specimens in Pinewood Lake, Columbia, South Carolina, gemmulate in late summer and early fall. This has enabled colonies to survive on logs at the lake margin at least since 1960 although the lake has been drained completely several times. It apparently has a similar seasonal cycle with maximum growth in summer, gemmulation in late summer or early fall, tissue degeneration in winter, and gemmule germination in late spring in other areas of the southeastern United States (Poirrier, 1969).

Summarizing the data presented by Old (1932a), Jewell (1935, 1939), Wurtz (1950), Moore (1953), Penney (1954), and Poirrier (1969), *T. pennsylvanica** is reported from waters with the range of conditions tabulated below. Unless otherwise noted, all data are given in parts per million.

Alkalinity	
Methyl orange	9.9–85.0
Phenolphthalein	0.0
Aluminum	0.06–0.27
Calcium	0.3–24.0
Chlorides	1.0–1.9
Chromium	0.004–0.057
CO_2	
Bound	0.5–21.0
Free	0.5–26.0
Color	160–180
Conductivity	7–163 micromhos/cm
Copper	0.09–0.45
Temporary hardness	1.0–10.0
Total hardness	40.0–80.0
Iron	0.004–0.033
Nickel	0.002–0.005
Residue	11.0–78.0
SiO_2	0.0–10.75
Sulfate	1.5–3.7
Temperature	11.5°–31°C

*See Poirrier (1969) for a discussion of possible synonomy with *T. horrida*.

V. Disturbed Ecology

Considering that this chapter is the initial treatment of the normal ecology of all the North American freshwater sponges, it is to be expected that their disturbed ecology, i.e., pollution ecology, is little understood. Most of our present knowledge concerns marine sponges. However, in that basic physiological processes are similar in marine and freshwater Porifera, large amounts of this informations are directly applicable to a treatment of the disturbed ecology of freshwater sponges. It must also be understood that, as opposed to the normal ecology of spongillids, considerable of the forthcoming dicussion is subjective in nature, acquired through correspondence with the exceedingly cooperative spongologists of the world.

A. Siltation

Potts (1887) noted that mud is the great enemy of sponges. Agreeing, Moore (1953) observed that the presence of abundant suspended material in the water limits or prevents the development of sponge fauna, except in those areas where water movement is slight. Limitation due to siltation is present throughout the phylum. In studies of benthic invertebrates of Fanning Island, central Pacific, Bakus (1968) found many sponges to be adversely affected by sediment deposition, primarily by burying and by clogging of canals and chambers. Additionally, V. E. Smith (personal communication), while participating in a benthic survey off California Edison's nuclear power plant at San Onofre, observed that the high turbidity created during offshore installation of the outfall line completely eliminated the sessile invertebrate community, including sponges. Within two years it was reestablished and seemed not to be influenced by the slightly warmer water (ca. 1°C).

R. E. Johannes (personal communication) suggested that (possibly) sedimentation in Kaneoh Bay, Hawaii may be favorable to presently unidentified sponges while killing other species of marine organisms. In view of de Laubenfel's (1950) report that the Hawaiian sponge *Laxosuberites* endures muddy or silt-filled waters with an endurance almost unique in the phylum Porifera, Johannes' observation seems very logical. It is possible that the normal sponge fauna was reduced or eliminated due to siltation and was replaced by silt-tolerant forms through larval invasion. The Kaneohe Bay situation appears to be rather significant and certainly warrants further investigation.

Although most sponge species are adversely affected by siltation, others have adapted to survive in highly silted waters. Potts (1887) observed that, in any body of water often charged with sedimentary material, natural selection favors those species growing on undersurfaces of supports and, thus, being protected.

Members of the genus *Ephydatia* are particularly tolerant of high levels of siltation. This, in part, can be related to an apparent generic tendency to prefer areas of low light intensity. *Ephydatia fluviatilis* may be found on either upper or lower surfaces of supports (Old, 1932b). *Ephydatia muelleri* attains its best development in waters high in color (Jewell, 1935). In waters of low transparency (Secchi disk 0.75–1.1 m) it grows at the water surface or at depths of a few centimeters in unshaded locations. In more transparent waters (Secchi disk 2.3–7.3 m) it is found near the surface in partly shaded locations or at depths of 1.0–1.5 m in less shaded areas (Jewell, 1935).

Eunapius fragilis, a species able to grow well on the underside of a substrate (Jewell, 1935), is frequently found in highly silted waters (Poirrier, 1969). Other factors than attachment preference are involved here, however, as Cheatum and Harris (1953) found healthy colonies growing completely buried in mud.

Radiospongilla crateriformis was collected in profusion by Cheatum and Harris (1953) in waters of 5300 ppm turbidity while Smith (1921) collected it in a stream "too muddy to permit development of an extensive sponge fauna." It usually grows on lower surfaces (Hoff, 1943; Cheatum and Harris, 1953) but was collected by Cheatum and Harris in a healthy condition buried in mud (see above). Eshleman (1950) considered it the only Florida species likely to be found in very stagnant or turbid water.

Members of the genus *Trochospongilla*, all normally light-shunning forms growing on lower surfaces of substraces, are tolerant of high silt levels. *Trochospongilla horrida* was found in waters of 5300 ppm turbidity by Cheatum and Harris (1953). *Trochospongilla leidyi* was often collected from very turbid waters (Poirrier, 1969). *Trochospongilla pennsylvanica*, a species favored by high color and organic content and tolerant of residue (Jewell, 1935), was abundant in Louisiana sloughs and ditches subject to silting (Poirrier, 1969).

In Australian habitats, all subject to severe flooding and concurrent enormous silt loads, most spongillid species have apparently fully adapted to prolonged exposure to muddy conditions (Racek, 1969). All species of *Eunapius*, *Ephydatia ramsayi*, and the *Heterorotula* complex successfully survive the effects of flood conditions in Australia without any perceptible structural changes or damages.

In addition to light tolerance and substrate selection, more subtle factors may operate in the siltation ecology of sponge species. Species with negatively phototactic larvae, attaching in cryptic locations, are often found colonizing lower surfaces of substrates (Wilson, 1935; Ali, 1956; Warburton, 1966;

Simpson, 1968; Bergquist *et al.* (1970). These habitats, in addition to being deeply shaded (Section III,B,2), also afford protection from sedimentation (Bakus, 1964; Simpson, 1968).

The architecture of the sponge itself may determine habitat selection in regard to siltation (Reiswig, 1971b). In more loosely structured sponges, a primary particle capture system involving amoebocyte cycling of particles greater than 2–5 μm operates efficiently. This system, involving amoebocyte capture of particles at the incurrent canal lining and migration to the exhalent canals for particle expulsion, has previously been suggested by van Tright (1919), van Weel (1949), and Kilian (1952). Harrison (1972) has, in addition, observed particle capture by cells of the basal pinacoderm of *Corvomeyenia carolinensis.* In view of the high levels of phagosomal acid phosphatase present in these cells (Harrison, 1972), their activities apparently supplement the amoebocyte capture system in removing materials from inhalent areas. The amoebocyte cycling system may also act in conjunction with canal reorganization in young sponges with few amoebocytes or in species with irregular or erratic pumping activities (Reiswig, 1971a).

The more densely organized sponges, such as *Verongia gigantea* (Hyatt, 1875), have an easily saturated amoebocyte cycle, less able to cope with sediment loads. The species is restricted to clean water habitats of the outer reef (Reiswig, 1971b). Loosely architectured forms, with more efficient amoebocyte cycling systems, are able to tolerate wide ranges of siltation and invade habitats that are inimical to more dense sponges (Reiswig, 1971b).

Harrison (1971, 1972) observed a complete population of amoebocytes migrating upon the outer surface of the upper pinacoderm of *C. carolinensis.* These amoebocytes, through phagocytic activities, no doubt cleanse the surface of the sponge and affect its tolerance to siltation.

Storr (personal communication) observed the interesting reaction of *Spheciospongia vesparia* to high siltation. Following placement of a handful of soft bottom material on the top edge of this sponge, a backwash along the sides of the sponge jetted material out of the side pores "much like a puff of smoke by a pipe smoker." This action apparently resulted from strong internal contractions at various points in the sponge.

The above observations, in addition to their relationship to habitat selection, are important in terms of the presence or absence of a nervous system in sponges. In view of reports of intrinsically generated patterns of pumping activities (Reiswig, 1971a), of histochemical localization of neurohumoral substances (Lentz, 1966), and of myofilament like thick and thin filaments in myocytes (Bagby, 1966), the possibility of a primitive nervous system (Jones, 1962) needs to be reinvestigated.

B. Organic Enrichment

It is generally considered that, due to their intimate relationship with the aquatic habitat, sponges are totally eliminated from polluted waters. This is not necessarily the case. The distribution of sponges in polluted waters depends upon the interacting factors of type and quantity of pollutant and individual species tolerances.

Potts (1887) found sponges growing occasionally in water "unfit for domestic uses." Annandale (1911) reported that Indian sponges apparently prefer waters polluted by human agencies and that some species may be entirely restricted to these domestically unfit habitats.

Old (1932a,b) repeatedly found sponges growing in Michigan waters polluted by organic and industrial wastes. In many cases he found the original fauna depleted while sponges, particularly *Ephydatia fluviatilis* and to a lesser extent *Heteromeyenia tubisperma*, grew normally. Old (1932a) considered polluted waters to form the optimal habitat for *E. fluviatilis*, while *H. tubisperma* was indifferent to the presence or absence of pollution.

The ability of *E. robusta* and *H. repens* to survive very low dissolved oxygen levels (Moore, 1953) suggests a capability of surviving, at least for a time, oxygen depletion due to eutrophication. *Ephydatia robusta* has been collected at dissolved oxygen tensions of 1.00 ppm while *H. repens* was found in waters with even lower levels (0.80 ppm).

Organic enrichment may actually favor certain species of sponges. Simon (personal communication) found *E. muelleri* occurred most often in polluted shore waters or in relatively more eutrophic waters of Lake Constance, Switzerland. This is in keeping with Jewell's (1935) observation that the species attained its highest development in waters relatively high in color. *Eunapius fragilis* (refer to Section IV,14) was found thriving in waters containing several wagon loads of decomposing potatoes and growing in waters with a coliform count of 24,500 colonies/ml (Author Anonymous, 1961). The above are interesting in view of M. Imlay's (personal communication) observation that dissolved oxygen concentrations of 2.5 ± 0.5 ppm were required for normal growth of *E. muelleri* and *E. fragilis* in the laboratory over a two-month period. Among the sponges of Florida, *R. crateriformis* is unique in the frequency of its occurrence in very stagnant or otherwise "naturally" polluted waters (Eshleman, 1950 and personal communication). In Kaneohe Bay, Hawaii (Johannes, personal communication), there are areas in the vicinity of sewage outfalls where the corals are all dead but where sponges not generally obvious on normal reefs are found growing luxuriantly.

A very interesting series of experiments demonstrated the capability of sponges to effectively purify water through filtration activities. Claus *et al.*

(1967), Madri *et al.* (1971), and Kumen *et al.* (1971) demonstrated the removal of microbial pollutants from waste effluents by *Microciona prolifera*. These authors found the bacterium *Esherichia coli* and the fungus *Candida albicans* to disappear rapidly from tanks containing specimens of *M. prolifera*. The elimination of these organisms was considered due to ingestion by the sponge and not due to an antibiotic elaborated by *M. prolifera* as the antibiotic is not diffusible. The authors feel that, as *M. prolifera* concentrates large quantities of bacteria, the sponge can be used in estuaries to combat microbial pollution from fecal contamination.

As a corollary, Racek (personal communication) notes that since the introduction of fluoridation into Sydney water supplies the incidence of sponge growths in large water mains is rapidly dropping. This in turn has brought about an increasing number of bacteria, which otherwise would have served as normal food for the sponges. Racek continues that it would be wiser to use fluoride at the outlet rather than the inlet.

Vacelet and Sarà (personal communication from manuscript in preparation for Grasse's "Traite de Zoologie") consider that the water purifying effects of sponges are of little practical interest in view of the sensitivity of most sponges to highly polluted waters.

C. Toxic Wastes

Casual observations (S. K. Eshleman, personal communication) suggest that spongillids are more sensitive to chemical pollution than to organic enrichment. This is very likely true but there is little more than subjective data to support this conclusion. Vacelet and Sarà (personal communication, see above) observe that Demosponges appear to be rather sensitive to pollution resulting from human activities. In highly polluted waters species diversity is low although particular species may be abundant (*Hymeniacidon sanguinea* and *Halichondria panicea* on quays, etc., at Naples).

Old (1932a) reported (see above, Section V,B) sponges from waters receiving industrial wastes. He also (1932b) found many lavender specimens of *E. fragilis*, *S. lacustris*, and *E. muelleri* growing on copper-bearing rocks in highly colored waters in the Keweenaw Peninsula, Michigan. He noted that almost all members of the genus *Heteromeyenia* are eliminated by the presence of pollution.

A possible indicator of the presence of toxic wastes has been reported by Potts (1887) and Old (1932b). Potts (1887) observed that the megascleres of specimens of *T. leidyi* growing in iron water pipes exhibited malformation of the axial canal. The canal occupied one half the width of the megascleres and opened at both extremities. Birotulates had lost their "entire" margins and appeared delicately rayed. Potts attributed the malformation to a chemical condition in the environment as the tissues of the sponges were strongly marked by iron rust.

Old (1932b) found specimens of *E. fluviatilis* with identical megascleral malformations in waters polluted with industrial wastes. However, numerous sponges at this same location were luxuriant and showed no spicular abnormalities. Moreover, Racek (1969) found specimens of *H. multiformis* with megascleres possessing "unusually wide" axial canals. He considered it a normal phenomenon in hot, arid climates, caused by the bacterial reduction of gypsum.

The phenomenon certainly warrants further investigation as the presence of modified spicules in sponges or in bottom samples might provide permanent indications of pollution if, indeed, there is a relationship present.

D. Oil Pollution

Johnson (personal communication from unpublished study) subjected specimens of *Clathrina coriacea* Montagu to the following substances in aerated seawater: Santa Barbara crude oil floating on the water surface, a sinking agent, oil plus sinking agent, sand of the same particle size as the sinking agent, and oil plus sand.

Clathrina coriacea, composed of numerous anastomosing tubes, was used in the experiments because the tube diameter decreases markedly in a short time period following injury. Specimens showed maximum tube diameter decrease in jars with oil floating on the water surface. The sponge recovered in running seawater. Apparently, water soluble components of the surface oil were toxic to the sponge. The recovery was likely due to the short exposure time (6 hours). The combination of oil plus sinking agent, although failing to elicit an immediate contractile response from the sponge, was fatal. This was more deleterious than oil alone because the sponge was brought into direct contact with the smothering nonsoluble components of the oil. Treatment with sinking agent alone had no effect, presumably due to the large particle size (too large to enter inhalent canals) of the agent employed. Treatment with sand alone killed the sponges, presumably due to clogging of inhalent canals by microscopic sand particles. The effects of oil plus sand were similar to those of oil plus sinking agent except the use of sand as a sinking agent was more deleterious, probably due to the additional clogging action of microscopic sand particles.

E. Disease

Sponges may survive eutrophication, toxic wastes, and oil spills only to be eliminated by disease. The occurrence of epidemic disease among sponges is never investigated unless, of course, economic factors are involved. Such epidemics do occur and because of this the possibility of an epidemic should never be discounted when seeking to determine the cause of a die-off of sponges, marine or freshwater.

Spongiophaga communis, a fungus, is the only adequately known disease organism of sponges (Storr, 1964). It was responsible for the 1938–1939 epidemic which destroyed 90–95% of all commercial sponges in most areas of the Caribbean and the Gulf of Mexico. The disease, as described by Smith (1941), attacked the sponge interior with maximum numbers of organisms in the narrow zone between healthy and necrotic tissue. Hyphae, appearing in groups as a number of short colorless unbranched filaments between 0.001 and 0.002 mm in diameter, attached by one end to the sponge tissue. The disease progressed rapidly toward the sponge periphery until, with the piercing of the outer surface, the entire sponge rotted away. A heavily diseased sponge, on being hooked by a sponge fisherman, was described as "disappearing in a cloud of dust" (Storr, 1964).

de Laubenfels (1950) notes that new areas were affected in sequence as they were down-current from infected areas, suggesting the contagious nature of the disease. No one made a study to investigate the spread of the disease to other noncommercial types of sponges. By 1940 there was little further spread of the disease, primarily because it was limited by mainland to the west, deep open ocean to the east, colder waters to the north, and, possibly, the freshwater Amazonian discharge to the south (de Laubenfels, 1950).

Acknowledgments

Because of the absence of published literature in the area of sponge pollution ecology, personal correspondence with the spongologists of the world was the only method of obtaining information about this field. The response of investigators throughout the world was gratifying.

I wish to express appreciation to M. Poirrier, Louisiana State University in New Orleans, for his careful criticism of the manuscript. I am grateful to the following persons for their contributions and suggestions: B. Afzelius, University of Stockholm; J. F. Allen, Environmental Protection Agency; M. Bacescu, Musee d'Histoire Naturelle "Grigore Antipa," Bucharest, Roumania; G. J. Bakus, University of Southern California; P. R. Bergquist, University of Auckland, New Zealand; H. V. Brøndsted, Copenhagen, Denmark; P. Brien, Université Libre de Bruxelles, Belgium; R. Desqueyroux, Universidad de Concepcion, Chile; R. H. Emson, King's College, London; S. K. Eshleman, Lancaster, Pennsylvania; C. W. Hart, Jr., Academy of Natural Sciences, Philadelphia; W. D. Hartman, Yale University; S. H. Hopkins, Texas Agricultural and Mechanical College; M. Imlay, Environmental Protection Agency; R. E. Johannes, University of Georgia; M. F. Johnson, University of Southern California; Sister Julita, SSpS, University of the Philippines; G. P. Korotkova, Leningrad State University, U.S.S.R.; M. Labate, Universita di Bari, Italy; H. Mergner, Ruhr Universität, West Germany; W. G. Moore, Loyola University; A. A. Racek, University of Sydney, Australia; R. E. H. Reid, The Queen's University of Belfast, Northern Ireland; V. Resh,

University of Louisville; M. Sarà, Istituto di Zoologica, Universita Degli Studi di Genova, Italy; R. T. Sawyer, College of Charleston; O. Sebestyen, Tihany, Hungary; L. K. Simon, Cyanamid GmbH, Munchen, West Germany; V. E. Smith, Cranbrook Institute of Science; J. F. Storr, State University of New York at Buffalo; J. Vacelet, Station Marine d'Endoume, Marseille, France; C. Volkmer-Ribeiro, Museu Rio-Grandense de Ciencias Naturais, Brasil.

I wish to especially acknowledge the support and encouragement of my wife, Marion, during the preparation of the manuscript.

References

Ali, M. A. (1956). Development of the monaxonid sponge *Lissodendoryx similis* Thiele. *J. Madras Univ.* **26**, 553–581.

Annandale, N. (1911). Freshwater sponges, hydroids, and polyzoa. In "*The Fauna of British India including Ceylon and Burma*," pp. 27–126, 241–245. Taylor and Francis, London.

Annandale, N. (1915). Fauna of the Chilka Lake. Sponges. *Mem. Indian Mus.* **5**, 1–54.

Author Anonymous (1961). Savannah River biological survey, South Carolina and Georgia, May-June and August-September 1960 for the E. I. Du Pont de Nemours and Company, Savannah River Plant (TID-14772. Order from Office of Technical Services, U. S. Dep. Of Commerce, Washington 25, D.C.). Biolog. Abstr. **41(2)**, 4833 (1963).

Bagby, R. (1966). The fine structure of myocytes in the sponges *Microciona prolifera* (Ellis and Solander) and *Tedania ignis* (Duchassaing and Michelotti). *J. Morphol.* **118**, 167–182.

Bakus, G. J. (1964). The effects of fish-grazing on invertebrate evolution in shallow tropical waters. *Allan Hancock Foundation Publ., Occasional Paper Number* **27**, 1–29.

Bakus, G. J. (1968). Sedimentation and benthic invertebrates of Fanning Island, Central Pacific. *Mar. Geol.* **6**, 45–51.

Bergquist, P. R., Sinclair, M. E., and Hogg, J. J. (1970). Adaptation to intertidal existence: Reproductive cycles and larval behaviour in Demospongiae. p. 247–271. *In* "The Biology of the Porifera," *Symp. Zoolog. Soc. London Number 25* (W. G. Fry, ed.), pp. 247–271. Academic Press, New York.

Cheatum, E. P., and Harris, J. P., Jr. (1953). Ecological observations upon the fresh-water sponges in Dallas County, Texas. *Field Lab.* **23**, 97–103.

Claus, G., Madri, P., and Kunen, S. (1967). Removal of microbial pollutants from waste effluents by the redbeard sponge. *Nature (London)* **216**, 712–714.

de Laubenfels, M. W. (1932). Physiology and morphology of Porifera, exemplified by *Iotrochota birotulata* Higgin. Carnegie Inst. of Washington, Washington, D. C., Publ. Number 435, pp. 37–66.

de Laubenfels, M. W. (1936). A discussion of the sponge fauna of the Dry Tortugas in particular and the West Indies in general, with material for a revision of the families and orders of the Porifera. Carnegie Inst. of Washington, Washington, D.C., Publ. Number 467. Papers Tortugas Lab. **30**, 1–225.

de Laubenfels, M. W. (1947). Ecology of the sponges of a brackish water environment at Beaufort, N. C. *Ecolog. Monogr.* **17**, 31–46.

de Laubenfels, M. W. (1950). An ecological discussion of the sponges of Bermuda. *Trans. Zoolog. Soc. London* **27**, 155–201.

Eshleman, S . K. (1950). A key to Florida's fresh-water sponges, with descriptive notes. *Quart. J. Florida Acad. Sci.* **12**, 35–44.

Hall, F. G. (1936). Physical and chemical survey of cenotes of Yucatan. *In* "The Cenotes of Yucatan, a Zoological and Hydrographic Survey" (A. S. Pearsé, E. P. Creaser, and F. G. Hall, eds.), pp. 5–16. Carnegie Inst. of Washington, Washington, D.C., Publ. Number 457.

Harrison, F. W. (1971). A taxonomical investigation of the genus *Corvomeyenia* Weltner (Spongillidae) with an introduction of *Corvomeyenia carolinensis sp. nov. Hydrobiologia* **38**, 123–140.

Harrison, F. W. (1972). The nature and role of the basal pinacoderm of *Corvomeyenia carolinensis* Harrison (Porifera: Spongillidae). A histochemical and developmental study. *Hydrobiologia* **39**, 495–508.

Hoff, C. C. (1943). Some records of sponges, branchiobdellids, and molluscs from the Reelfoot Lake region. *J. Tennessee Acad. Sci.* **18**, 223–227.

Hopkins, S. H. (1956a). Notes on the boring sponges in Gulf Coast estuaries and their relation to salinity. *Bull. Mar. Sci. Gulf Caribbean* **6**, 44–58.

Hopkins, S. H. (1956b). The boring sponges which attack South Carolina oysters, with notes on some associated organisms. Contributions from Bears Bluff Lab., Number 23, 1–30.

Hutchinson, G. E. (1957). A treatise on limnology. "Geography, Physics and Chemistry," Vol. 1. Wiley, New York.

Jewell, M. E. (1935). An ecological study of the fresh-water sponges of north-eastern Wisconsin. *Ecolog. Monogr.* **5**, 463–504.

This study provides the foundation for any investigation of the species ecology of the Spongillidae.

Jewell, M. E. (1939). An ecological study of the fresh-water sponges of Wisconsin, II. The influence of calcium. *Ecology* **20**, 11–28.

Jewell, M. E., and Brown, H. W. (1929). Studies on Northern Michigan bog lakes. *Ecology* **10**, 427–475.

Jones, W. C. (1962). Is there a nervous system in sponges? *Biolog. Rev.* **37**, 1–50.

Jorgensen, C. B. (1944). On the spicule-formation of *Spongilla lacustris* (L.). I. The dependence of the spicule-formation on the content of dissolved and solid silicic acid of the milieu. *Kgl. Dan. Vidensk. Selsk. Biol. Meddelelser* **19**, 1–45.

Juday, C., and Birge, E. A. (1933). The transparency, the color, and the specific conductance of the lake waters of northeastern Wisconsin. *Trans. Wisconsin Acad. Sci. Arts Lett.* **25**, 337–352.

Kilian, E. F. (1952). Wasserströmung und Nahrungsaufnahme beim Susswasserschwamm *Ephydatia fluviatilis Z. Vergl. Physiol.* **34**, 407–447.

Korotkova, G. P. (1970). Regeneration and somatic embryogenesis in sponges, *In* "The Biology of the Porifera," Symp. Zoolog. Soc. London Number 25 (W. G. Fry, ed.), pp. 423–436. Academic Press, New York.

Kunen, S., Claus, G., Madri, P., and Peyser, L. (1971). The ingestion and digestion of yeast-like fungi by the sponge, *Microciona prolifera. Hydrobiologia* **38**, 565–576.

Landacre, F. L. (1901). Sponges and bryozoans of Sandusky Bay. *Ohio Natur.* **1**, 96–97.

Lentz, T. L. (1966). Histochemical localization of neurohumors in a sponge. *J. Exp. Zool.* **162**, 171–180.

Madri, P., Hermel, M., and Claus, G. (1971). The microbial flora of the sponge *Microciona prolifera* Verrill and its ecological implications. *Bot. Marina* **14**, 1–5.

McDougall, K. D. (1943). Sessile marine invertebrates at Beaufort, North Carolina. *Ecol. Monogr.* **13**, 321–374.

Moore, W. G. (1953). Louisiana fresh-water sponges, with ecological observations on certain sponges of the New Orleans area. *Trans. Amer. Microsc. Soc.* **32**, 24–32.

Neidhoefer, J. R. (1940). The fresh-water sponges of Wisconsin. *Trans. Wisconsin Acad.* Sci. *Arts Lett.* **32**, 177–197.

Old, M. C. (1932a). Environmental selection of the fresh-water sponges (Spongillidae) of Michigan. *Trans. Amer. Microsc. Soc.* **51**, 129–136.

Old, M. C. (1932b). Taxonomy and distribution of the fresh-water sponges (Spongillidae) of Michigan. *Papers Michigan Acad. Sci. Arts Lett.* **15**, 439–476.

Old, M. C. (1936). Yucatan fresh-water sponges. *In* "The Cenotes of Yucatan, a Zoological and Hydrographic Survey" (A. S. Pearse, E. P. Creaser, and F. H. Hall, ed.), pp. 29–31. Carnegie Inst. Washington, D. C., Publ. Number 457.

Old, M. C. (1941). The taxonomy and distribution of the boring sponges (Clionidae) along the Atlantic coast of North America. *Chesapeake Biolog. Lab. Publ.* **44**, 1–30.

Ott, E., and Volkheimer, W. (1972). *Palaeospongilla chubutensis n.g.* et *n.sp.*—ein Susswasserschwamm aus der Kreide Patagoniens. *Neues Jahrb. Geol. Palaontol. Abh.* **140**, 49–63.

Patrick, R., Cairns, J., Jr., and Roback, S. S. (1966). An ecosystematic study of the fauna and flora of the Savannah River. *Proc. Acad. Natur. Sci. Philadelphia* **118**, 109–407.

Penney, J. T. (1933). Reduction and regeneration in fresh-water sponges (*Spongilla discoides*). *J. Exp. Zool.* **65**, 475–495.

Penney, J. T. (1954). Ecological observations on the fresh-water sponges of the Savannah River Project Area. *Univ. South Carolina Publ. Ser. 3, Biol.* **1**, 156–172.

Penney, J. T. (1957). A new species of fresh-water sponge from South Carolina, *Univ. South Carolina Publ. Ser. 3, Biol.* **2**, 112–115.

Penney, J. T. (1960). Distribution and bibliography (1892–1957) of the fresh-water sponges. *Univ. South Carolina Publ. Ser. 3 Biol.* **3**, 1–97.

Penney, J. T., and Racek, A. A. (1968). Comprehensive revision of a worldwide collection of fresh-water sponges (Porifera: Spongillidae). *U.S. Nat. Mus. Bull.* **272**, 1–184.

This pioneering revisionary work has brought, for the first time, spongillid systematics into some order. It must be considered the base-line from which all studies, systematic and otherwise, of fresh-water sponges must begin. The monograph may be ordered from the Superintendent of Documents; U.S. Government Printing Office; Washington, D. C. 20402. Price $1.50 (Paper Cover).

Poirrier, M. A. (1969). Louisiana fresh-water sponges: Taxonomy, ecology, and distribution. PhD. Thesis. Louisiana State Univ. Univ. Microfilms Inc., Ann Arbor, Michigan, Number 70–9083.

Poirrier, M. A. (1972). Additional records of Texas fresh-water sponges (Spongillidae) with the first record of *Radiospongilla cerebellata* (Bowerbank, 1863) from the Western Hemisphere. *Southwest. Natur.* **16**, 434–435.

Potts, E. (1887). Contributions towards a synopsis of the American forms of fresh-water sponges with descriptions of those named by other authors and from all parts of the world, *Proc. Acad. Natur. Sci. Philadelphia* **39**, 158–279.

Potts, E. (1889). Report upon some fresh-water sponges collected in Florida by Jos. Willcox, Esq. *Trans. Wagner Free Inst. Sci. Philadelphia* **2**, 5–7.

Racek, A. A. (1969). The fresh-water sponges of Australia (Porifera: Spongillidae). *Aust. J. Mar. Freshwater Res.* **20**, 267–310.

Rasmont, R. (1970). Some new aspects of the physiology of fresh-water sponges. *In* "The Biology of the Porifera," Symp. Zool. Soc. of London Number 25 (W. G. Fry, ed.), pp. 415–422. Academic Press, New York.

Redeke, H. C. (1932). Ueber den jetzigen Stand unserer Kenntnisse der Flora und Fauna des Brackwassers. *Verh. Int. Vereinigung Theor. Angew. Limnol. Sechsten Mitgleiderversammlung, Amsterdam, 4–11 September 1932* **6**, 46–61.

Reiswig, H. M. (1971a). *In situ* pumping activities of tropical Demospongiae. *Mar. Biol.* **9**, 38–50.

Reiswig, H. M. (1971b). Particle feeding in natural populations of three marine Demosponges. *Biolog. Bull.* **141**, 568–591.

Sarà, M. (1960). Osservazioni sulla composizione ecologia e differenziamento della fauna di Porifera di acqua salmastra. *Annu. Instituto Museo Zool. Univ. Napoli* **12**, 1–10.

Simpson, T. L. (1968). The biology of the marine sponge *Microciona prolifera* (Ellis and Solander) II. Temperature-related, annual changes in functional and reproductive elements with a description of larval metamorphosis. *J. Exp. Mar. Biol. Ecol.* **2**, 252–277.

Smith, F. (1921). Distribution of the fresh-water sponges of North America. *State Ill. Dept. Registration Educ. Div. Nat. History Survey* **14**, 9–22.

Smith, F. G. W. (1941). Sponge disease in British Honduras, and its transmission by water currents. *Ecology* **22**, 415–421.

Stephens, J. (1912). Clare Island survey. Freshwater Porifera. *Proc. Roy. Irish Acad.* **31**, 1–15.

Storr, J. F. (1964). Ecology of the Gulf of Mexico commercial sponges and its relation to the fishery. U. S. Fish and Wildlife Serv. Spec. Sci. Rep., Fisheries Number **466**, pp. 1–73.

Tendal, O. S. (1967). On the fresh-water sponges of Denmark. *Videnskabelige Meddelelser Fra Dansk Naturhistorisk Forening* **130**, 173–178.

van Tright, H. (1919). A contribution to the physiology of the fresh-water sponges. *Tijdschr. Diergeneesk. Ser. 2,* **17**, 1–220.

Warburton, F. E. (1966). The behavior of sponge larvae. *Ecology* **47**, 672–674.

Weel, P. B. van (1949). On the physiology of the tropical fresh-water sponge *Spongilla proliferens* Annand. I. Ingestion, digestion, and excretion. *Physiol. Comparata Oecol.* **1**, 110–126.

Wells, H. W. (1959). Boring sponges (Clionidae) of Newport River, North Carolina. *J. Elisha Mitchell Sci. Soc.* **75**, 168–173.

Wells, H. W. (1961). The fauna of oyster beds, with special reference to the salinity factor. *Ecolog. Monogr.* **31**, 239–266.

Wells, H. W., Wells, M. J., and Gray, I. E. (1960). Marine sponges of North Carolina. *J. Elisha Mitchell Sci. Soc.* **76**, 200–245.

Weltner, W. (1895). Spongillidenstudien III. Katalog und Verbreitung der bekannten Süsswasserschwämme. *Arch. Naturges.* **61**, 114–144.

Wilson, H. V. (1935). Some critical points in the metamorphosis of the halichondrine sponge larva. *J. Morphol.* **58**, 285–345.

Wurtz, C. B. (1950). Fresh-water sponges of Pennsylvania and adjacent states. *Not. Natur. Acad. Natur. Sci. Philadelphia* **228**, 1–10.

CHAPTER 3

Flatworms (Platyhelminthes: Tricladida)

ROMAN KENK

I. Ecology

Freshwater triclads or planarians are regularly found in aquatic habitats that fulfill certain requirements which may vary from species to species. Some species are characteristic of running waters (springs, brooks, rivers), others occur in standing water bodies (lakes, ponds, ditches) in both aboveground (epigean) and subterranean (hypogean) habitats. Among the important parameters determining the suitability of a habitat for planarians are temperature, dissolved oxygen, the chemical constitution of the water, the nature of the substrate, water current, turbidity, availability of food, and interspecific food competition.

Some species are confined to cold waters with little diurnal and seasonal temperature fluctuations (stenothermic species). These will be found in springs, cold creeks, and in lakes at high altitudes. Other species tolerate higher and more variable temperatures (eurythermic species).

Following a stream from its headwaters to its mouth, one frequently encounters a succession of species such as has been observed in Europe: In an ideal situation, the spring and upper reaches are populated by the stenothermic *Crenobia alpina* (Dana) which farther downstream is joined by *Polycelis felina* (Dalyell); then follows a section in which *P. felina* alone occurs, to be gradually replaced by the more eurythermic *Dugesia gonocephala* (Dugès) (Voigt, 1905). Similar gradations are reported from Japan where *Polycelis*, *Phagocata* and *Dugesia* succeed each other in a regular fashion (Kawakatsu, 1965). Little is known about the succession of planarian species in North America. Kenk (1944) observed that in Michigan *Phagocata velata* (Stringer) and *P. morgani* (Stevens and Boring) inhabit the upper reaches of streams, to be supplanted by *Cura foremanii* (Girard) and finally by *Dugesia tigrina* (Girard). Some data are given by Chandler (1966) for streams near Bloomington, Indiana, where *Phagocata gracilis*, *Cura foremanii*, *Dugesia dorotocephala* (Woodworth), and *D. tigrina* appear in sequence. Whitehead (1965) reports analogous conditions for Cattaraugus County, New York (*Phagocata morgani*, *Cura foremanii*, *Dugesia tigrina*), and Darlington and Chandler (1972) analyze the associations of seven planarian species in Tennessee.

The study of planarian successions is of some importance to the analysis of water pollution, since they occur in clean, unpolluted waters, chiefly on account of the temperature gradient in the course of a stream. The replacement of one planarian species by another in waters suspected of pollution must be weighed against the possibility of a natural succession.

Planarians require well-oxygenated water although some species tolerate temporary oxygen depletion to less than 2 ppm (Abbott, 1960). The role of dissolved carbon dioxide is not well understood; it may be in part responsible for the natural succession of planarian species in a river (Alause, 1962). Most chemical constituents found in natural waters have little effect on the distribution of triclads, with the possible exception of the calcium content or hardness (Leloup, 1944; Reynoldson, 1958). Acid waters are, in general, unfavorable, and planarians are absent from bog water containing humic acids (Harnisch, 1925; Tucker, 1958).

The sensitivity of planarians to poisons introduced into their habitats varies somewhat with the different species. Little is known about the differential susceptibility of the individual American forms, since most observers recorded only the presence or absence of unidentified "planarians" in their field work, and since there are hardly any laboratory studies with well-identified American species reported in the literature.

Planarians are photonegative and are more active at night than during the day. They are usually collected by examining the undersurfaces of stones and other objects where they rest to avoid bright daylight. Some species may be attracted by bait (liver, meat, a dead fish, etc.) placed in a shaded location in their habitats. Because of these collecting methods, which depend to a great extent on the experience of the investigator, and of the great variety of substrates to be inspected, precise quantitative studies of the population density of planarians are often difficult.

II. Effects of Pollution

A. Organic Pollution

Organic pollution is, in general, detrimental to planarians, chiefly on account of two factors: the depression of the content of dissolved oxygen in the water caused by oxidative processes and the metabolism of microorganisms and the presence of toxic intermediary and end products of putrefaction. In the saprobic system of Kolkwitz and Marsson (1909) and its later refinements, common freshwater planarians are listed as indicators of the various saprobiotic zones. Among the European species, *Crenobia alpina* is considered oligosaprobic and has been widely used as indicator of clean, unpolluted water; *Polycelis felina* and *Dugesia gonocephala* are oligo- to mesosaprobic, and *Dendrocoelum lacteum* (Müller) mesosaprobic. Besch (1967), who studied the effects of unspecified pollutants on the fauna of streams in southern Württemberg, Germany, observed that *Crenobia alpina* and *Dugesia gonocephala* tolerated very mild pollution, while *Dendrocoelum lacteum* and *Dugesia lugubris* (O. Schmidt) proved to be less sensitive. In more heavily polluted streams planarians are absent though other faunal constituents (tubificid oligochaetes, Ephemeroptera larvae, gammarid amphipods) may still be found. An extensive investigation of the effects of organic pollution on the animal life, including planarians, in streams of Switzerland was carried out by Steinmann and Surbeck (1918).

The American species, *Dugesia tigrina*, introduced to Europe in this century, proved to be beta- to alpha-mesosaprobic. It will be noticed that the pollution tolerances of these species parallel their temperature tolerances, the stenothermic species being more sensitive to pollution than the eurythermic species. No attempts have been made to classify the American planarians according to their pollution resistance. Rivers receiving mild organic waste generally recover their suitability as habitats for triclads some distance below the intake of the sewage, by self-purification and by dilution of the pollutants.

Wiebe (1928) considered "*Planaria*" to be a clean-water form as it was absent in polluted sections of the upper Mississippi River. Another example of the effects of organic river pollution on planarians was given by Hynes (1960) who observed that *Polycelis felina* was eliminated in the river Dee in Wales below the inflow of pollutants for at least 16 miles. Whitehead (1965) commented on the lack or depletion of four species of triclads in several streams in Cattaraugus County, New York, polluted by dairy wastes or human sewage. The first species to reappear downstream from the inflow of the wastes was *Dugesia tigrina*. Animal dung present in the water in streams which flow through pastures had no deleterious effect on the triclads. Pickavance (1971) recorded the scarcity or absence of planarians in sections of Rennies River in St. John's, Newfoundland, which were polluted by domestic waste and road drainage. Mettrick *et al.* (1970) reported that flatworms were absent in streams flowing through areas of heavy urbanization in southern Ontario, Canada, while they were present in rural sections of the same area. On the other hand, mild organic pollution may increase the abundance of planarians if other physicochemical parameters are favorable. Macan (1962) reports that in a stream in the English Lake District the numbers of *Polycelis felina* increased spectacularly when effluents of an overloaded farm septic tank entered the stream. Similarly, Holsinger (1966) observed great numbers of *Phagocata subterranea* Hyman (*P. gracilis*) in pools in Banners Corner Cave, Virginia, which were highly contaminated with leakage from septic tanks and contained many isopods (*Asellus*), oligochaetes (*Tubifex*), and considerable bacterial growth, including coliform bacteria. In these instances, the presence of abundant food organisms may explain the great population density of the planarians. It is a well-known fact that planarians may be quite numerous in the effluents from fish hatcheries which presumably contain remnants of fish food. Oliff (1960), who studied the effects of organic pollution on the biology of Bushmans River, Natal, states that "*Planaria* sp." and "*Dugesia* sp." were present in unpolluted and lightly polluted sections of the stream, but were replaced by great numbers of "*Sorocelis* sp." in sections with heavy pollution. He considers *Sorocelis* (apparently a misidentification) to be an indicator of polluted water. Gaufin (1958) also lists an unidentified planarian species among the organisms tolerant to pollution in Mad River, Ohio. Mettrick *et al.* (1970) state that *Dugesia polychroa* (O. Schmidt), a European species recently introduced to the St. Lawrence River system by shipping, is abundant in the heavily polluted harbor of Toronto, Ontario, Canada.

The effects of raw sewage and of some poisonous constituents of organic wastes and products of their degradation were studied in the laboratory by determining the survival times of planarians exposed to various concentrations of the compounds.

Hydrogen sulfide and free ammonia are very toxic to planarians. Stammer (1954) determined the lethal limits for *Crenobia alpina*, *Dugesia gonocephala*, and *Dendrocoelum lacteum* to be 3 mg/liter for H_2S and 0.2–0.4 mg/liter for NH_3. Van Oye (1941), however, reported that *Crenobia alpina* occurred in a mineral water spring in Belgium, containing 4.5 mg hydrogen sulfide per liter.

Seibold (1956), using the same experimental species as Stammer, performed a series of experiments with various concentrations of raw and of aerated domestic sewage from the city of Munich, Germany, at various temperatures. The most sensitive species was the oligosaprobic *Crenobia alpina*, followed by the oligo- to mesosaprobic *Dugesia gonocephala* and the mesosaprobic *Dendrocoelum lacteum*. Toxic effects were stronger at elevated temperatures. Seibold also tested the sensitivity of the planarians to individual poisonous components of the sewage. Of the six putrefactive compounds tested, indole was the most toxic, damaging *Crenobia* and *Dugesia* at a concentration of 1 mg/liter and *Dendrocoelum* at 3 mg/liter. The other compounds, arranged by decreasing toxicity, were skatole (1, 3, and 5 mg/liter, respectively), hydroxylamine (2, 5, and 5 mg/liter), histamine (25, 20, and 20 mg/liter), putrescine (30, 20, and 20 mg/liter), and cadaverine (10, 40, and 10 mg/liter). Various combinations of the toxicants frequently resulted in toxicities which exceeded the sum of the toxic actions of the individual components. This was particularly true with mixtures which contained histamine. Thus histamine appears to have an activating or superadditive role in the toxic action of the sewage. The toxicity of the various mixtures of the six substances tested did not, however, approach the toxicity of the raw sewage, indicating that still other components must be active in the sewage.

The toxicity of 11 quinones for *Dugesia gonocephala* was investigated by Schreier (1949) with particular reference to their physicochemical action on planarian tissues. As quinones are not likely to appear in the putrefactive processes, no analysis of his results is needed here.

B. Industrial Pollution

Data on the deleterious effects of effluents from industrial plants on planarians are given by Thienemann (1911) for a chemical factory, by Hynes (1960) for wastes containing ammonia and cyanides, by Whitehead (1965) for effluents from glue and tanning factories in Cattaraugus County, New York, and by Mettrick *et al.* (1970) for the outflow of tanneries below Newmarket, Ontario. Oliff (1963) observed that the acidic effluents from coal mines in Natal, South Africa, impoverished the fauna of rivers while in neutralized sections of the streams planarians generally increased in numbers.

The range of acid and alkali tolerance of planarians is rather wide. *Dugesia dorotocephala*, *D. tigrina*, and *Phagocata velata* withstand hydrogen ion concentrations roughly from pH 5 to pH 9, with slight differences between the species and between young and adult animals (MacArthur, 1920).

Chlorine, in a concentration of 1 mg/liter, kills *Polycelis felina* in 5 hours, in higher concentrations in a shorter time (Birrer, 1932).

The density of *Polycelis felina* in streams in Cornwall, England, was significantly reduced at stations polluted by effluents of china-clay refineries (Nuttall and Bielby, 1973).

1. Oil

Neel (1953) reports that "planarians" were absent in stretches of North Platte River, Wyoming, which were polluted by effluents of oil refineries. Above the inflow of the pollutants and 110 miles below it planarians were common. The effects of fuel oil pollution in Muddy River near Boston, Massachusetts, were studied by McCauley (1966), who found that *Dugesia* sp. was absent in the polluted section of the stream. Whitehead (1965) states that no planarians occurred in many streams that received wastes from oil fields in Cattaraugus County, New York.

2. Metallic Salts

Carpenter (1924) observed that "Turbellaria" were absent in streams in Cardiganshire, west Wales, which were polluted by effluents of lead mines, while they were present in unpolluted waters of the same area. According to Kawakatsu and Ito (1963), no planarians were found in localities polluted by discharges of an abandoned copper mine in Japan, but occurred regularly in neighboring unpolluted streams.

According to Laurie and Jones (1938), *Phagocata vitta* (Dugès) survives in lead nitrate solution (containing 3 ppm lead) 40 hours and *Polycelis nigra* (Müller) 70 hours.

Newton (1944) reports than, according to Davies, *Polycelis nigra* can withstand zinc sulfate solutions in concentrations up to 30 ppm.

Intensive studies of the toxicity of metallic salts for the European planarian, *Polycelis nigra*, were carried out by J. R. E. Jones (see references). Both the cations and anions of the salts may exhibit toxic action. Jones (1940b) determined the toxicity thresholds of the individual metals at which a mean survival time of 48 hours was attained. The most poisonous cations, which were active in concentrations below 50 mg/liter, were, in descending order: silver (0.15 mg/liter), mercury (0.2), copper (0.47), gold (0.6), cadmium (2.7), iron (>20), zinc (30), arsenic (40), and nickel (45). Of the anions tested (Jones, 1941b) the following had toxicity thresholds below molar concentrations of 0.01: hydroxide (0.00004), sulfide (0.00045), cyanide

(0.0006), nitrate (0.0006), nitroprusside (0.0008), ferrocyanide (0.0008), fluoride (0.0011), iodate (0.0013), chromate (0.0028), and arsenate (0.0048).

3. Soap and Detergents

Brown (1966) states that in Mexican streams polluted by the washing of clothes on their banks, "planarians" seem to be tolerant of soapy water. Hynes and Roberts (1962) observed that a degradable synthetic detergent contained in municipal sewage had little effect on triclads in the river Lee, Hertfordshire, England.

C. Pesticides and Fish Poisons

DDT seems to have little effect on planarians. Frey (1961) reports that "*Planaria*" were eliminated in a stream in Georgia, the watershed area of which had been sprayed with DDT dissolved in oil. His results are, however, not significant statistically. Hynes and Williams (1962), who studied the effects of DDT applied to a stream in Uganda, state that *Dugesia* was not affected by it in the concentration used (1 ppm).

The toxicity of two insecticides, DDT and Sevin (= Arylam, 1-naphthyl-*N*-methylcarbamate) for planarians was investigated in the laboratory by An der Lan (1959, 1962), Aspöck (1962), and Aspöck and An der Lan (1963), who also analyzed the histological alterations produced by these poisons. DDT in a concentration of 0.5 ppm was tolerated by *Crenobia alpina* up to 20 days and 5 ppm Sevin up to 8 days. Some other insecticides, Methoxychlor, Rhotane, and Kelthane, induced similar histological injuries, but no data on individual toxicities are given (An der Lan, 1962).

Crenobia alpina and *Polycelis felina* were not affected by the outflow of a sheep-dip containing the insecticide BHC (hexachlorocyclohexane), although amphipods (*Gammarus*) and some insect larvae were almost entirely eliminated (Hynes, 1961).

The concentrations of Derris root preparations, including Rotenone, lethal to unidentified "*Planaria*" are variously given as 2 mg/liter (Scheuring and Heuschmann, 1935) and as 10 and 0.5 mg/liter (Hamilton, 1940, 1941; see also Lindgren, 1960). Rotenone in a concentration of 4.5 ppm killed "*Planaria* sp." within 7 hours (Almquist, 1959).

Van Jaarsveld (1970) studied the toxicity of Dieldrin to various aquatic invertebrates and found that a South African species of *Euplanaria* (*Dugesia*) is more resistant to this insecticide than are most insect larvae.

Nicotine extract, diluted 1:1000, is recommended by Remkes (1941) as a means of ridding aquaria of undesirable "planaria."

Spraying of water bodies with gas oil to control mosquito larvae in the vicinity of Copenhagen, Denmark, reduced the numbers of "Turbellaria" without eliminating them (Arevad, 1961).

Smith (1967) reports that TFM (3-trifluoromethyl-4-nitrophenol), used for the control of the sea lamprey (*Petromyzon marinus*) in the upper Great Lakes, United States and Canada, killed "triclads" in concentrations above 2–6 ppm.

D. Herbicides

A study of the effects of 14 herbicides on various freshwater animals was published by Pravda (1973). *Planaria gonocephala* (*Dugesia gonocephala*) proved to be, in general, less sensitive to herbicide pollution than the majority of the other organisms studied (*Tubifex*, *Daphnia*, *Asellus*, *Gammarus*, etc.). The greatest toxicity was recorded for Dinoseb (4,6-dinitro-*o*-*s*-butylphenol), followed by Liro-CIPC (isopropyl-*m*-chlorocarbanilate), Alipur (mixture of two active ingredients, OMU = *N*-cyclooctyl-*N'N'*-dimethylurea and BiPC = butinol-*N*-[3-chlorophenyl] carbamate), and Rafex (2-methyl-4,6-dinitrophenol). Dinoseb in a concentration of 1000 mg/liter killed the planarian in 7 minutes, at 100 mg/liter in 37 minutes, and at 10 mg/liter in 12 hours.

E. Thermal Pollution

No field observations on the effects of high-temperature effluents of industrial plants on planarians have been reported. Presumably raising the temperature of a stream will interfere with the natural succession of the planarian species as discussed in the section on ecology. A cold-stenothermic species may be eliminated and replaced by a more eurythermic species which had normally been present only in the lower, warmer, reaches of the stream.

Wurtz and Dolan (1961) observed the presence of the warm-water planarian *Dugesia tigrina* in Schuylkill River, Pennsylvania, both above and below the inflow of thermal discharge in waters that sometimes exceeded 95°F (35°C).

In Yellowstone National Park, Wyoming, outflows from hot springs enter cooler rivers in several places. In the Upper Geyser Basin, for example, above the Morning Glory pool, water from a hot spring flows into Firestone River. The river, of about 17°C temperature, is inhabited by the cold-stenothermic *Polycelis* sp. and a small population of *Dugesia dorotocephala*. In the zone where the two waters mix, at a temperature of 21.9° to 26.3°C, enormous numbers of the eurythermic *D. dorotocephala* were observed under stones covering the river bed. Thus *Dugesia* was attracted to a portion of the river which had a temperature close to the optimum temperature for the species (25°C).

F. Radioactive Pollution

From laboratory experiments (see review by Brøndsted, 1969), we may conclude that ionizing radiation, unless exceptionally heavy, would not kill planarians outright. It would, however, destroy certain cellular elements in the mesenchyme of the animals, the neoblasts, which play an important role in the growth, physiological replacement of tissues, and regeneration of the planarians. Radioactive pollution would, therefore, eliminate planarians in the course of time, perhaps after several weeks.

III. Systematics

Of the approximately 50 species of freshwater triclads described from North America, the majority is either, as far as we know, confined to narrow geographic areas or to subterranean habitats. Recent reviews and keys for their identification have been published by Hyman (1953, 1959) and Kenk (1972). Only a few species are likely to be encountered in epigean springs, rivers, and lakes and may play a role in the evaluation of the quality of the water. These are listed here with their outstanding differential characteristics.

Phylum Platyhelminthes, class Turbellaria, order Tricladida, suborder Paludicola or Probursalia

FAMILY PLANARIIDAE

Cura foremanii (Girard) (Fig. 1A). Anterior end bluntly triangular, with rounded auricles which protrude only slightly laterally. Color uniformly brown, with a light oblique dash (auricular sense organ) behind each auricle. Pharynx unpigmented. Inhabitant of cool streams in the eastern half of North America.

Dugesia tigrina (Girard) (Fig. 1B). Head triangular, with bluntly pointed or rounded anterior tip and short lateral auricles. Pigmentation very variable, usually spotted, more rarely uniformly brown, frequently with a pair of dark longitudinal stripes on the dorsal side of the body. Ventral side generally without pigment. Pharynx pigmented. A eurythermic species occurring in stagnant waters and the lower stretches of rivers, widely distributed in the Americas. Has been introduced to Europe and is considered there a mesoaprobic species which tolerates moderate organic pollution (Heuss, 1971). Hunt's (1962) statement, that *D. tigrina* is oligosaprobic, is apparently erroneous.

Dugesia dorotocephala (Woodworth) (Fig. 1C). Head triangular, similar to that of *D. tigrina* but with longer, slender, pointed auricles. Color almost uniformly brown to black to the naked eye, composed of colored and white patches under magnification. Ventral side lighter, but also with pigment. Pharynx pigmented.

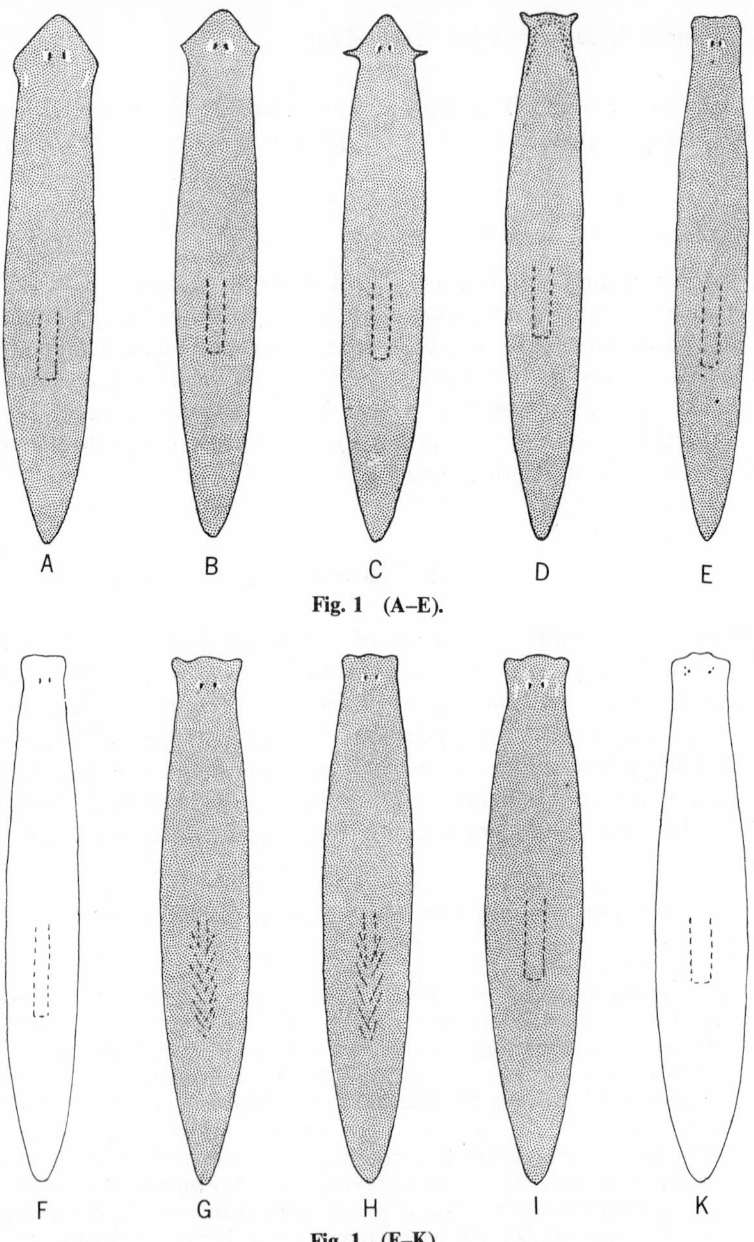

Fig. 1 (A–E).

Fig. 1 (F–K).

Fig. 1. Sketches of common North American planarians. Pigmented species are shaded, unpigmented species white. A, *Cura foremanii*, up to 15 mm long; B, *Dugesia tigrina*, 18 mm; C, *Dugesia dorotocephala*, 30 mm; D, *Polycelis sierrensis*, 22 mm (*P. coronata* is almost identical); E, *Phagocata velata*, 20 mm (*P. crenophila* and *Planaria dactyligera* are very similar); F, *Phagocata morgani*, 14 mm; G, *Phagocata gracilis*, 30 mm; H, *Phagocata woodworthi*, 30 mm; I, *Dendrocoelopsis vaginata*, 22 mm; K, *Procotyla fluviatilis*, 20 mm.

Eurythermic species, found usually in running waters, more rarely (in the western states) in lakes. Distributed all across the North American continent.

Polycelis coronata (Girard) and *P. sierrensis* Kenk (Fig. 1D). These two species are not distinguishable externally but differ in their anatomy. Head truncate, with convex frontal margin and a pair of triangular auricles with rounded tips extending anterolaterally. Eyes, many, arranged in a band paralleling the margins of the anterior region. Uniformly pigmented, brown. Both species inhabit cold springs, streams, and high altitude lakes in the western half of North America. Apparently very sensitive to pollution (hard to keep in laboratory cultures).

Planaria dactyligera Kenk (see Fig. 1E). This is one of the species which cannot be identified with certainly in life. Head, truncate, with almost straight frontal margin and rounded lateral edges. Eyes, usually two. Color, a variable shade of brown. In springs and streams in the eastern United States.

Phagocata velata (Stringer) (Fig. 1E). Externally similar to *Planaria dactyligera*. Color, gray to almost black. Reproduces sexually or by fragmentation and encystment. Inhabits springs and spring-fed ponds, apparently across the continent.

Phagocata crenophila Carpenter (see Fig. 1E). Resembles *P. velata* and *Planaria dactyligera* in external characters. Gray to black, ventral side lighter. A cold stenothermic species found in cold springs and streams at high altitudes in the western parts of the United States.

Phagocata morgani (Stevens and Boring) (Fig. 1F). An unpigmented species, white (the color of the intestinal contents may be visible through the body wall, but the head, the body margins, and places above the pharynx and copulatory complex are always white). Eyes, normally two, close together and removed from the frontal margin (in a subspecies, the eyes are multiple and arranged in a pair of parallel longitudinal rows on the head). Lives in springs and cool creeks in the eastern half of North America.

Phagocata gracilis (Haldeman) (Fig. 1G). Head truncate, with straight or slightly convex frontal margin and rounded lateral edges which usually project laterally. Normally one pair of eyes. Color, gray or brown. Characteristic is the multiplicity of the pharynges which may be recognized in living specimens (this is also seen in *P. woodworthi*). Inhabits cool springs and streams in the eastern part of the United States, south of the Delaware River. Tolerates and may even be attracted to mild organic pollution.

Phagocata woodworthi Hyman (Fig. 1H). Externally similar to *P. gracilis*. The lateral edges of the head protrude laterally less than in that species. Eyes, normally two. Polypharyngeal. Gray or brown, to almost black. In streams in eastern North America north of Delaware River.

FAMILY DENDROCOELIDAE

Dendrocoelopsis vaginata Hyman (Fig. 1I). Head truncate, with slightly convex frontal margin bearing an adhesive organ, and rounded lateral corners. Eyes, two. Color, gray or brown dorsally, somewhat lighter ventrally. A pair of light oblique streaks on the dorsal side of the head laterally to the eyes is very characteristic. Occurs in springs, streams, and lakes in the northwestern part of the United States.

Procotyla fluviatilis Leidy (Fig. 1K). A white species with truncate head, with a conspicuous adhesive organ in the center of the frontal margin and rounded, somewhat protruding, lateral lobes. Eyes, usually more than two, situated in a pair of groups rather far apart and close to the frontal end (only two eyes in young specimens and in worms in the southern part of the geographic range). A eurytopic species inhabiting springs, brooks, rivers, ponds, and lakes in the eastern half of North America.

References

Abbott, B. J. (1960). A note on the oxygen and temperature tolerances of the triclads *Phagocata gracilis* (Haldeman) and *Dugesia tigrina* (Girard). *Va. J. Sci. New Ser.* **11**, 1–8.

Alause, P. (1962). Gradient d'acidité carbonique dans deux rivières du Département de l'Hérault et écologie de *Polycelis felina* Dalyell (= *Polycelis cornuta* Johnson). *Vie Milieu* **13**, 341–358.

Almquist, E. (1959). Observations on the effect of Rotenone emulsives on fish food organisms. *Inst. Freshwater Res. Drottningholm Rep.* **40**, 146–160.

An der Lan, H. (1959). Physiologische Besonderheiten moderner Pflanzenschutzmittel. *Schri. Vereines Verbreit. naturwissensch. Kenntnisse Wien* **99**, 29–65.

An der Lan, H. (1962). Histopathologische Auswirkungen von Insektiziden (DDT und Sevin) bei Wirbellosen und ihre cancerogene Beurteilung. *Mirroskopie* **17**, 85–112.

Arevad, K. (1961). Om den øvrige faunas reaktion på myggebekaempelse med olie (On the effect on other organisms of mosquito control by the oiling method). *Entomol. Meddelelser* **31**, 27–50.

Aspöck, H. (1962). Untersuchungen über biologische Eigenschaften des Sevin (1-Naphthyl-*N*-Methylcarbamat). Dissertation, Univ. of Innsbruck, Austria (unpublished).

Aspöck, H., and An der Lan, H. (1963). Ökologische Auswirkungen und physiologische Besonderheiten des Pflanzenschutzmittels Sevin (= 1-Naphthyl-*N*-methylcarbamat). *Z. angew. Zool.* **50**, 343–380.

Besch, W. (1967). Biologischer Zustand und Abwasserbelastung der Fliessgewässer Südwürttembergs. *Veröff. Landesstelle Natur Schutz Landsch. Baden-Württemberg* **35**, 111–128.

Birrer, A. (1932). Aktives Chlor und seine Einwirkung auf niedere Wasserorganismen. *Z. Hydrol.* **6**, 64–104.

Brøndsted, H. V. (1969). "Planarian Regeneration." Pergamon, Oxford.

Brown, H. P. (1966). Effects of soap pollution upon stream invertebrates. *Trans. Amer. Microsc. Soc.* **85**, 167.

Carpenter, K. E. (1924). A study of the fauna of rivers polluted by lead mining in the Aberystwyth district of Cardiganshire. *Ann. Appl. Biol.* **11**, 1–23.

Chandler, C. M. (1966). Environmental factors affecting the local distribution and abundance of four species of stream-dwelling triclads. *Invest. Indiana Lakes Streams* **7**, 1–56.

Darlington, J. T., and Chandler, C. M. (1972). A survey of the epigean triclad turbellarians of Tennessee. *Amer. Midl. Natur.* **88**, 158–166.

Frey, P. J. (1961). Effect of DDT spray on stream bottom organisms in two mountain streams in Georgia. *U. S. Fish Wildl. Serv. Spec. Sci. Rep. Fish.* **392**, 1–11.

Gaufin, A. R. (1958). The effects of pollution on a midwestern stream. *Ohio J. Sci.* **58**, 197–208.

Götz, R. (1959). Zur Wirkung moderner Schädlingsbekämpfungsmittel auf *Planaria gonocephala* Dugès und *Planaria alpina* Dana. Dissertation, Univ. of Innsbruck, Austria (Cited by An der Lan, 1962).

Hamilton, H. L. (1940). The biological action of Rotenone on lake fauna. *Proc. Iowa Acad. Sci.* **46**, 457–458.

Hamilton, H. L. (1941). The biological action of Rotenone on freshwater animals. *Proc. Iowa Acad. Sci.* **48**, 467–479.

Harnisch, O. (1925). Studien zur Ökologie und Tiergeographie der Moore. *Zool. Jahrb. Abt. Syst.* **51**, 1–166.

Heuss, K. (1971). Neufunde von *Dugesia tigrina* (Girard) (Turbell., Tricladida) im Gebiet des Niederrheines und der unteren Maas. *Decheniana* **123**, 53–57.

Holsinger, J. R. (1966). A preliminary study on the effects of organic pollution of Banners Corner Cave, Virginia. *Int. J. Speleol.* **2**, 75–89.

Hunt, G. S. (1962). Water pollution and ecology of some aquatic invertebrates in the lower Detroit River. *Univ. of Michigan, Great Lakes Res. Div. Publ.* **9**, 29–49.

Hyman, L. H. (1953). Turbellaria (Flatworms). *In* "Fresh-water Invertebrates of the United States" (R. W. Pennak, ed.), pp. 114–141. Ronald Press, New York.

Hyman, L. H. (1959). Order Tricladida. *In* H. B. Ward and G. C. Whipple, "Fresh-water Biology", 2nd edition, pp. 326–334. Edited by W. T. Edmonson. Wiley, New York.

Hynes, H. B. N. (1960). "The Biology of Polluted Waters." Liverpool Univ. Press, Liverpool.

Hynes, H. B. N. (1961). The effect of sheep-dip containing the insecticide BHC on the fauna of a small stream, including *Simulium* and its predators. *Annals Trop. Med. Parasitol.*, **55**, 192–196.

Hynes, H. B. N., and Roberts, F. W. (1962). The biological effects of synthetic detergents in the River Lee, Herfordshire. *Ann. Appl. Biol.* **50**, 779–790.

Hynes, H. B. N., and Williams, T. R. (1962). The effect of DDT on the fauna of a Central African stream. *Ann. Trop. Med. Parasitol.* **56**, 78–91.

Jaarsveld, J. H. van (1970). A laboratory study on the toxicity of Dieldrin to fresh water invertebrates. *Phytophylactica* **2**, 269–273.

Jones, J. R. E. (1937). The toxicity of dissolved metallic salts to *Polycelis nigra* (Muller) and *Gammarus pulex* (L.). *J. Exp. Biol.* **14**, 351–363.

Jones, J. R. E. (1939). Antagonism between two heavy metals in their toxic action on freshwater animals. *Proc. Zool. Soc. London. Ser. A* **108**, 481–499.

Jones, J. R. E. (1940a). The fauna of the river Melindwr, a lead-polluted tributary of the River Rheidol in North Cardiganshire, Wales. *J. Anim. Ecol.* **9**, 188–201.

Jones, J. R. E. (1940b). A further study of the relation between toxicity and solution pressure, with *Polycelis nigra* as test animal. *J. Exp. Biol.* **17**, 408–415.

Jones, J. R. E. (1940c). A study of the zinc-polluted river Ystwyth in North Cardiganshire, Wales. *Ann. Appl. Biol.* **27**, 368–378.

Jones, J. R. E. (1941a). The effect of ionic copper on the oxygen consumption of *Gammarus pulex* and *Polycelis nigra. J. Exp. Biol.* **18**, 153–161.

Jones, J. R. E. (1941b). A study of the relative toxicity of anions, with *Polycelis nigra* as test animal. *J. Exp. Biol.* **18**, 170–181.

Kawakatsu, M. (1965). On the ecology and distribution of freshwater planarians in the Japanese Islands, with special reference to their vertical distribution. *Hydrobiologia* **26**, 349–408.

Kawakatsu, M., and Itô, T. (1963). Report on the ecological survey of freshwater planarians in the Ishizuchi mountain range, Shikoku. *Nihon Seitai Gakkai Shi (Japan. J. Ecol.)* **13**, 231–234 (In Japanese).

Kenk, R. (1944). The fresh-water triclads of Michigan. *Misc. Publ. Mus. Zool. Univ. Michigan* **60**, 1–44, 7 plates.

Kenk, R. (1972). Freshwater planarians (Turbellaria) of North America. *Biota Freshwater Ecosyst. Identification Manual* **1**. U. S. Environmental Protection Agency, Washington.

Kolkwitz, R., and Marsson, M. (1909). Ökologie der tierischen Saprobien: Beiträge zur Lehre von der biologischen Gewässerbeurteilung. *Int. Rev. ges. Hydrobiol. Hydrogr.* **2**, 126–152.

Laurie, R. D., and Jones, J. R. E. (1938). The faunistic recovery of a lead-polluted river in North Cardiganshire, Wales. *J. Anim. Ecol.* **7**, 272–289.

Leloup, E. (1944). Recherches sur les Triclades dulcicoles épigés de la Forêt de Soignes, *Mém. Musée Roy. Hist. Natur. Belg.* **102**, 1–112, 3 plates.

Lindgren, P. E. (1960). About the effect of Rotenone upon benthonic animals in lakes. *Inst. Freshwater Res. Drottningholm, Rep.* **41**, 172–184.

Macan, T. T. (1962). Biotic factors in running water. *Schweiz. Z. Hydrol.* **24**, 386–407.

MacArthur, J. W. (1920). Changes in acid and alkali tolerance with age in planarians. *Amer. J. Physiol.* **54**, 138–146.

McCauley, R. N. (1966). The biological effects of oil pollution in a river. *Limnol. Oceanogr.* **11**, 475–486.

Metrick, D. F., Boddington, M. J., and Gelder, S. R. (1970). Distribution of freshwater triclads (Platyhelminthes: Turbellaria) in central-southern Ontario. *Proc. Conf. Great Lakes Res.*, 13th, pp. 71–81.

Neel, J. K. (1953). Certain limnological features of a polluted irrigation stream. *Trans. Amer. Microsc. Soc.* **72**, 119–135.

Newton, L. (1944). Pollution of the rivers of West Wales by lead and zinc mine effluent. *Ann. Appl. Biol.* **31**, 1–11.

Nuttall, P. M., and Bielby, G. H. (1973). The effect of china-clay wastes on stream invertebrates. *Envir. Pollution*, **5**, 77–86.

Oliff, W. D. (1960). Hydrobiological studies on the Tugela River system. II. Organic pollution in the Bushmans River. *Hydrobiologia* **16**, 137–196.

Oliff, W. D. (1963). Hydrobiological studies on the Tugela River system. III. The Buffalo River. *Hydrobiologia* **21**, 355–379, 5 unpaged leaves.

Oye, E. L. van (1941). Verbreitung und Ökologie der paludicolen Tricladen in Belgien. *Arch. Hydrobiol.* **38**, 110–147, plates 1–2.

Pickavance, J. R. (1971). Pollution of a stream in Newfoundland: Effects on invertebrate fauna. *Biolog. Conservation* **3**, 264–268.

Pravda, O. (1973). Über de Einfluss der Herbizide auf einige Süswassertiere. *Hydrobiologia*, **42**, 97–142.

Remkes, M. C. (1941). Planaria. *Aquarium* (*Philadelphia*) **10(1)**, 15–16.

Reynoldson, T. B. (1958). Observations on the comparative ecology of lake-dwelling triclads in southern Sweden, Finland and northern Britain. *Hydrobiologia* **12**, 129–141.

Scheuring, L., and Heuschmann, O. (1935). Ueber die Giftwirkung von Derrispräparaten auf Fische. *Allg. Fischerei-Z.* **60**, 370–378.

Schreier, O. (1949). Die schädigende Wirkung verschiedener Chinone auf *Planaria gonocephala* Dug. und ihre Beziehung zur Child'schen Gradiententheorie. *Österreich. Zool. Z.* **2**, 70–116.

Seibold, A. (1956). Die Einwirkung von organischen Fäulnisstoffen auf tierische Leitformen des Saprobiensystems. *Vom Wasser* **22**, 90–166.

Smith, A. J. (1967). The effect of the lamprey larvicide, 3-trifluoromethyl-4-nitrophenol, on selected aquatic invertebrates. *Trans. Amer. Fish. Soc.* **96**, 410–413.

Stammer, H. A. (1954). Der Einfluss von Schwefelwasserstoff und Ammoniak auf tierische Leitformen des Saprobiensystems. *Vom Wasser*, **22**, 34–71.

Steinmann, P., and Surbeck, G. (1918). "Die Wirkung organischer Verunreinigungen auf die Fauna schweizerischer fliessender Gewässer." Schweizerisches Departement des Innern, Inspektion für Forstwesen, Jagd und Fischerei, Bern.

Thienemann, A. (1911). Die biologische Untersuchung der Abwässer. *In* "Die Untersuchung landwirtschaftlich und gewerblich wichtiger stoffe" (J. König, ed.), pp. 1032–1050. Parey, Berlin.

Tucker, D. S. (1958). The distribution of some fresh-water invertebrates in ponds in relation to annual fluctuations in the chemical composition of the water. *J. Anim. Ecol.* **27**, 105–123, 1 unpaged table.

Voigt, W. (1905). Über die Wanderungen der Strudelwürmer in unseren Gebirgsbächen. *Verh. Naturhist. Vereins Preussischen Rheinlande, Westfalens Regierungsbezirks Osnabrück* **61**, 103–178.

Whitehead, M. M. (1963). The Triclads of Cattaraugus County, New York. Dissertation, St. Bonaventure Univ., St. Bonaventure, New York. (unpublished).

Wiebe, A. H. (1928). Biological survey of the upper Mississippi River, with special reference to pollution. *Bull. Bur. Fish.* (*U. S.*) **43(2)**, 137–167.

Wurtz, C. B., and Dolan, T. (1961). A biological method used in the evaluation of effects of thermal discharge in the Schuylkill River. *In Proc. Ind. Waste Conf., 15th Lafayette, Indiana, 3–5 May 1960. Eng. Bull. Purdue Univ.* **45(2)**, 461–472.

CHAPTER 4
Leeches (Annelida: Hirudinea)

ROY T. SAWYER

I. Introduction

Most of our understanding of the ecology of freshwater leeches is based on work on European species which are basically similar to those of North America. The two regions have in common a large number of genera (e.g., *Helobdella*, *Erpobdella*, *Glossiphonia*, *Batracobdella*, and *Piscicola*) and even several species (e.g., *Helobdella stagnalis*, *Glossiphonia complanata*, and *Glossiphonia heteroclita*) (Table I). Such European workers as Herter (1937) and Kalbe (1966) in Germany; Pawlowski (1936) and Sandner (1951) in Poland; Bennike (1943) and Berg (1938, 1948), in Denmark; Ökland (1964) in Norway; and Mann (1955a,b) and Tucker (1958) in England have given us insight into the physical, chemical, and biological factors which affect the distribution of leeches. There have been countless biological and faunistic accounts of European leeches by such workers as Herter (1929), Mann (1953; 1957a,b; 1961), Pawlowski (1955), Hatto (1968), and Gruffydd (1965). Our knowledge of the ecology of North American leeches is much more sketchy, but some thorough accounts, such as Herrmann (1970a), Kenk (1949), Klemm (1972a), Kopenski (1972), and Sapkarev (1968), have been recently published. In addition there have been a number of biological and faunistic studies by such workers as J. P. Moore (1901, 1906, 1912, 1924, 1936), Meyer (1937, 1940, 1946), J. E. Moore (1964, 1966), and Sawyer (1967, 1968, 1970, 1971, 1973). On the whole, however, North American leeches have received relatively little attention, primarily because of the problems of identification.

Even though leeches constitute a significant portion of the fauna of most kinds of freshwater habitats, almost nothing has been specifically written on the pollutional ecology of North American freshwater leeches. What little information that is available must be gleaned from many diverse sources, most of which are not primarily concerned with leeches. It is strange that this is so because so often certain leech species, especially *Helobdella stagnalis* and *Erpobdella punctata*, are associated with polluted conditions. Presented below is a preliminary analysis of the relationships of leeches to normal factors of the environment and their responses to various kinds of polluted conditions.

Until recently the keys and aids to identification of the freshwater leeches of all or parts of the United States and Canada were to be found in J. P. Moore (1912, 1918, 1959), Miller (1929, 1937), Meyer (1940, 1946), Pennak (1953), Mann (1961), J. E. Moore (1964, 1966), Soos (1965–1969), Hoffman (1967), Davies (1971), and Klemm (1972b), but most of these are outdated and incomplete. The recent monograph on North American leeches by Sawyer (1972) summarizes our current understanding of the systematics as well as the ecology of three of the four major families of leeches. For each species a complete American synonymy is included, along with an illustrated

TABLE I

Ecosystematic Classification of All North American (North of Mexico) Freshwater Leeches and Their Approximate Ecological Counterparts in Europe, Based on the Normal Hosts of Each Species[a]

Food Organism	Family	Leech: North America	Leech: Europe
Oligochaetes and insect larvae	Glossiphoniidae	****Helobdella stagnalis (L.)**	****Helobdella stagnalis (L.)**
		****Helobdella elongata** (Castle, 1900)	....
	Erpobdellidae	****Erpobdella punctata** (Leidy, 1870)	****Erpobdella octoculata (L.)**
		***Dina parva* Moore, 1912	***Erpobdella testacea* (Savigny, 1820)
		**Dina dubia* Moore and Meyer, 1951	**Dina lineata* (O. F. Müller, 1774)
		***Mooreobdella microstoma** (Moore, 1901) (W) }	....
		Morreobdella fervida (Verrill, 1874) (C) }	
		Mooreobdella bucera (Moore, 1949) }	
		Nephelopsis obscura Verrill, 1872 (C)	{ *Trocheta subviridis* Dutrochet, 1817 { *Trocheta bykowskii* Gedroyc, 1913
	Hirudinidae Aquatic species	***Haemopis (Percymoorensis) marmorata** (Say, 1824) (C)	**Haemopis (Haemopis) sanguisuga (L.)**
		Haemopis (Mollibdella) grandis (Verrill, 1874) (C) }	....
		Haemopis (*Percymoorensis*) *lateromaculata* Mathers, 1963 (C) }	
		Haemopis (*Percymoorensis*) *kingi* Mathers, 1954 (C) }	
		Haemopis (*Bdellarogatis*) *plumbea* Moore, 1912 (C) }	

TABLE I (*continued*)

Food Organism	Family	Leech: North America	Leech: Europe
	Terrestrial species	*Haemopis (Percymoorensis) terrestris* (Forbes, 1890)	*Xerobdella annulata* Autrum, 1958 (Haemadipsidae)
Snails	Glossiphoniidae	***Glossiphonia complanata** (L.)	***Glossiphonia complanata** (L.)
		***Glossiphonia heteroclita** (L.)	***Glossiphonia heteroclita** (L.)
		***Helobdella lineata** (Verrill, 1874) (W) *Helobdella fusca* (Castle, 1900) (C) *Helobdella papillata* (Moore, 1906) *Helobdella transversa* Sawyer, 1972 *Marvinmeyeria lucida* (Meyer and Moore, 1954) (C)	....
Fish	Glossiphoniidae	*Placobdella pediculata* Hemingway, 1908	*Hemiclepsis marginata* (O. F. Müller, 1774)
		Placobdella montifera Moore, 1906	....
		Actinobdella inequiannulata Moore, 1901 *Actinobdella annectens* Moore, 1906 *Actinobdella triannulata* Moore, 1924	....
	Piscicolidae	....	*Acanthobdella peledina* Grube, 1851 (C)
		***Piscicola punctata** (Verrill, 1871) ***Piscicola salmositica** Meyer, 1946 *Piscicola milneri* (Verrill, 1871)	***Piscicola geometra** (L.)
		***Illinobdella moorei** Meyer, 1940	....
		Cystobranchus verrilli Meyer, 1940 *Cystobranchus virginica* Hoffman, 1964	*Cystobranchus mammillatus* (Malm, 1863) *Cystobranchus respirans* (Troschel, 1859) *Cystobranchus fasciatus* (Kollar, 1842)
		Piscicolaria reducta Meyer, 1940	....

TABLE I *(continued)*

Food Organism	Family	Leech: North America	Leech: Europe
Amphibians	Glossiphoniidae	*Oligobdella biannulata* (Moore, 1900) (C)	....
		Batracobdella picta (Verrill, 1872) (C)	*Batracobdella paludosa* (Carena, 1824)
		Batracobdella phalera (Graf, 1899) (W)	*Batracobdella algira* (Moquin-Tandon, 1846) (W)
		Batracobdella michiganensis Sawyer, 1972	*Batracobdella verrucata* (Fr. Müller, 1884)
Reptiles	Glossiphoniidae	**Placobdella ornata** (Verrill, 1872) (C) **Placobdella multilineata** Moore, 1953 (W)	*Placobdella costata* (Fr. Müller, 1846)
		Placobdella parasitica (Say, 1824) *Placobdella hollensis* (Whitman, 1892) (C) *Placobdella papillifera* (Verrill, 1872)	....
Birds	Glossiphoniidae	**Theromyzon rude** (Baird, 1863) (C) *Theromyzon meyeri* (Livanow, 1902) (C)	**Theromyzon tessulatum** (O. F. Müller, 1774) (C) *Theromyzon maculosum* (Rathke, 1862) (C)
Mammals (and other vertebrate hosts)	Hirudinidae	**Macrobdella decora** (Say, 1824) (C) **Macrobdella sestertia** Whitman, 1886	*Hirudo medicinalis* (L.)
		Macrobdella ditetra Moore, 1953 (W) **Philobdella gracilis** Moore, 1901 (W) *Philobdella floridana* (Verrill, 1874) (W)	*Limnatis nilotica* Savigny, 1822 (W)

[a]Those species in bold print are common, widely distributed species and those in normal print are either uncommon or geographically restricted species. Those prefixed with one or two asterisks are occasionally (*) or commonly (**) associated with polluted water; and species thought to be temperature sensitive are suffixed with a (C) or (W) for cold-water or warm-water species, respectively. A dotted line indicates there is no ecological counterpart.

key and an exhaustive bibliography of the primary American literature on leeches. In addition information is also given on the relative abundance, distribution, and general biology of each species. The remaining family, the Piscicolidae, has been similarly treated by Meyer (1940, 1946) and Hoffman (1967).

II. Systematics

In the United States and Canada there are about 50 nominal species of freshwater leeches which belong to the four major families of the Hirudinea: the Glossiphoniidae, Piscicolidae, Erpobdellidae, and Hirudinidae. The North American leeches fall ecologically into four groups: a group of widely distributed, widely tolerant species; a group of predominantly northern species; a group of predominantly southern species; and a group of geographically restricted species. Most of the dozen or so species commonly found in polluted water belong to the first group of almost ubiquitous species. Many of these same species also live in Europe or have close European affinities. The northern or southern groups listed above are not so directly involved in our discussion of pollutional ecology of leeches. However, in some regions some of the early signs of pollution, especially thermal pollution, may be the replacement of certain predominantly northern species with warm-water species. In the brief systematic account that follows the focus will be on the forms which are directly or indirectly associated with disturbed environmental conditions. No significance should be attached to the order of presentation.

A. Glossiphoniidae

1. *Helobdella stagnalis* (L.)

Helobdella stagnalis, which is readily recognized by the small chitinous scute on its dorsal surface, is the most common leech in North America and is almost always the dominant leech in polluted water (Fig. 1). For this reason the ecology of this species deserves special mention in any discussion of pollution ecology. In addition to occurring on every continent except Australia, *Helobdella stagnalis* occurs abundantly throughout the northern United States and Canada, but is less common in the southeastern states.

Although most members of the genus *Helobdella* feed almost exclusively upon snails, *Helobdella stagnalis* feeds predominantly upon tubificids and related oligochaetes (e.g., *Ilyodrilus prespaenis*), tendipedid larvae (e.g., *Tendipes plumosus*), and various small crustaceans (e.g., *Asellus aquaticus* and *Hyalella azteca*) (Moore, 1912; Bennike, 1943; Sapkarev, 1963). That

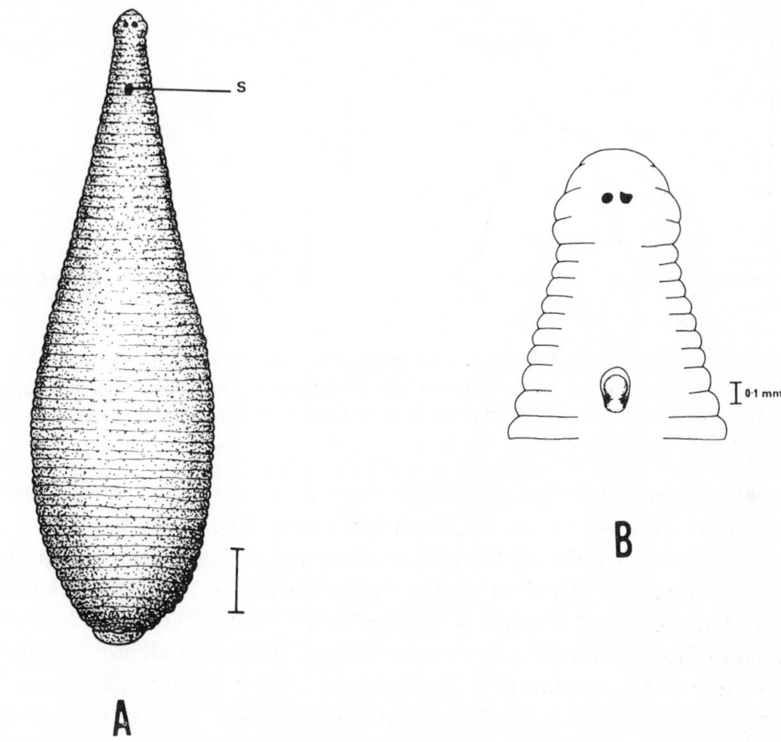

Fig. 1. *Helobdella stagnalis.* A, dorsal view (scale = 1 mm); B, close-up of head. s, scute (unique characteristic of this species). (Sawyer, 1972).

many of these host organisms often thrive in various kinds of polluted conditions probably accounts for the abundance of *Helobdella stagnalis* in many disturbed environments. While *Helobdella stagnalis* has been known to feed occasionally on the small snails, *Promenetus exacuous* and *Planorbis albus*, it apparently does not feed at all upon the larger snails *Helisoma trivolvis*, *Lymnaea emarginata*, and *Physa heterostropha* (Moore, 1966; Sawyer, 1972). Like other glossiphoniid leeches *Helobdella stagnalis* possesses an eversible proboscis which can be used to suck out the soft portions of the bodies of these animals. An increase in temperature caused an increase in the rate of predation of *Helobdella stagnalis* on the larvae of *Tendipes plumosus*, the increase being greatest between 10° and 20°C (Hilsenhoff, 1963). After 38 days the average number of larvae fed upon by the leech was 1.2 at 5°C and 11.1 at 25°C. Similarly the average number of days between feeding was 31.7 at 5°C and 3.4 at 25°C. The rate of predation and the interval between feeding also depended a great deal upon the density of the leeches.

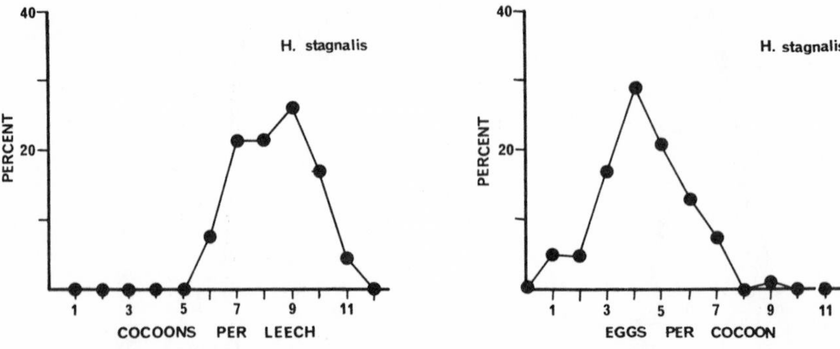

Fig. 2. *Helobdella stagnalis* from a small permanent pond in Michigan. Left, distribution of the number of cocoons carried by 23 individuals (average number of cocoons per individual = 8.35; right, distribution of the number of eggs contained in 193 cocoons (average number of eggs per cocoon = 4.23). (Sawyer, 1972).

Hilsenhoff concluded that *Helobdella stagnalis* exerts a considerable influence on the *Tendipes plumosus* population in any lake where both occur.

Like other *Helobdella*, *Helobdella stagnalis* lays its eggs in thin transparent cocoons which are attached to the ventral surface of the parent (Sawyer, 1971). The eggs and the newly hatched embryos and young are carried around in this manner for a few weeks. In Michigan each individual deposits an average of 8.4 cocoons (Fig. 2), and each cocoon contains an average of 4.2 eggs (Sawyer, 1972). The larger individuals deposit even more cocoons. Brooding individuals are encountered in Michigan from early April to August. In England each individual broods an average of 13–17 eggs (Mann, 1957). The eggs are deposited predominantly in April, and the parents are dead by June soon after the young have been liberated. In Denmark the length of its breeding season, i.e., from May 15 to the end of August, is similar to, if somewhat earlier than, other portions of Europe (Bennike, 1943). Breeding starts when the water is about 12°–13°C; each individual deposits an average of 20 (7–37) eggs. Only in the Alps at an altitude of 2500 m is the breeding season later, from the end of July to early August.

In certain circumstances a population of *Helobdella stagnalis* may produce two generations a year (Mann, 1967). In England a proportion of the offspring hatched in April and May are breeding in July and August and the others reproduce the next spring. The leeches which hatch in August also reproduce the next spring. Few, if any, individuals of this species live for more than one year and many die after only three months of life. It is very doubtful if one individual could produce two broods in one year. In Denmark more than one generation may also be produced in a season. In Lake Washington *Helobdella stagnalis* takes one year to complete its life cycle (Thut, 1969). Egg-laying starts as early as January, but occurs predominantly in early May. The breeding season lasts until late September. Each adult

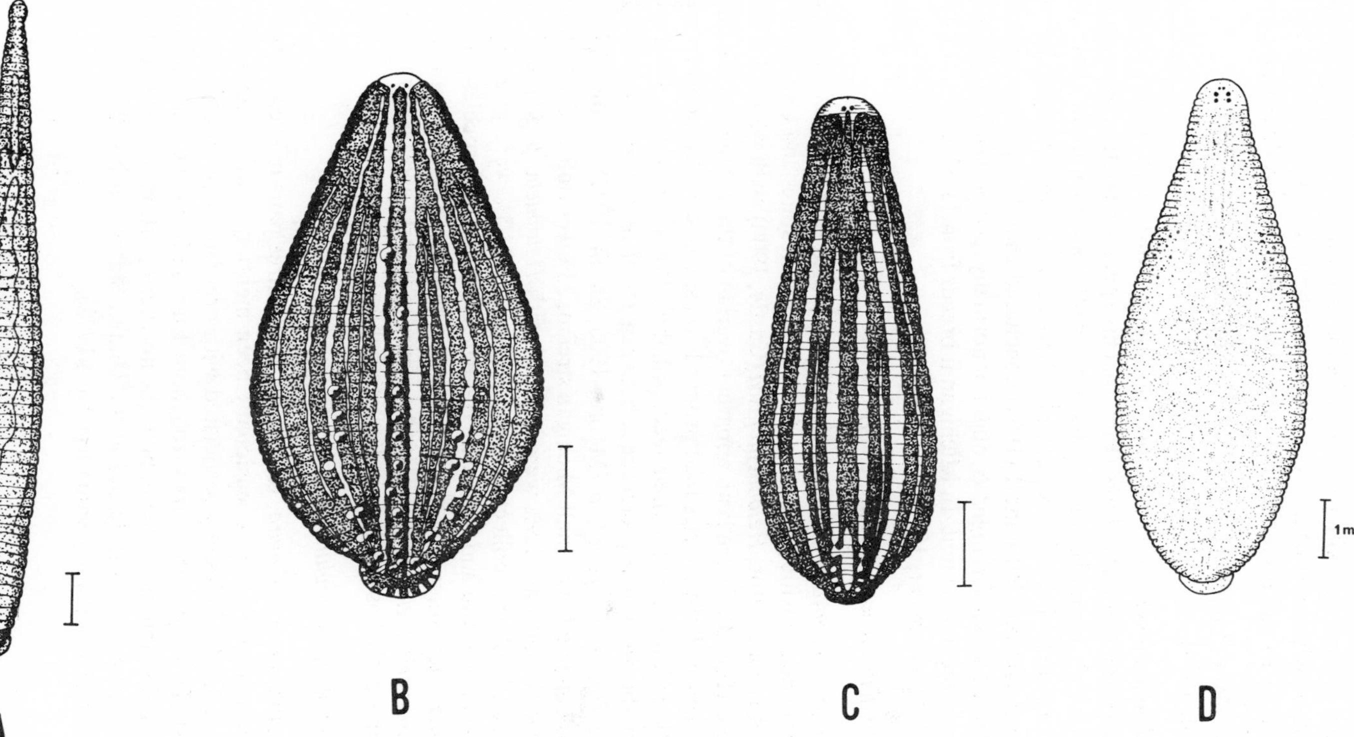

Fig. 3. Dorsal views of common North American glossiphoniids. A, *Helobdella elongata*; B, *Helobdella lineata*; C, *Helobdella fusca*; D, *Glossiphonia heteroclita*. (Sawyer, 1972).

carries an average of 14.5 (6–36) young and the adults die at the end of the breeding season. In Alberta the breeding season is from May to early September, with most of the eggs deposited about mid-May (Moore, 1964). Moore also found evidence that two generations are produced in a year.

2. *Helobdella elongata* (Castle, 1900) = *Helobdella nepheloidea* (Graf, 1899)

Helobdella elongata is widely, but sporadically, distributed from southern Canada to the southern states (Fig. 3). This species has often been associated with high levels of organic pollution, but little is known about its biology. It appears likely that it feeds on snails, oligochaetes, or insects. After laboratory investigations Hilsenhoff (1964) concluded that this species, unlike its congenitor *Helobdella stagnalis*, does not normally feed on *Tendipes plumosus* to any appreciable extent. In Michigan it breeds from late May to early June (water 21°C) (Sawyer, 1972).

3. *Helobdella lineata* (Verrill, 1874)

Helobdella lineata, which may also be associated with polluted water, is a southern species which is replaced in the nothern states and Canada by its cold water congenitor, *Helobdella fusca* (Castle, 1900), with which it is often confused (Fig. 3). There is an apparent overlap of their ranges in the Great Lakes region. Both species feed primarily on snails, e.g., *Helisoma trivolvis*, but little is known about their feeding habits. Some interesting differences exist in the snail host preferences between *Helobdella lineata* and its near relative *Helobdella papillata* (Moore, 1906; Sarah, 1971). Of the *Helisoma trivolvis* collected in a small Michigan stream, 23% were infected with *Helobdella lineata* and only 1.9% were infected with *Helobdella papillata*. On the other hand, of the collected *Helisoma anceps* 24.5% were infected with *Helobdella papillata* and 0.9% with *Helobdella lineata*. Both species preferred *Helisoma* over *Stagnicola palustris* or *Physa gyrina*. The uncommon leech *Marvinmeyeria lucida* (Moore, 1954) occurs in central and western Canada and apparently has similar snail-eating habits.

4. *Glossiphonia complanata* (L.)

Glossiphonia complanata, recognized by its three pairs of eyes and pair of longitudinal stripes, has occasionally been mentioned as a species directly or, more likely, indirectly associated with organically polluted conditions (Fig. 4). This species occurs throughout Eurasia from Western Europe to Japan, and is abundant throughout the northern portion of North America. To date it has not been found in the United States south of Missouri, the Ohio River, and Pennsylvania (Sawyer, 1972).

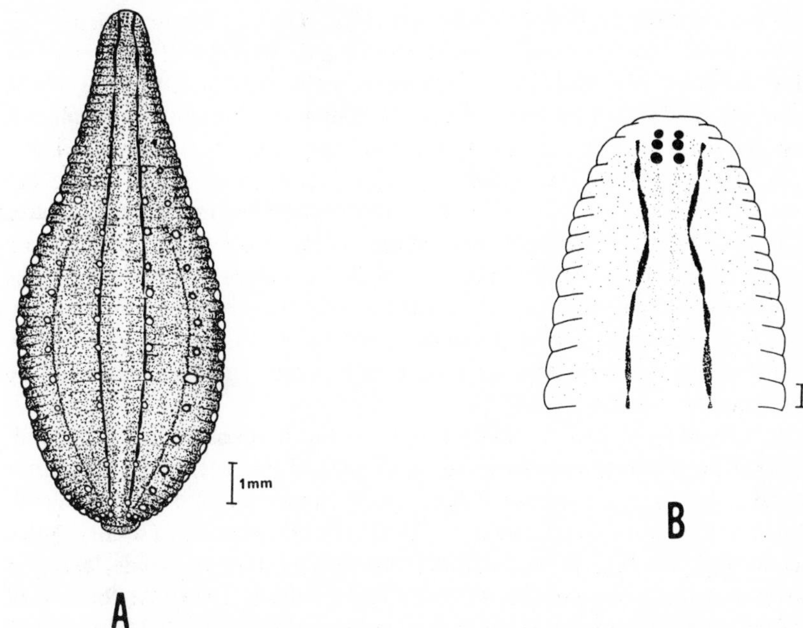

Fig. 4. *Glossiphonia complanata.* A, dorsal view; B, close-up of head (scale = 0.1 mm) (Sawyer, 1972).

By means of an eversible proboscis, *Glossiphonia complanata* can suck out most of the body fluids from a variety of invertebrate hosts, especially snails. This species feeds predominantly upon snails. (e.g., *Ancylus fluviatilis, Bithynia tentaculata, Hydrobia jenkini, Lymnaea reflexa, Lampsilis siliquoides, Promenetus exacuous, Physa fontinalis, Physa gyrina, Physa heterostropha, Physa integra, Planorbis corneus, Planorbis parvus,* and *Planorbis vortex*), and to a less extent upon aquatic oligochaetes (e.g., *Limnodrilus, Ilyodrilus hammoniensis,* and *Tubifex ochridanus*) and possibly insect larvae (Bennike, 1943; Mann, 1955; Moore, 1964; Sapkarev, 1963, 1968; Klemm, 1972a).

The reproductive biology of *Glossiphonia complanata* is fairly well known. The eggs are tightly enveloped in delicate membranous sacs (cocoons) containing little albumen. The sacs are attached to rocks or other hard objects while the parent covers them with its body for the 5–6 days prior to hatching (Sawyer, 1971). On hatching, the young (about 1.5 mm or less) attach to the ventral surface of the parent and are carried about in this manner for 2–3 weeks. In any given locality this species has a well-defined, temperature-dependent, breeding season. In Michigan most of the adult individuals deposit their eggs around April 30 (April 5 to June 6) when the daytime water temperature is about 15°C (Sawyer, 1967). On the average each individual produces 6.24 cocoons, each containing an average of 20.6 eggs. In Lake Mendota, Wisconsin, the breeding season is from the end of May to the beginning of July (Sapkarev, 1968). In Denmark breeding *Glossiphonia complanata* are first encountered between May 5 and 25, when the water is 12.5°C. The number of young is about 45 (25–67) (Bennike, 1943). In England most of the breeding occurs when the minimum and maximum temperatures are 8° and 14°–15°C, respectively (Mann, 1955). Each adult produces an average of 26 young, which hatch from 2–3 cocoons. Individuals in the warmer portions of the stream (warmed by the sun) breed earlier, and conversely those in the colder portions breed later than the average. Also, leeches in more dense aggregations breed earlier than do those not so aggregated.

Glossiphonia complanata takes one or two years to reach maturity, depending upon local conditions. In either case most of the leeches die soon after breeding. In a small permanent pond in Michigan this species has a well-defined annual life cycle (Sawyer, 1972). On the other hand in a small English stream eggs are laid in two distinct broods, the two-year-olds laying in March and the one-year-olds in April–May (Mann, 1955). At the end of the first year 100% of the March brood are breeding and only about 40% of the April–May brood are breeding, but practically all have bred by the end of the second year. Most of the leeches die soon after breeding. The mortality rate is 97% during the first 6–8 months, 33% during the next year, and 84–88% during the next year. Only 5–6% of the adult population are three years old. The rates of growth of the two broods are similar except that the April–May brood grows more slowly during the first few months of life.

Glossiphonia complanata is most commonly found in lakes and running water, especially in those with a stone bottom and with a high calcium content. It occurs only in the oligotrophic and, especially, the eutrophic lakes and is not found in dystrophic or acidotrophic lakes. This species has been found at pH values as low as 5.5, bicarbonate contents not higher than 3.6 ppm, and in water with only 0.69 ml 0_2/liter (Bennike, 1943; Mann, 1955a,b).

In Lake Mendota *Glossiphonia complanata* occurs in all types of habitats but reaches its highest density, up to 266 individuals/m^2, at a depth of 0–1 m (rarely lower) where the bottom is covered with rocks (Sapkarev, 1968). Like many other leeches, *Glossiphonia complanata* displays seasonal changes in its average population density. It has its lowest density (11 individuals/m^2) in the winter and spring months and its highest density (100 individuals/m^2) in the summer and fall months. The biomass of *Glossiphonia complanata* reaches its maximum in the fall (0.93 gm wet weight/m^2 or 0.20 gm dry

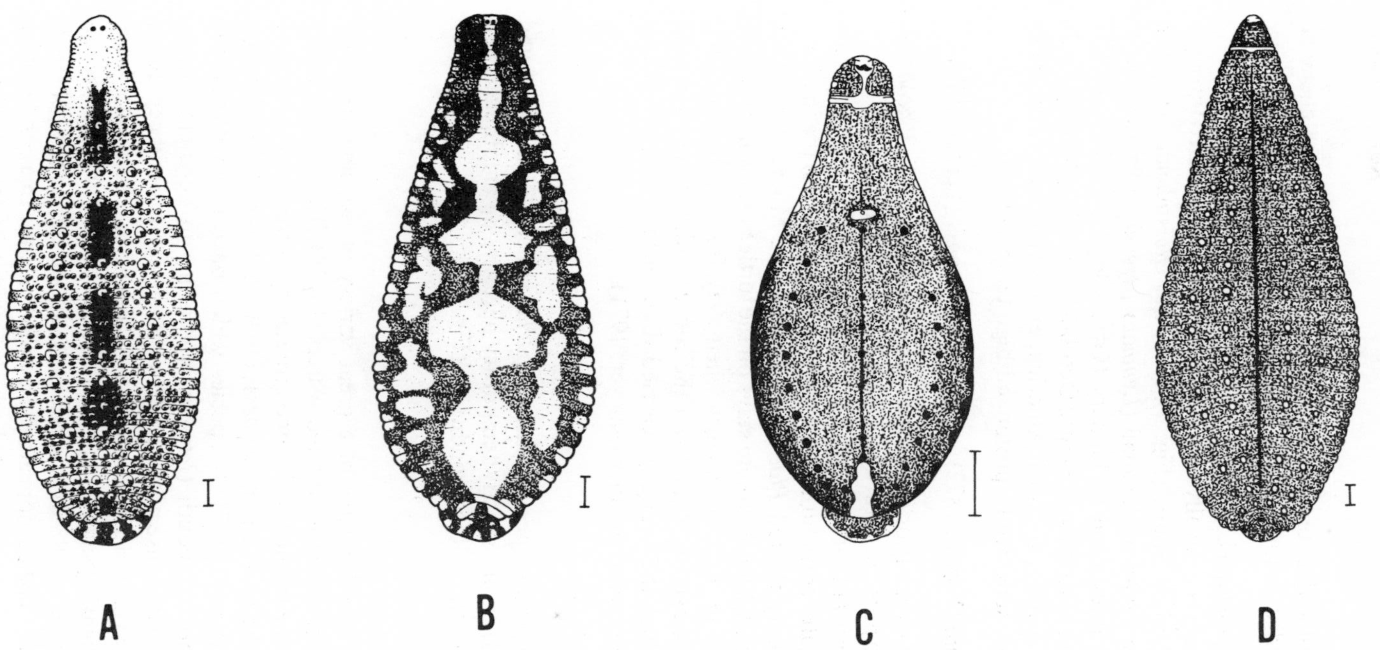

Fig. 5. Dorsal views of common North American glossiphoniids. A, *Placobdella ornata*; B, *Placobdella parasitica*; C, *Batracobdella phalera*; D, *Batracobdella picta*. Scale = 1 mm (Sawyer, 1972).

weight/m^2). The roles that temperature and other physical and chemical factors play in the distribution of this species are discussed below.

5. *Glossiphonia heteroclita* (L.)

This holarctic species is locally abundant and widely distributed throughout the northern United States (Fig. 3). In Wales *Glossiphonia heteroclita* inhabits the mantle cavity of the snail (*Lymnaea pereger*) from October to May, with a peak infestation in January (Gruffydd, 1965; Hatto, 1968). During the breeding season from May to October, they are free-living. Little is known about its ecology in North America, except that it feeds on snails. This species appears sensitive to prolonged high temperatures.

6. *Placobdella ornata* (Verrill, 1872) = *Placobdella rugosa* (Verrill, 1874)

Like its common congenitor *Placobdella parasitica* (Say, 1824), *Placobdella ornata* is primarily a parasite of turtles (Fig. 5). Any association of this species with polluted water is at best indirect. It is a common species throughout the northern United States and southern Canada, and is replaced in the southern states with the closely related *Placobdella multilineata* Moore, 1953. Like most species of *Placobdella*, *P. ornata* stays on the turtle host until the breeding season, usually in June to August when they leave to deposit their cocoons on a solid substrate. They cover the cocoons with their bodies until they hatch, after which the young are carried about on their ventral surfaces in the typical glossiphoniid manner (Sawyer, 1971).

7. *Batracobdella*

The two most common species, *Batracobdella phalera* (Graf, 1899) and *B. picta* (Verrill, 1872), are strictly temporary parasites on amphibians and are not associated with polluted conditions (Fig. 5).

B. Piscicolidae

The Piscicolidae are parasites of a great variety of fish and are rarely associated with most types of polluted waters (Fig. 6). This family, which includes many marine members, has been treated by meyer (1940, 1946) and Hoffman (1967). The most common species in the nothern United States is *Piscicola punctata* (Verrill, 1871), a species which once reached epidemic proportions on the red-mouth (*Ictiobus cyprinellus*) in northern Illinois (Thompson, 1927). The cause of this epidemic is unclear. A similar epidemic is described by Sawyer and Hammond (1973) for the brackish water leech *Calliobdella carolinensis* (Sawyer and Chamberlain, 1972) which was epidemic on the menhaden, *Brevoortia tyrannus*, during the winters of 1970–1971 and 1971–1972. The authors attributed the epidemic to a combination of unusually low temperatures and the high turbidity of the water. In the

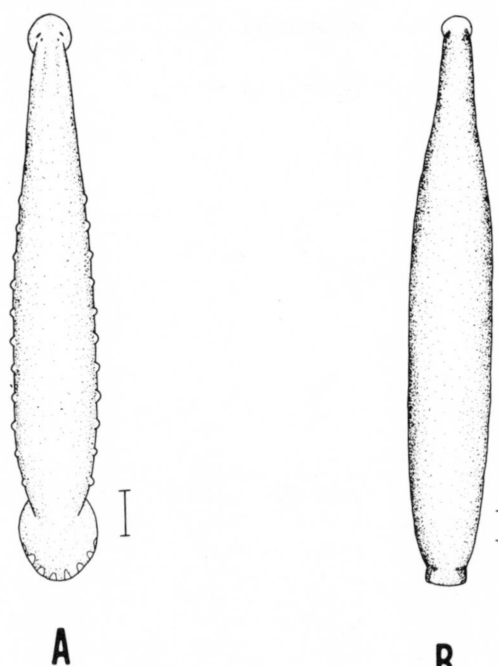

Fig. 6. Dorsal views of common North American piscicolids. A, *Piscicola punctata*; B, *Illinobdella moorei*. Scale = 1 mm. (Sawyer, 1972).

northwestern states the most common piscicolid is *Piscicola salmositica* Meyer, 1946, a parasite on salmonid and other fish. In the central and southern states the most common species is *Illinobdella moorei* Meyer, 1940, which occurs on a great variety of fish hosts. In the brackish water areas of the southern states it is especially common on the mullet (*Mugil cephalus*) and the catfish (*Ictalurus catus*), on which they can reach near epidemic proportions.

C. Erpobdellidae

1. *Erpobdella punctata* (Leidy, 1870)

Erpobdella punctata, recognized by the paramedial rows of pigmentation and a two-annulus separation of the gonopores, is one of the most commonly encountered and widely distributed species of freshwater leeches in North America (Fig. 7). The ecology of this species deserved special attention because no other American leech, with the possible exception of *Helobdella stagnalis*, has been so consistently associated with polluted water. It occurs throughout Canada and over most of the United States as far south as

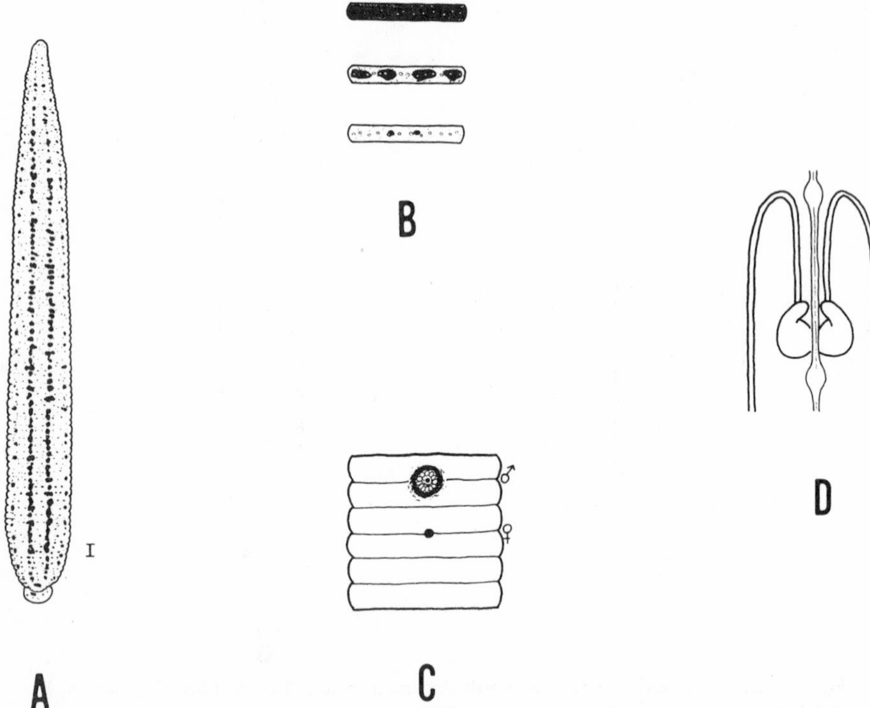

Fig. 7. *Erpobdella punctata*. A, dorsal view (scale = 1 mm); B, three annuli showing color variants; C, ventral view showing the two annuli separation of the male and female gonopores; D, dorsal view of male reproductive system and a portion of the ventral nerve cord. (Sawyer, 1972).

Mexico. Like several other leech species which are common in the northern United States, *Erpobdella punctata* is scarce in the southeastern states. *Erpobdella punctata* is typical of other erpobdellids in being strictly a scavenger and a predator; there is no evidence that it is ever a parasite. It feeds primarily upon oligochaetes and the larvae of Odonata, Coleoptera, Trichoptera, Ephemeroptera, and Diptera (Simuliidae and Tendipedidae) (e.g., *Caenis*, *Hexagenia*, and *Polycentropus*) (Muttkowski, 1918; Moore, 1920; Moore, 1966; Sapkarev, 1968; Sawyer, 1970). That many of these organisms are commonly associated with polluted conditions probably accounts for the abundance of *Erpobdella punctata* in many polluted environments.

Our understanding of the reproductive biology and life cycle of *Erpobdella punctata* comes mainly from a study by Sawyer (1970) on two populations in southeastern Michigan. One population occurs in a small, permanent pond, whereas the other occurs in a small stream that dries up in the summers.

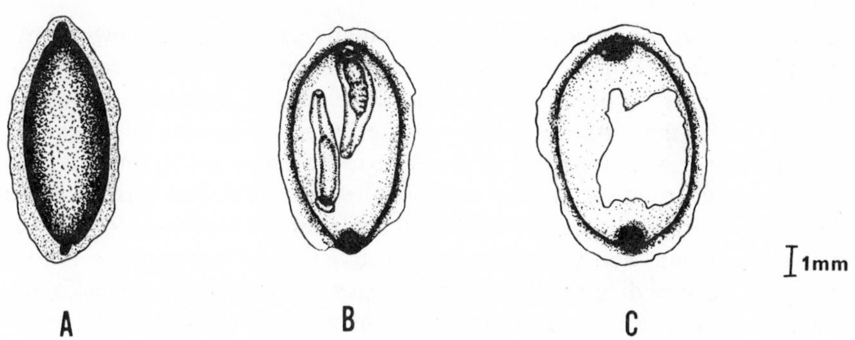

Fig. 8. A, dorsal view of cocoon of *Erpobdella punctata*; B, young of *Erpobdella octoculata* ready to leave the cocoon via the terminal plug; C, cocoon of *Erpobdella octoculata* emptied by a snail. (A: Sawyer, 1972; B–C: Bennike, 1943).

Erpobdella punctata is especially active on damp spring nights when they will briefly leave the water in search of food. Like most leeches, this species feeds most voraciously just prior to the reproductive season. This species is known to make rather remarkable upstream migrations prior to breeding. The small eggs of *Erpobdella punctata* are contained in brown, helmet-shaped cocoons which are attached to pebbles, vegetation, and other suitable hard substrates (Fig. 8). In Michigan cocoon deposition begins in April, soon after the water has suddenly warmed on the melting of the winter ice from the winter level of 0.3°C to about 12°C (Fig. 9). Peak numbers are found in May after which egg-laying gradually declines. The high temperature of the water after mid-May, which precedes the drying of the stream at the end of June, probably plays a role in this decline. On the average each individual

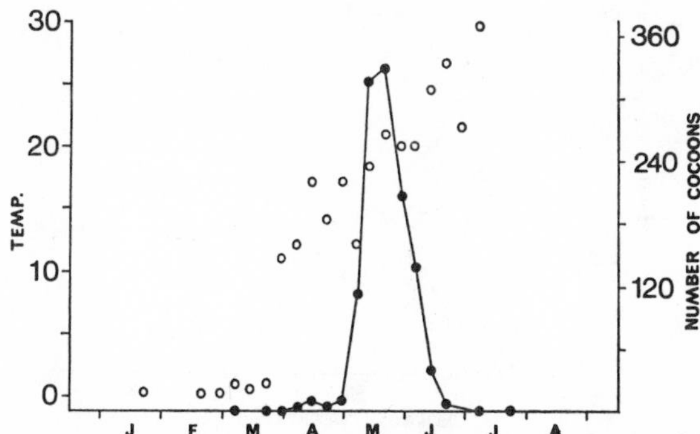

Fig. 9. Relationship between water temperature and cocoon deposition by *Erpobdella punctata* in a small Michigan stream. Open circles, water temperature (°C) from January to June; solid circles, the number of cocoons of *Erpobdella punctata* collected at weekly intervals. (Sawyer, 1970).

deposits about ten cocoons and each initially contains an average of five eggs. There is a steady decline in the number of eggs per cocoon each week. The young hatch in about 3–4 weeks after being laid. The size distributions within each of the two populations indicate that growth to maturity takes one year in the permanent pond, but two years in the temporary stream, presumably because growth is interrupted when the stream dries up in the summer (Fig. 10). Mortality in the stream is estimated at about 93, 73, and 93% during the first, second, and third years, respectively. In either case few survive to a second breeding season. A major cause of mortality, at least in the laboratory, is a cannibalistic devouring of many of the freshly deposited cocoons. In nature the cocoons are also attacked by some predator, presum-

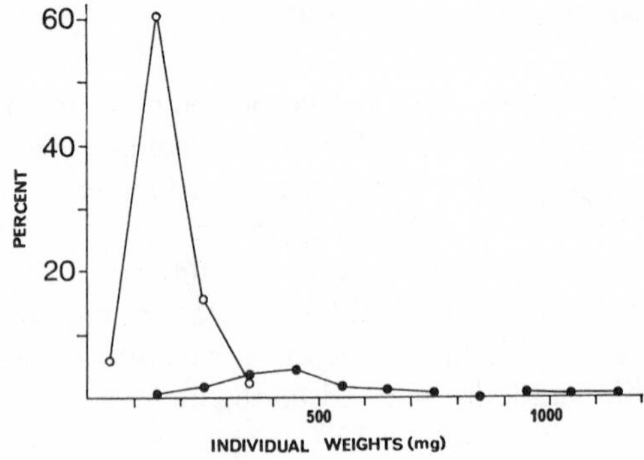

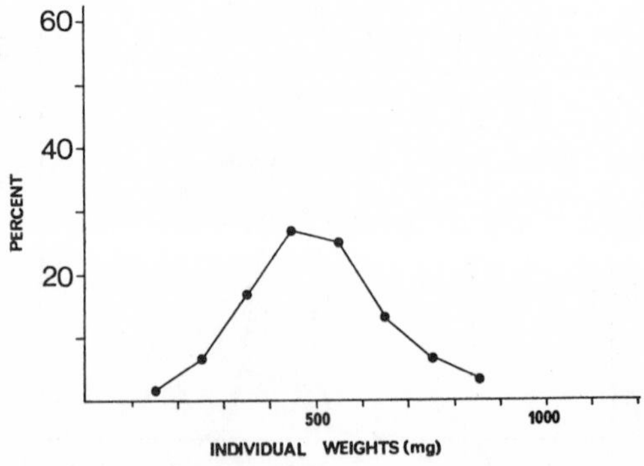

Fig. 10. Size distributions of *Erpobdella punctata* from Michigan from a temporary stream (above) and a permanent pond (below). Above, three distinct groups were found on 16 June 1967, one immature and two mature groups (N = 503); below, a single group was found on 3 June 1967 (N = 60), (Sawyer, 1970).

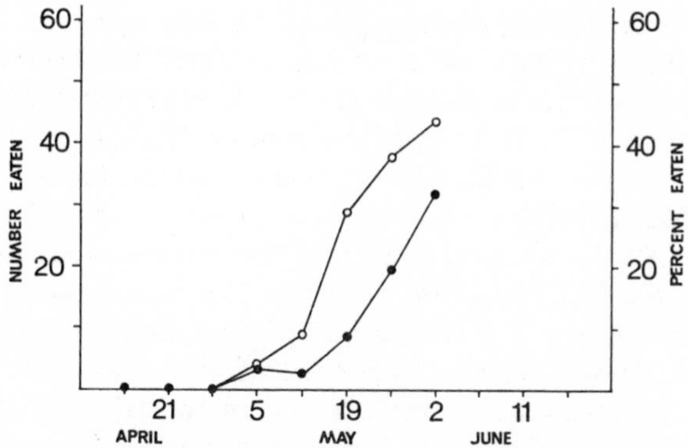

Fig. 11. *Erpobdella punctata.* The number (open circles) and percentage (solid circles) of the cocoons eaten by snails in a small Michigan stream. (Sawyer, 1970).

ably snails, which create irregular holes completely through the cocoon wall (Fig. 8). In each succeeding week increasing numbers of cocoons are thus destroyed, up to about 32% toward the end of the breeding season (Fig. 11).

In Lake Mendota optimal conditions for this species are along the shoreline where coarse gravel and stones are intermixed (Sapkarev, 1968). In such rocky conditions the density of adults can reach 1332 individuals/m². Upon the emergence of the newly hatched young in late summer the density can be considerably greater (Fig. 12). The number of cocoons deposited along the shoreline are between 888 and 2220/m² (Muttkowski, 1918). In Lake Mendota the greatest number of individuals are found at a depth of 0–0.5 m.

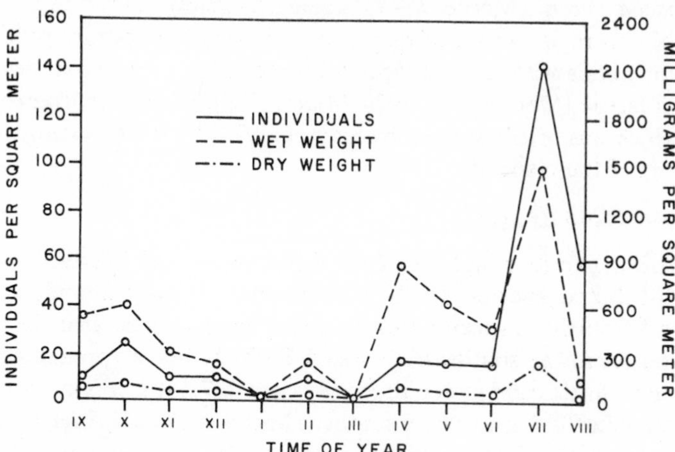

Fig. 12. Average seasonal changes in the littoral zone of Lake Mendota, Wisconsin, during 1964–1965 (Sapkarev, 1968).

The life history of *Erpobdella punctata* closely resembles that of its Eurasian counterpart *Erpobdella octoculata*, which has been thoroughly studied by such workers as Kulajew (1929), Bennike (1943), Mann (1953), and Pawlowski (1953). In England cocoons of *Erpobdella octoculata* are deposited in June and July, and the young emerge during August and September (Mann, 1953). Each leech produces an average of 23.5 young of which about 89% do not survive the first four months of life. Most of the leeches breed the first year of life, but about 13% require two or even three years to reach maturity. The related European species, *Erpobdella testacea*, is usually found in small temporary ponds and, unlike *Erpobdella octoculata*, in dystrophic waters, even those containing humic acids (pH 4.6–5.1) (Mann, 1961; Bennike, 1943). *Erpobdella testacea* would appear to be a pollution-tolerant species. In North America there is no systematic counterpart to this species, but its ecological counterparts may be *Mooreobdella bucera*, *M. microstoma*, or *Dina parva*. The physical and chemical factors which influence the distributions of *Erpobdella punctata*, *E. octoculata*, and *E. testacea*, and the tolerance of these species to various forms of pollution are discussed below. The excellent papers by Elliott (1973a,b) arrived too late to be included in the account.

2. *Mooreobdella microstoma* (Moore, 1901)

Mooreobdella microstoma is especially abundant in the southern states and extends up the Mississippi and Ohio River systems and into southwestern Lake Erie. In the northern states and in Canada this species is replaced by its cold water congenitor, *Mooreobdella fervida* (Smith and Verrill, 1871). *Mooreobdella bucera* (Moore, 1949) is geographically restricted to southeastern Michigan where it occurs primarily in small temporary ponds and permanent ponds without drainage. All three species feed upon tubificids and insect larvae (Moore, 1912, 1920; Miller, 1929). Except for *Mooreobdella bucera*, which was briefly studied by Sawyer (1972), little else is known about the biology of these species.

3. *Dina parva* (Moore, 1912)

Dina parva, like its congenitor *Dina dubia* Moore and Meyer, 1951, is a widely distributed species which is occasionally locally abundant. Both appear to be northern species, but there also appears to be an undescribed related species in the southeastern states. Both species are predaceous and feed on oligochaetes and insect larvae (Sawyer, 1972). Little is known about reproduction in *Dina parva*, but in southern Michigan *D. dubia* breeds in early May (Sawyer, 1972). Each individual deposits 7.9 (5–13) cocoons containing an average of 4.15 (1–9) eggs.

4. *Nephelopsis obscura* (Verrill, 1872)

Nephelopsis obscura is a large (5–6 cm) species restricted to the cold waters of the extreme northern United States, the Rocky Mountains region, and Canada. This species feeds primarily upon oligochaetes and insect larvae, but has been reported feeding on the wastes of a fish-packing station (Moore, 1912, 1924). *Nephelopsis obscura* has a summer breeding season and each cocoon contains 5–10 eggs, but otherwise little is known about reproduction of this species.

D. Hirudinidae

The large (6–10 cm) hirudinid leeches, which are probably the best known of all American leeches, are the so-called "bloodsuckers" which are so dramatically represented in films such as "African Queen." If the truth were known, only a very few suck mammalian blood. The most common hirudinid leeches in the northern United States and Canada are the American medicinal leech, *Macrobdella decora* (Say, 1824) and the horse leeches, *Haemopis* (*Percymoorensis*) *marmorata* (Say, 1824) and *Haemopis* (*Mollibdella*) *grandis* (Verrill, 1874) (Fig. 13). *Haemopis* (*Percymoorensis*) *terrestris* (Forbes, 1890) is a truly terrestrial species and is primarily restricted to

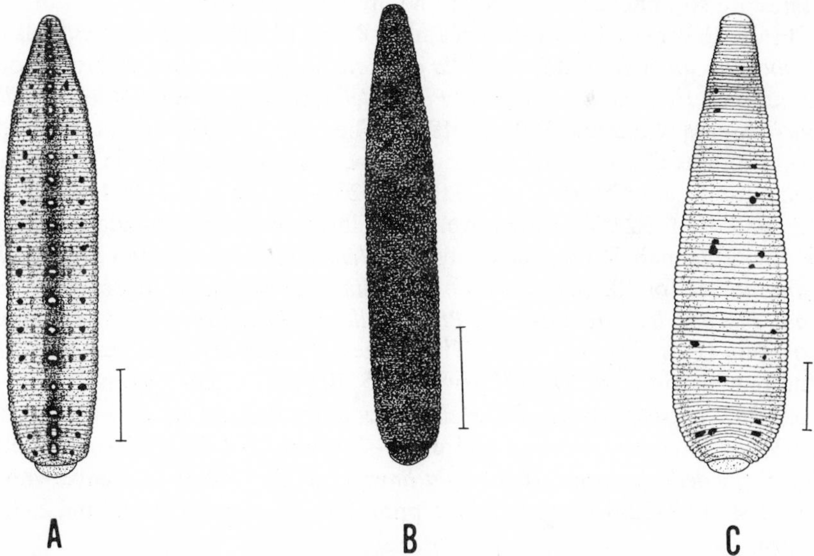

Fig. 13. Dorsal view of common North American hirudinids. A, *Macrobdella decora*; B, *Haemopis* (*Percymoorensis*) *marmorata* (color variable); C, *Haemopis* (*Mollibdella*) *grandis* (color variable). Scale = 1 cm. (Sawyer, 1972).

Illinois, Indiana, and Ohio. In the southern United States there are only two common hirudinids, *Philobdella gracilis* Moore, 1901 and *Macrobdella ditetra* Moore, 1953. The only European hirudinid species are *Haemopis sanguisuga* (L.), *Limnatis nilotica* Savigny, 1822, and *Hirudo medicinalis* (L.). In spite of the fact that these species commonly eat many kinds of macroinvertebrates such as oligochaetes, insect larvae, and snails, they are hardly ever associated with polluted conditions and, therefore, will not be treated further in this discussion.

E. Ecology of the Leeches of the Southern United States

Throughout the northern United States and Canada the dominant leech species are *Helobdella stagnalis*, *Glossiphonia complanata*, and *Erpobdella punctata*, all three of which are commonly associated with polluted water. In the southern United States *Glossiphonia complanata* is completely absent, and *Helobdella stagnalis* and *Erpobdella punctata* are either absent or uncommon in broad expanses of this region (Sawyer, 1972). At first glance the restricted distributions of these species are unexpected because of their seemingly broad tolerances of environmental extremes in the northern states or in Europe. The reasons for their ecological restrictions in the southern states are still unclear and need to be investigated further.

In the lower reaches of the Escambia River, Florida, a few individuals of *Helobdella stagnalis* and *Erpobdella punctata* are found, along with the more numerous *Helobdella elongata*, *Placobdella parasitica*, and *Mooreobdella microstoma* (Wurtz and Roback, 1955). The ranges of alkalinity, total hardness (both as $CaCO_3$), Ca, pH, and temperature of the water in the spring and autumn months are 21.1–29.0 ppm, 20.6–160 ppm, 6.4–15.4 ppm, 6.7–7.1, and 19.0°–22.0°C, respectively. Similarly, in the intermediate regions of the Savannah River, relatively few *Helobdella stagnalis* and *Erpobdella punctata* are found, whereas *Helobdella lineata* (= *fusca*), *Helobdella elongata*, *Placobdella parasitica*, and *Placobdella multilineata* (= *rugosa*) are more common (Patrick *et al.*, 1967). The ranges of alkalinity, hardness, Ca, pH and temperature are 20.0–30.0 ppm, $>$ 5–10 ppm, $<$ 3.0–3.0 ppm, 6.5–7.0, 23.0–28.0°C, respectively for *Helobdella stagnalis*, and $>$ 10.0–20.0 ppm, 10.0– $<$ 50.0 ppm, $<$ 3.0–3.0 ppm, 6.5–7.0, and 12.0°–17.0°C, respectively, for *Erpobdella punctata*. *Helobdella lineata* has the widest tolerance with a range of 14.7–33.6 ppm, 8.0–21.9 ppm, 2.0–5.8 ppm, 6.5–7.0, and 23.0°–29.0°C.

As in most southern states, *Helobdella lineata* is by far the most common leech in South Carolina. In this state both *Helobdella stagnalis* and *Erpobdella punctata* are restricted to the basins of a few major rivers or lakes: *Helobdella stagnalis* is known only from around Lake Marion (limestone area) and from the lower Black River; and *Erpobdella punctata* is known only from the upper to middle portions of the Savannah River. On the whole the state can be divided into two faunistic zones: the Coastal Plain, which has many leeches; and the Piedmont, which is relatively devoid of both numbers and kinds of leeches, especially erpobdellids. The thorough summary of the quality of the surface waters of South Carolina by Cummings (1969) sheds little light on the unusual ecological distributions of the leeches in this state, except that the state does have unusually soft water (1–60 ppm, $CaCO_3$). In Louisiana *Helobdella stagnalis* is absent and *Erpobdella punctata* is uncommon (Sawyer, 1967). In central and southern Florida, a region of surface limestone, *Helobdella stagnalis* is a common species, but *Erpobdella punctata* is absent. In the light of the above discussions, it is tempting to speculate that in the southern United States, *Erpobdella punctata* and *Glossiphonia complanata* are limited by the high water temperatures, whereas *Helobdella stagnalis* is limited by the hardness of the water. As a group, leeches can easily withstand alkaline humus material but cannot tolerate acid humus material, thus accounting for the relative scarcity of leeches in sphagnum bogs and dystrophic lakes in Denmark and possibly in many of the swamps of the southeastern United States.

III. Factors Which Influence the Distribution of Freshwater Leeches

For the most part it is the quantitative, rather than the qualitative, composition of the leech fauna which characterizes the different types of habitats. Few leech species, if any, are so ecologically restricted that a single environmental factor can determine their distributions. As with most freshwater organisms, the ecological distributions of leeches are usually determined by two or more physical, chemical, or biological characteristics of the environment. The most important of these environmental factors, in approximate order of significance, are (1) availability of food organisms, (2) the nature of the substrate, (3) the depth of the water, (4) water currents (lentic *vs*. lotic), (5) size and nature of the body of water, (6–8) hardness, pH, and temperature of the water, (9) minimum concentration of dissolved oxygen, (10) siltation and turbidity of the water, and (11) the salinity of the water. The following account is a summary of our relatively sketchy understanding of the roles that these and other environmental factors play in the distribution of leeches. It must be stressed that much work remains to be done before we can have a clear picture of this problem.

A. Food Organisms

No other single factor is more important in restricting the distribution of freshwater leeches than the availability of food organisms. However, no definitive work has ever been done on this difficult problem. It is well established that most leeches are not host specific, but they tend to restrict their diets to certain groups of animals. For example, piscicolids feed on fish; *Batracobdella* on amphibians; *Placobdella* on turtles; *Theromyzon* on aquatic birds; *Glossiphonia* and most *Helobdella* on snails; and *Helobdella stagnalis* and erpobdellids on aquatic insect larvae and oligochaetes. Space does not allow here for further discussion of the very important role that the presence and relative abundance of these food organisms play in the distribution of the various leech species, except to say that any factor which disrupts the distributional patterns of the host would directly affect the leeches which are so dependent upon them.

B. Substrate

A solid substrate is necessary for the proper functioning of the leech sucker, a characteristic shared by all leeches. The suckers are used for locomotion, feeding and reproduction and cannot function well in pure mud, sand or in other substances in which strong suction cannot be attained. In addition, a solid substrate is required by most leeches for the deposition of cocoons. In Lake Wigry, Poland, leeches prefer the rocky bottom (68.4 individuals/m^2), the *Chara* bottom (52.0 individuals/m^2), the calcareous soil bottom ("Kalkboden") (9.9 individual/m^2), and the peat-soil bottom ("Torfboden") (5.0 individuals/m^2), in that order (Pawlowski, 1936). On the rocky bottom the most common species are *Helobdella stagnalis* (27.3 individuals/m^2), *Erpobdella octoculata* (14.0 individuals/m^2), *Glossiphonia complanata* (6.0 individuals/m^2), *Erpobdella testacea* (5.3 individuals/m^2), *Glossiphonia heteroclita* (3.9 individuals/m^2), and *Theromyzon tessulatum* (1.3 individuals/m^2). On the *Chara* bottom the species are *Erpobdella octoculata* (24.0 individuals/m^2), *Helobdella stagnalis* (18.0 individuals/m^2), *Glossiphonia complanata* (6.0 individuals/m^2), and *Erpobdella testacea* (4 individuals/m^2). On the calcareous soil bottom the species are *Helobdella stagnalis* (6.8 individuals/m^2), *Erpobdella octoculata* (1.3 individuals/m^2), *Glossiphonia heteroclita* (0.5 individuals/m^2), and *Glossiphonia complanata* (0.3 individuals/m^2). On the peat-soil bottom the species are *Erpobdella testacea* (2.0 individuals/m^2), *Erpobdella octoculata* (2.0 individuals/m^2), and *Helobdella stagnalis* (1 individuals/m^2). *Glossiphonia complanata* is absent from the peat-soil bottom and *Erpobdella testacea* from the calcareous soil bottom. *Glossiphonia heteroclita* is absent from the peat-soil and the calcarous soil bottoms.

In Lake Mendota, Wisconsin, leeches prefer stone, gravel, mud and detritus with vegetation, sand, sand and mud, mud with detritus, sand and mud with shells, and deep lake mud, in that order (Sapkarev, 1967). On the stone substrate the dominant species are *Helobdella stagnalis* (maximum density 3108.0 individuals/m^2), *Erpobdella punctata* (1332.0 individuals/m^2), *Nephelopsis obscura* (1110.0 individuals/m^2), *Glossiphonia complanata* (266.4 individuals/m^2), *Dina parva* (266.4 individuals/m^2), and *Helobdella elongata* (222.0 individuals/m^2). On the mud-and-detritus-with-vegetation substrate the major species are *Helobdella stagnalis* (2175.6 individuals/m^2), *Haemopis marmorata* (266.4 individuals/m^2), *Helobdella elongata* (133.2 individuals/m^2), *Erpobdella punctata* (133.2 individuals/m^2), and *Nephelopsis obscura* (133.2 individuals/m^2). Muttkowski (1918) found in Lake Mendota that leeches preferred rock, gravel, sand, and mud bottoms, in that order. In southern Michigan, few leeches occur in areas with only a mud or sand bottom (Klemm, 1972a).

In Colorado the following species are most commonly found under rocks, boards, and other submerged objects: *Helobdella stagnalis*, *Glossiphonia complanata*, *Helobdella fusca* (? *lineata*), *Batracobdella phalera*, *Batracobdella picta*, and *Placobdella ornata* (Herrmann, 1970a). *Helobdella stagnalis* is common on most substrates except sand. *Erpobdella punctata*, *Dina dubia*, and *Nephelopsis obscura* are most common in semidrainage conditions with an abundance of gyttja and forna. In Denmark, leeches are scarce in a purely soft mud substrate, but are unaffected by a mixed hard and soft bottom (Berg, 1948). Similarly, in the cleaner, open lake regions of western Lake Erie the leeches which occur on a purely mud substrate have a mean density of 40.5 individuals/m^2, whereas on a mixed substrate the density is 74.5 individuals/m^2; in the more polluted regions near the mouths of the Raisin and Detroit Rivers, the densities are 12.4 and 16.3 individuals/m^2, respectively (Carr and Hiltunen, 1965).

The bottom of a relatively unpolluted portion of the Mississippi River, near Keokuk, Iowa, consists primarily of silt-loam which is rich in organic matter. The leeches in this area, primarily *Helobdella stagnalis*, *Erpobdella punctata*, and *Helobdella elongata*, have an average density of 20–40 individuals/m^2 (Carlson, 1968).

C. Depth

Leeches are maximally concentrated in submerged vegetation, more or less independent of the depth at which it is found (usually in the littoral zone) (Fig. 14). This vegetation offers protection from predation as well as a solid substrate for locomotion and cocoon deposition. In Lake Mendota, Wisconsin, the following species occur only at 0–1 m depths, *Placobdella parasitica*, *Placobdella montifera*, *Placobdella ornata*, and *Batracobdella picta*; to 2 m only, *Glossiphonia complanata*, *Dina parva*, *Haemopis marmorata*, *Erpobdella punctata*, *Nephelopsis obscura*, *Helobdella lineata*, *Batracobdella phalera*; to 4 m only, *Glossiphonia heteroclita* and *Helobdella elongata*; and to

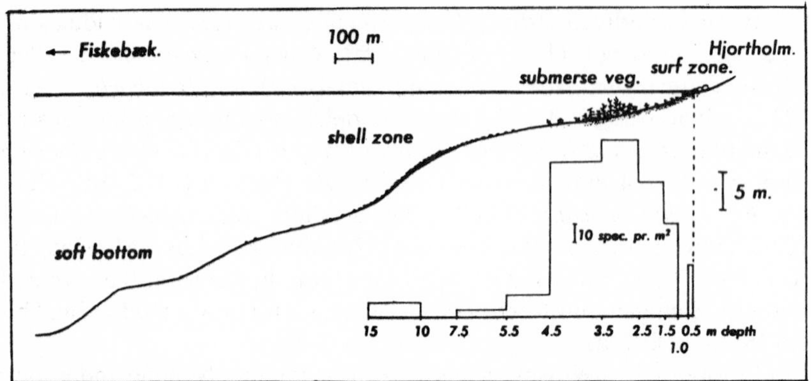

Fig. 14. Schematic section of the shore line of Lake Furesø, Denmark. The inset gives the distribution of all leeches according to depth zones. Vertical scale magnified 20 times. (Bennike, 1943).

12 m only, *Helobdella stagnalis* (Sapkarev, 1968). Unlike the Glossiphoniidae, the Erpobdellidae have little vertical distribution in Lake Mendota (Fig. 15). The latter are restricted to 0–1.5 m with a maximum density of 0–0.5 m. In the same lake, Muttkowski (1918) calculated that for depths of 0–1, 1–2, 2–3, 3–5, and 5–7 m, the densities of all leeches are 11.2, 1.93, 0.75, 14.0, and 0.24 individuals/m², respectively. Between the shoreline and a depth of 7 m (an area of 12.9 km²) he estimated a total of 103,200,000 individuals. In Lake Mendota, *Helobdella stagnalis* has its greatest density in the littoral zone (421.8 individuals/m² at 0 m), an intermediate density in the

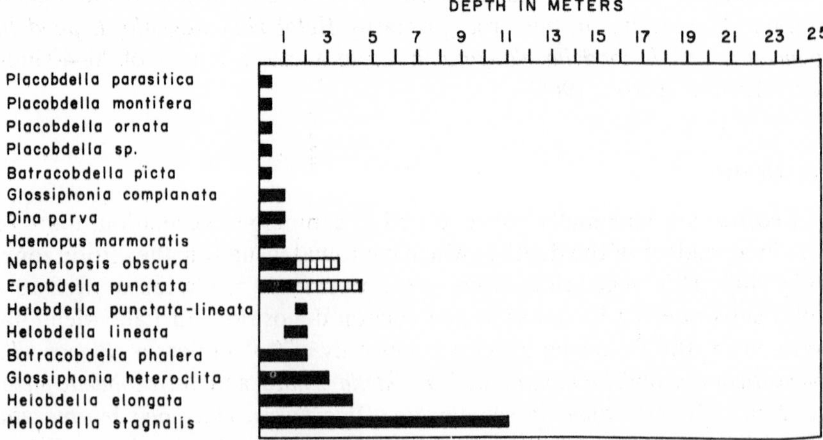

Fig. 15. Depth distribution of leeches in Lake Mendota, Wisconsin. (Sapkarev, 1968).

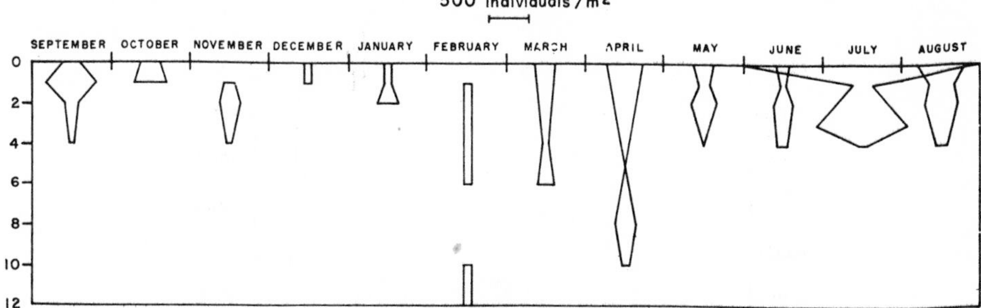

Fig. 16. Depth distribution and seasonal changes in the density of *Helobdella stagnalis* in Lake Mendota, Wisconsin, during 1964–1965. (Sapkarev, 1968).

sublittoral zone (14.8 individuals/m^2 at about 6 m), and a minimum density in the profondal zone (3.7 individuals/m^2 at 12 m) (Sapkarev, 1968) (Fig. 16). In Lake Washington, *Helobdella stagnalis* has a definite maximum density (150 individuals/m^2) at 40 m and occurs in large numbers from 20 to 45 m (Thut, 1969). The maximum temperatures at depths of 20 and 30 m are about 13° and 10°C, respectively. The minimum dissolved oxygen at 30 m is 5 ppm. In Lake Superior, no leeches occur deeper than 30 m, and only *Glossiphonia complanata* and *Dina parva* are found below 15 m (Thomas, 1966). In Michigan ponds most leeches occur at depths ranging from 0 to 1.5 m (Klemm, 1972a).

In Lake Mendota, the leech densities are subject to seasonal fluctuation, the densities greatest in the littoral zone in July, and greatest in the profondal zone in February (Figs. 16 and 17). In Lake Fures, a large eutrophic lake in Denmark, *Helobdella stagnalis* is commonly encountered in the surf zone (0–1.5 m) only from May to September (Bennike, 1943). Bennike felt that in this zone, *Helobdella stagnalis* are killed by the winter ice rather than migrate to greater depths, because the density of this species in deeper water is not greater in the winter than in the summer.

The greatest depths at which various leech species have been reported in European lakes are: *Helobdella stagnalis* (55 m), *Glossiphonia complanata* (35 m, with a questionable record of 120 m), *Glossiphonia heteroclita* (5 m), *Erpobdella octoculata* (30m), and *Haemopis sanguisuga* (0.5m). In Lake Fures *Helobdella stagnalis*, *Glossiphonia complanata*, *Glossiphonia heteroclita*, *Erpobdella octoculata*, *Erpobdella testacea*, and *Piscicola geometra* occur at the maximum depths, 12, 2.5, 2.5, 3.0, 1.75, and 20 m, respectively (Bennike, 1943). In this lake, they have their maximum concentration in the submerged vegetation located at a depth of 1.5 to 4.5 m. In Lake Prespa, Macedonia, the following species occur at more than 14 m, *Helobdella stagnalis* (30 m) *Erpobdella octoculata* (30 m), *Glossiphonia complanata* (20 m), and *Hemiclepsis marginata* (16 m) (Sapkarev, 1963).

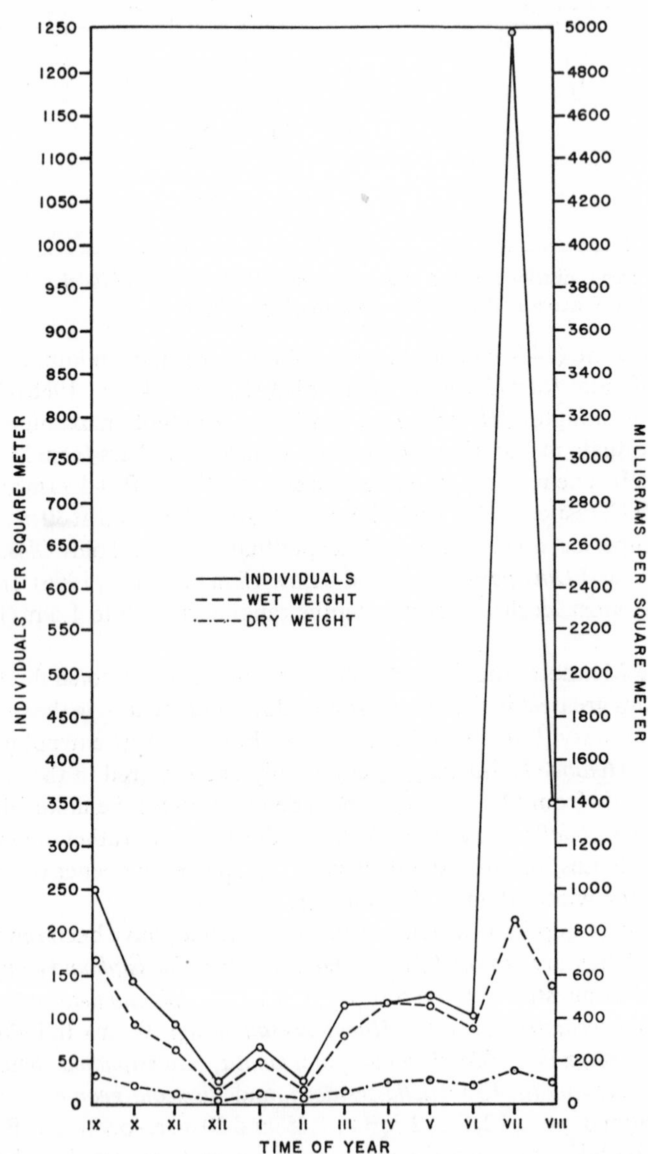

Fig. 17. Average seasonal changes of the population density and biomass of *Helobdella stagnalis* in the littoral zone of Lake Mendota, Wisconsin, during 1964–1965. (Sapkarev, 1968).

D. Lentic Habitats (Standing Water)

The size of ponds and lakes as a factor influencing the distribution of leeches is not so striking as the difference between standing and running water. In Denmark, *Helobdella stagnalis* and *Erpobdella octoculata* are encountered equally frequently in lakes, ponds, marshes, and running water (Bennike, 1943) (Fig. 18). *Glossiphonia complanata* is also abundant in all

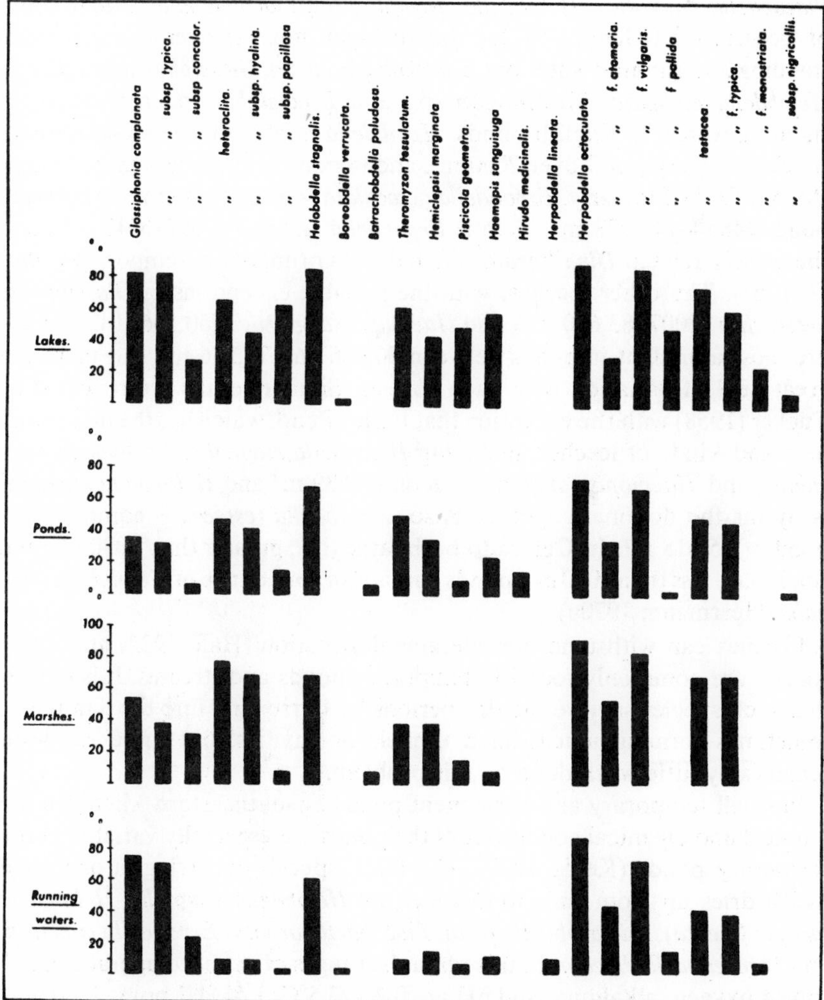

Fig. 18. Graphic representation of the percentage of occurrence in Denmark of leech species in the four types of fresh water. (Bennike, 1943).

types of water, but is more frequent in lakes and running water than in ponds and marshes. *Glossiphonia heteroclita, Theromyzon tessulatum,* and *Hemiclepsis marginata* are stagnant water species which are dependent upon the presence of vegetation. *Piscicola geometra* is characteristic of lakes. In Colorado the most common species in standing water are *Helobdella stagnalis*, *Glossiphonia complanata*, *Erpobdella punctata*, *Dina dubia*, and *Nephelopsis obscura* (Herrmann, 1970a). In southern Michigan, the most common species in standing water are *Erpobdella punctata*, *Glossiphonia complanata*, *Placobdella parasitica*, *Helobdella stagnalis*, and *Placobdella ornata*, in order of occurrence (Klemm, 1972a). In Michigan most species occur in both standing and running water but *Batracobdella picta*, *Glossiphonia heteroclita*, *Helobdella elongata*, *Theromyzon* sp., and *Dina dubia* occur primarily in standing water. In English ponds, *Helobdella stagnalis* is the most common species in every pond over 2090 m^2, and in none of those less than this size (Mann, 1955). Similarly, *Erpobdella octoculata* is more frequent in the larger ponds (4850–142,973 m^2) than is *Erpobdella testacea* (1087–5518 m^2), and the closely related *Dina lineata* which occurs primarily in temporary ponds (753 m^2). The other leeches, with the possible exceptions of *Glossiphonia heteroclita* (2007–83,610 m^2) and *Haemopis sanguisuga* (50,166–142,973 m^2) are most abundant in ponds between 30,936 and 76,336 m^2, but there is a great deal of variation with each species. Similar results are reported by Tucker (1958) with the exception that Farley Pond, which has the most numbers and kinds of leeches, including *Helobdella stagnalis*, *Erpobdella octoculata*, and *Haemopis sanguisuga*, is only 1589 m^2 and *Helobdella stagnalis* is by far the dominant species. Also, *Erpobdella testacea* is common in a pond of 35,618 m^2. In Colorado both large (i.e., greater than 2090 m^2) and small (i.e., less than 2090 m^2) ponds contain an abundance of *Helobdella stagnalis* (Herrmann, 1970a).

Leeches can withstand considerable desiccation (Hall, 1922) and many species are commonly found in temporary ponds and streams. It is known that such species survive the dry periods by burrowing into the damp soil, sometimes forming mucus-lined tunnels or cavities, but there has been remarkably little work done on this problem.

In small temporary and permanent ponds in southeastern Michigan the physical and chemical conditions of the ponds are especially variable in the temporary ponds (Kenk, 1949). The leech species occurring in one pond which dries up from June to October are *Helobdella stagnalis*, *Helobdella fusca* (? *lineata*), *Batracobdella picta*, *Placobdella ornata*, *Erpobdella punctata*, and *Mooreobdella bucera*. In this pond the ranges of water temperature, dissolved oxygen, alkalinity, and pH are 0.3°–23.5°C, 1.5–11.1 ppm, 3–80 ppm and 5.8–7.7, respectively. In a nearby permanent pond the leech species commonly encountered are *Helobdella stagnalis*, *Helobdella fusca* (? *lineata*),

Batracobdella picta, *Placobdella parasitica*, *Glossiphonia heteroclita*, *Erpobdella punctata*, and *Mooreobdella bucera*. The ranges of water temperature, dissolved oxygen, alkalinity, and pH in this pond are 0.8°–27.0°C, 4.1–17.0 ppm, 26–157 ppm, and 6.8–8.2, respectively. An impressive list of the macroinvertebrates associated with the leeches in each of these ponds includes oligochaetes, snails, crustaceans, and many types of aquatic insect larvae.

On the George Reserve, a wooded area of about 518 hectares in southeastern Michigan, the leech species which occur in temporary or semipermanent ponds or streams are, in order of tolerance to the temporary conditions, *Mooreobdella bucera*, *Marvinmeyeria lucida*, *Erpobdella punctata*, *Helobdella stagnalis*, and *Glossiphonia heteroclita*. The dominant species in the permanent ponds are, in order of abundance, *Glossiphonia complanata*, *Helobdella stagnalis*, *Erpobdella punctata*, *Placobdella hollensis*, *Batracobdella picta*, *Placobdella ornata*, *Macrobdella decora*, *Placobdella parasitica*, *Helobdella lineata*, *Mooreobdella bucera*, and *Marvinmeyeria lucida*.

E. Lotic Habitats (Running Water)

In Denmark only *Erpobdella octoculata*, *Glossiphonia complanata*, *Helobdella stagnalis*, and possibly *Erpobdella testacea* are classified as characteristic of running water (Bennike, 1943). These are the same species commonly found in the surf zone of lakes. On the other hand, in Colorado no species can be considered an ecological indicator of running water (Herrmann, 1970a). Almost all leech species occur in running water in Colorado but the following species are the most commonly found: *Helobdella stagnalis*, *Glossiphonia complanata*, *Erpobdella punctata*, and *Haemopis marmorata*. The species less common in running water are *Placobdella ornata*, *Mooreobdella microstoma*, *Dina parva*, *Dina dubia*, and *Macrobdella decora*. Similarly, in Michigan the most common species in running water are *Glossiphonia complanata*, *Erpobdella punctata* and *Placobdella ornata*, in order of occurrence (Klemm, 1972a). In England *Erpobdella octoculata*, *Erpobdella testacea*, *Glossiphonia complanata*, and *Piscicola geometra* are more frequent in running water than in standing water (Mann, 1955). *Glossiphonia heteroclita* is absent from running water. The leech fauna of relatively fast water is scarce and consists primarily of *Erpobdella octoculata*, *Glossiphonia complanata*, and *Helobdella stagnalis*.

The distribution of leeches in relation to the current and accompanying substrate in the River Susaa, Denmark, has been thoroughly studied (Berg, 1948). At 25–50 cm/sec the most common species are *Erpobdella octoculata* and *Glossiphonia complanata*; at 10–25 cm/sec *Glossiphonia complanata*, *Erpobdella octoculata*, and *Piscicola geometra*; and at less than 10 cm/sec

Piscicola geometra, *Helobdella stagnalis*, *Erpobdella octoculata*, and *Glossiphonia complanata*, in that order. At 25–50 cm/sec the substrate is basically stony, which gives up to submerged vegetation and detritus as the current reduces to less than 10 cm/sec. *Haemopis sanguisuga* prefers a stony bottom and can live where the current is occasionally 50 cm/sec or more. *Piscicola geometra* requires a vegetated substrate in the breeding season for the attachment of the cocoons.

A small, unpolluted brown water stream in central Alberta has water temperatures near 0°C for six months (completely iced over for five months), but approaches 20°C in the summer (Clifford, 1969). The water has a pH of 7.0–7.5 and a discharge of less than 1 m^3/sec. The phosphate phosphorus varies from 0.10 to 0.60 ppm and the oxygen content varies from 5.0 to 11.5 ppm. The leech species which are found in any numbers in this subarctic stream are *Glossiphonia complanata* and *Helobdella stagnalis*. The associated macroinvertebrates include *Leptophlebia cupida* (Say), *Baetis tricaudatus* Dodds, *Callibaetis coloradensis* Banks, *Hydropsyche*, *Cheumatopsyche analis* (Banks), and various simuliid larvae. Unfortunately, neither the tendipedid larvae nor the oligochaetes are quantified.

F. Hardness and pH

The distribution of leeches in relation to hardness, total alkalinity, and pH of the water has received considerable attention with equivocal results. Except perhaps for the extremely low levels of these and other functions of calcium, these factors have little or no direct influence on the distribution or relative abundance of leeches. However, such factors can profoundly affect the occurrence of their various food organisms and, thus, indirectly determine the leech density. Almost all leech species are most abundant in water with a total alkalinity above 60 ppm $CaCO_3$, and few species occur in an alkalinity below 18 ppm. Similarly, all leech species are most abundant in water with a pH of 7.0 or above, and few species occur at pH 6.0 or below. The following accounts of the distribution of freshwater leeches in relation to alkalinity and pH are important contributions to our understanding of leech ecology, but the reader is cautioned that these observations are at best indirect functions of alkalinity or pH.

In England *Erpobdella octoculata*, *Helobdella stagnalis*, *Glossiphonia*, and *Theromyzon tessulatum* occur over a remarkably wide range of alkalinities ($CaCO_3$) (Mann, 1955a,b). For the larger bodies of water *Erpobdella octoculata* is the most numerous species in soft water (0–17 ppm), and *Helobdella stagnalis* is the most numerous in intermediate (18–59 ppm) and hard water (60–242 ppm). The relative abundance of these two species may be a rough approximation of hardness of the water in the larger ponds and lakes. Other species, which are absent from the smaller soft water ponds, are most common in water of intermediate hardness. The following species occur in alkalinities over 239 ppm: *Erpobdella octoculata*, *Helobdella stagnalis*, *Glossiphonia complanata*, *Glossiphonia heteroclita*, and *Dina lineata*. *Glossiphonia complanata* is the most numerous species only in the hardest water ponds, where the bicarbonate content is as high as 242 ppm.

In Denmark few species occur in water with alkalinities ($CaCO_3$) below 9 ppm, and all common species can live at 100 ppm or higher (Bennike, 1943). Of 34 Danish ponds only the acidic ponds containing 9 ppm or less have a severely limited leech fauna. In these ponds only *Helobdella stagnalis*, *Erpobdella octoculata*, and *Erpobdella testacea* normally occur. *Erpobdella octoculata*, *Helobdella stagnalis*, and *Haemopis sanguisuga* rarely occur in 1 ppm and *Erpobdella testacea* in 4–6 ppm. Other species, including *Glossiphonia complanata* (9.0 ppm), *Glossiphonia heteroclita* (10.0 ppm), and *Theromyzon tessulatum* (10.0 ppm), become more abundant as the bicarbonate level rises. *Piscicola geometra* never occurs below a level of 20 ppm in Denmark or 33.9 ppm in England.

In Colorado *Helobdella stagnalis* is not an indicator of hardness (Herrmann, 1970a). In fact, total alkalinity does not appear to limit Colorado leeches at all. Almost all species occur in water with alkalinities (bound carbon dioxide) below 50 ppm (Fig. 19). Only *Glossiphonia complanata*, *Helobdella stagnalis*, and *Dina dubia* are regularly taken in oligotrophic water with

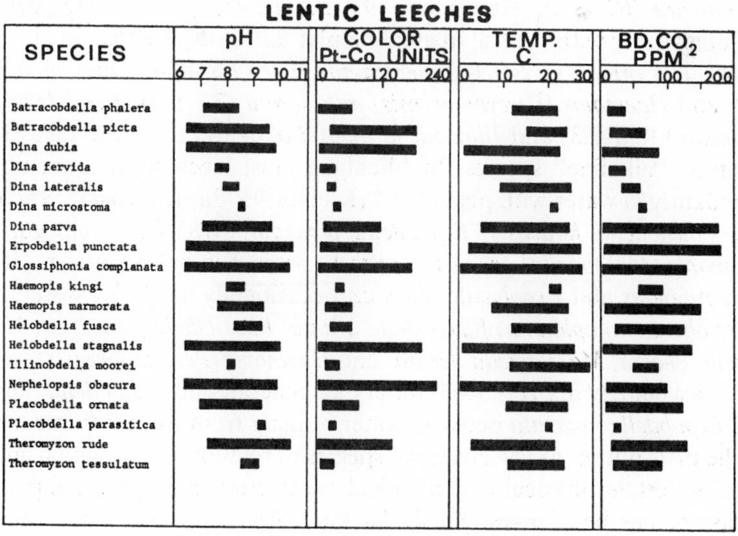

Fig. 19. Observed ranges of lentic leech species from Colorado for pH, color, temperature, and bound carbon dioxide. (Herrmann, 1970a).

total alkalinity below 10.0 ppm. *Glossiphonia complanata* has even been found at 3.6 ppm. In addition *Erpobdella punctata* (5.6–184.6 ppm) and *Dina parva* (6.0–179.2 ppm) enjoy a wide range of alkalinities. In Michigan most species are most abundant where total alkalinity is above 60 ppm (Klemm, 1972a; Kopenski, 1972).

In Denmark few leeches can live in localities with pH values lower than 6.0, but all species occur at pH 9.0 (Bennike, 1943). The species living at the lowest pH values are *Helobdella stagnalis* (pH 4.0), *Erpobdella testacea* (pH 4.3), *Erpobdella octoculata* (pH 4.6), and *Haemopis sanguisuga* (pH 4.6). The lowest pH values for these same and other species in England are 5.3 for *Helobdella stagnalis*, 6.9 for *Erpobdella testacea*, 5.3 for *Erpobdella octoculata*, 6.8 for *Haemopis sanguisuga*, 5.5 for *Glossiphonia heteroclita*, and 5.9 for *Theromyzon tessulatum* (Tucker, 1958). Laurie and Jones (1938) reported one specimen of *Haemopis sanguisuga* at pH 6.6 in a Welsh stream polluted with lead. The maximum pH values of English species which occur at Ph 9.0 or higher are 9.4 for *Erpobdella octoculata*, 9.6 for *Helobdella stagnalis*, 9.4 for *Glossiphonia complanata*, and 9.6 for *Glossiphonia heteroclita* (Mann, 1955a). In northern Norway *Helobdella stagnalis* and *Theromyzon maculosum* can be found at pH 9.0 (Fjeldsa, 1972). *Erpobdella octoculata* can live experimentally for over a month at pH 4.5–10.1 (Gresens, 1928).

In Colorado the following species have a wide range of pH: *Helobdella stagnalis* (6.3–9.9), *Glossiphonia complanata* (6.3–10.2), *Batracobdella picta* (6.4–9.7), *Erpobdella punctata* (6.3–10.3), *Dina dubia* (6.4–9.9), and *Nephelopsis obscura* (6.3–9.8) (Herrmann, 1970a). The following species occur in standing water with neutral to alkaline pH: *Batracobdella phalera* (7.1–9.4), *Placobdella ornata* (7.2–9.7), *Theromyzon rude* (7.6–10.4), *Dina parva* (7.0–9.8), and *Haemopis (Percymoorensis) marmorata* (7.7–9.4). *Helobdella fusca* (? *lineata*) (8.0–9.3) and *Illinobdella moorei* (8.0–8.1) are classified by Herrmann as "alkaline" species. In Michigan most leech species occur most abundantly in water with pH 7.2–7.5 (Klemm, 1972a; Kopenski, 1972) (Fig. 20). *Placobdella hollensis, Erpobdella punctata, Nephelopsis obscura,* and *Haemopis (Mollibdella) grandis* are rarely present in pH 4.5 waters; *Helobdella stagnalis* and *Erpobdella punctata* occasionally in pH 5.0 waters; and, *Glossiphonia complanata*, *Placobdella ornata*, *Placobdella parasitica*, *Mooreobdella bucera*, *Macrobdella decora*, and *Haemopis (Percymoorensis) marmorata* occasionally in pH 5.5–6.0 waters. In general, only *Helobdella stagnalis* and *Erpobdella punctata* occur in water ranging from pH 5.0 to pH 8.0.

The distribution of various leech species in relation to the annual fluctuations of certain physical and chemical parameters in ten selected ponds in Berkshire has been investigated (Tucker, 1958). The seasonal changes of many of the chemical factors, such as alkalinity, hardness, and pH, can be attributed to the annual fluctuations of the calcium content of the water. In general the calcium content of the water reaches a maximum value in

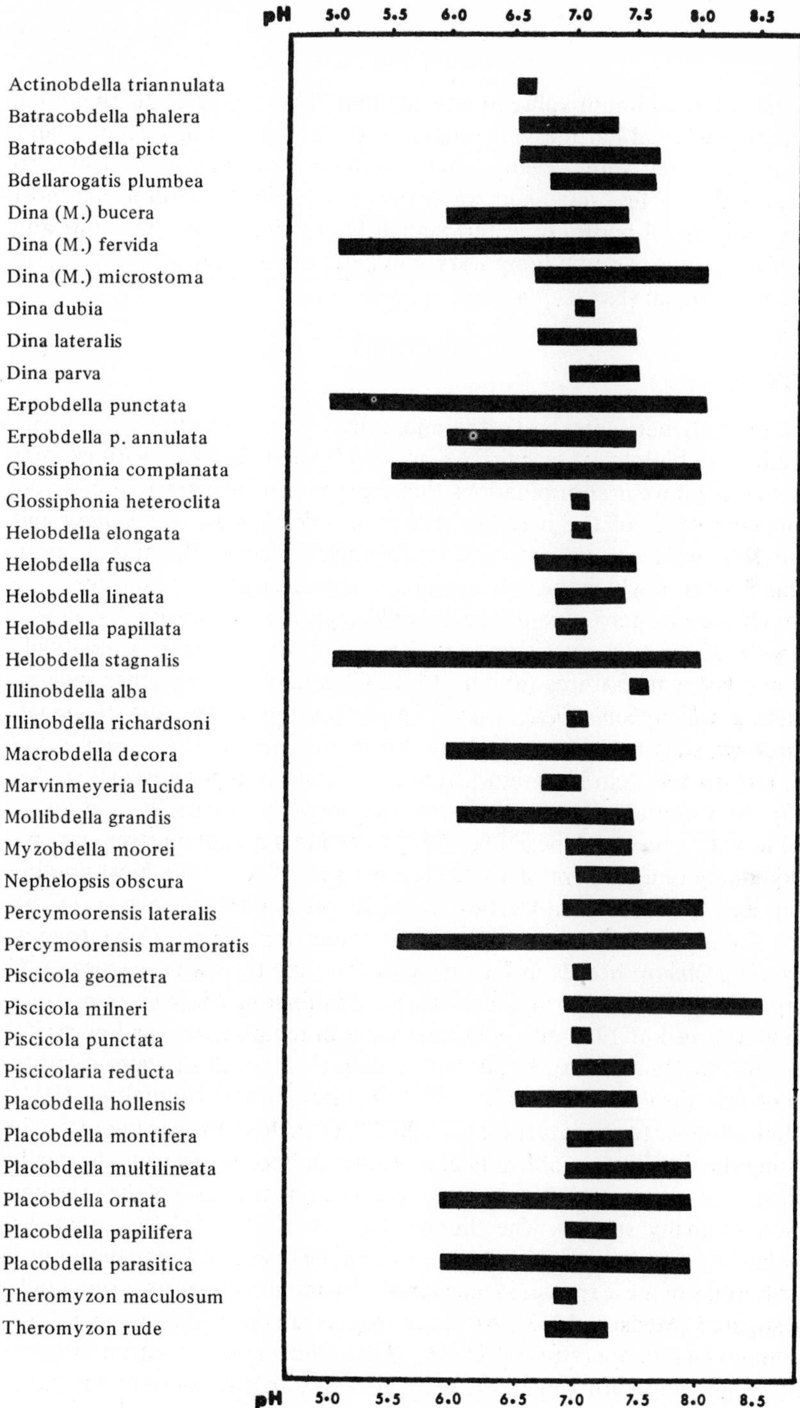

Fig. 20. Observed pH ranges of leech species from Michigan. (Klemm, 1972).

winter and a minimum value in late summer. The degree of fluctuation of calcium (and total hardness) is related to the amount of dissolved organic matter, but other things being equal the fluctuations are greatest in hard waters and in a few small soft water ponds. Erratic fluctuations are most common in small bodies of water. Similar fluctuations in the alkalinity and pH of the water of small temporary and permanent ponds occur in southeastern Michigan (Kenk, 1949).

G. Temperature

All the evidence indicates that temperature plays an important role in the reproductive biology of leeches. Mann (1957b) and Cristae (1970) present evidence from natural populations that temperature is important in determining the onset of the breeding season in *Glossiphonia complanata* and *Erpobdella testacea*, respectively. In Denmark leeches do not occur in running water in which the temperature has not been above 11.0°C for some time. Of eleven species examined only *Glossiphonia complanata*, *Helobdella stagnalis*, *Erpobdella octoculata*, and *Erpobdella testacea* start at relatively low midday temperatures (about 10°–11°C), whereas seven other species, including *Glossiphonia heteroclita*, *Theromyzon tessulatum*, and *Haemopis sanguisuga*, start breeding only when the midday temperatures reach about 14°C and do not occur in water with lower maximum temperatures (Bennike, 1943). In Colorado, however, *Helobdella stagnalis* occurs in a spring at 2820 m with a June temperature of 5.0°C, and in a mountain stream with a midday July temperature of 10.0°C (Herrmann, 1970a). The most notable exceptions to the rule that leeches propagate only in water above 11.0°C are to be found in the *Piscicola-Calliobdella* group of piscicolid (fish) leeches. *Piscicola geometra* breeds in Denmark by the time the water reaches 9.0°C (Berg, 1948). Similarly, in the state of Washington *Piscicola salmositica* begins to breed at 12°C and continues even in temperatures as low as 5°C (Becker and Katz, 1965). In South Carolina their brackish water relative, *Calliobdella carolinensis*, breeds at 9°–10°C (Sawyer and Hammond, 1973).

High summer temperatures (about 30.0°C or higher) are also important in limiting the distribution of freshwater leeches in both natural and thermally polluted environments. Heated effluents accelerate the onset of the breeding season of many species. The thermal tolerances of *Helobdella stagnalis*, *Erpobdella octoculata*, and *Glossiphonia complanata* in relation to the natural distributions of these species in ten Danish streams have been experimentally investigated (Madsen, 1963). All three species are optimal in a stream with an annual high temperature of 21.1°C. *Helobdella stagnalis* occurs in streams with a high temperature of 21.1°C or higher and is absent from streams with a high of 16.5°C or lower. *Glossiphonia complanata* occurs in streams with a high of 14.6°–21.1°C, and is absent from streams with a high of 7.6°–13.1°C.

For *Helobdella stagnalis*, *Erpobdella octoculata*, and *Glossiphonia complanata* no lethal effects occur below experimental temperatures of 24.7°, 30.0°, and 30.0°C, respectively, and 100% lethalities occur at 35.1°, 35.1°, and 33.0°C, respectively. *Helobdella stagnalis* probably has a somewhat lower lethal temperature than *Erpobdella octoculata*. It is clear that at 33.9°C the mortality of the latter species depends upon prior acclimation. Relatively few other laboratory investigations on the reactions of various leech species to temperature changes have been conducted (Baal, 1928). In response to a hollow glass rod which contains circulating hot water, *Helobdella stagnalis*, *Glossiphonia complanata*, and *Erpobdella* sp. show definite withdrawal reactions at temperatures of 27°, 26°, and 29.5°C, respectively; coiling reactions at 31°–36.5°C; and moribund reactions at 40.0°C (Herter, 1929). Both *Erpobdella punctata* and *Helobdella stagnalis* die at 38.9°C (Wurtz and Bridges, 1961).

In Colorado the following species are considered by Herrmann (1970a) to be eurythermal: *Helobdella stagnalis* (0.5°–26.0°C), *Glossiphonia complanata* (0.5°–28.0°C), *Theromyzon rude* (3.0°–22.0°C), *Erpobdella punctata* (3.0°–28.0°C), *Dina dubia* (2.0°–21.0°C), *Dina parva* (5.0°–24.0°C), and *Nephelopsis obscura* (0.5°–24.0°C). In Michigan *Nephelopsis obscura* is a cold water species and does not occur in water above 16°C (Kopenski, 1972), thus accounting for its northern restriction in that state.

Closely related to this question of temperature is the altitudinal limits of the various species. In Colorado all leech species have an upper limit of 3000 m or less, except *Glossiphonia complanata* (3610 m), *Helobdella stagnalis* (3550 m), *Nephelopsis obscura* (3500 m), *Dina dubia* (3300 m), *Erpobdella punctata* (3234 m), and *Batracobdella picta* (3224 m) (Herrmann, 1970a).

H. Oxygen

Most leeches can withstand anaerobic conditions for unusually long periods. Under depleted oxygen conditions *Hirudo medicinalis* can live for three days, *Haemopis sanguisuga* for two days, *Erpobdella* sp. for 45 hours, and two unidentified glossiphoniids for one and five days (Bunge, 1888). Similarly, *Erpobdella testacea* and *Macrobdella decora* are little affected by oxygen depletion (Mann, 1961; Moore, 1923). This being the case most adult leeches are probably little restricted, if at all, by temporary low oxygen levels in both natural and polluted environments. In Denmark *Helobdella stagnalis*, *Erpobdella octoculata*, and *Glossiphonia complanata* occur in small to moderate numbers in polluted regions with oxygen concentrations as low as 0.69 ml/liter (25.3°C) (Bennike, 1943).

The oxygen consumptions by five common species of British leeches have been investigated in relation to their ecology (Mann, 1956, 1961). The oxygen consumptions by *Helobdella stagnalis*, *Glossiphonia complanata*,

Erpobdella octoculata, *Erpobdella testacea*, and *Piscicola geometra* (all 30 mg at 20°C) are 6.10, 4.95, 4.0, 5.98, and 10.0 μl/hour, respectively. The rate of oxygen uptake is roughly proportional to surface area rather than weight of the leeches. The consumption is higher in the smaller individuals than in the larger ones. For example, the rate for *Glossiphonia complanata* is about 286 μl/gm/hr for small individuals (about 5 mg) and 102 μl/gm/hr for large individuals (about 100 mg). Developing embryos are probably very sensitive to oxygen levels.

The oxygen consumptions by *Erpobdella octoculata* and *Piscicola geometra* are dependent upon the ambient oxygen tension throughout the range 1.0 to 6.0 ml/liter. The unusually high consumption by *Piscicola geometra*, which normally lives in fast flowing streams and other oxygenated regions, is probably facilitated by lateral pulsatile vesicles which are characteristic of this species. The consumption by *Helobdella stagnalis*, which normally lives in eutrophic water where the summer concentration can become very low, is basically constant over the range of 2.0 to 4.0 ml/liter, but is dependent on the ambient oxygen tension outside this range. The relationship between oxygen consumption and ambient concentration in *Glossiphonia complanata* is unclear but appears to constant over the range 3.5 to 6.0 ml/liter. Prior acclimation has little effect on the oxygen consumptions of any species except *Erpobdella testacea.* If the latter species is acclimated overnight, its oxygen consumption is basically constant over the range 3.0 to 6.0 ml/liter. Without prior acclimation the consumption is dependent upon the ambient concentration. The constant consumption of oxygen in the face of declining oxygen tension by *Erpobdella testacea*, and probably *Helobdella stagnalis*, is at least partially due to ventilatory movements of the body. While this may be the case, it should be pointed out that these ventilatory movements are characteristic of most species of leeches.

These laboratory data agree well with ecological observations that *Erpobdella testacea* is better adapted to life in low oxygen than is *Erpobdella octoculata* (Mann, 1961). While these two species do tend to live in different optimal habitats, the role that oxygen concentration plays in this ecological difference is still obscure. One optimal habitat (Farley Pond) for *Erpobdella testacea* has an oxygen level of 1.8 (0.9–2.4) ml/liter and yet contains a fair number of *Erpobdella octoculata* as well as six other leech species, including a few *Piscicola geometra* (Tucker, 1958).

I. Siltation and Turbidity

Changes in the levels of siltation and turbidity constitute real forms of pollution which appear to have profound, but undetermined, effects on the ecology of leeches. Unfortunately, this problem has not received the attention it deserves. My own impression has been that ectoparasites and external commensals may be indicators of silt pollution because the turbidity of the water reduces natural predation on these animals. The epidemic of the brackish water leech *Calliobdella carolinensis* on the menhaden *Brevoortia tyrannus* was attributed partially to the highly turbid (Secchi disk typically visible at 0.50 m or less) waters of the estuaries of southern South Carolina (Sawyer and Hammond, 1973; Federal Water Pollution Control Administration, 1966). In addition, the population densities of the leech, *Illinobdella moorei*, found on the catfish, *Ictalurus catus*; the leech, *Myzobdella lugubris*, found on the blue crab, *Callinectes sapidus*; and the triclad, *Bdelloura rustica* Verrill, are much higher in areas of high turbidity than in similar regions of South Carolina where the water is much clearer.

The role that turbidity plays in predation by fish on leeches was examined in my laboratory with the following preliminary experiments. Three small *Lepomis* (SL, 23–29 mm), which had not eaten for at least 24 hours prior, were maintained at room temperature in a 1-gal glass aquarium with no substrate, under three different conditions: clear water in subdued light, clear water in total darkness, and turbid (humus and clay suspension) water (visibility about 1–1.5 cm) in subdued light. Five fish leeches *Illinobdella moorei* (10–15 mm) were dropped into each of these aquaria with the following results (Fig. 21). In the aquarium with clear water in subdued light all the leeches were eaten in an average (based on two trials) of 6.5 seconds.

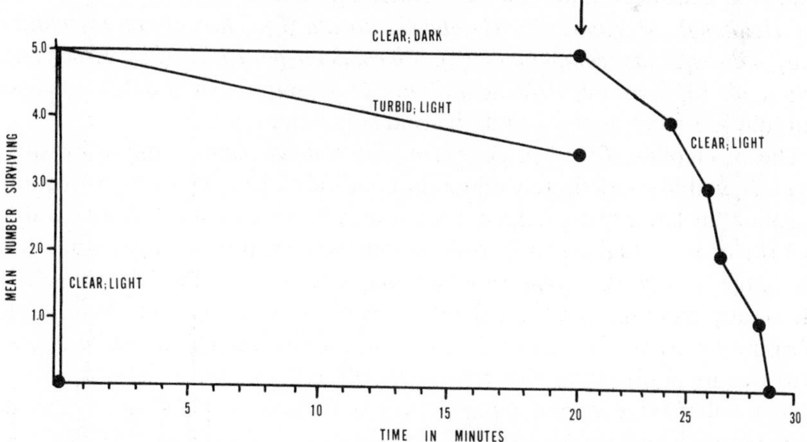

Fig. 21. Predation rate by the fish *Lepomis* on the leech *Illinobdella moorei* in relation to the turbidity of the water and to clear water under light and dark conditions. All leeches were dropped into the aquaria from the top at time 0. The arrow indicates that the aquarium maintained in the dark was continuously exposed to the light at this point. Note the subsequent decline in the number of leeches. See text for further explanation. (Original).

To insure constant conditions, the latter two aquaria were checked only after 20 minutes. After the 20 minutes the aquarium with turbid water in subdued light still had an average of 3.5 leeches and the aquarium with clear water in total darkness had all five leeches remaining. When the aquarium in total darkness was uncovered all the leeches were eaten in an average of 7.5 minutes rather than the 6.5 seconds reported above. This simple experiment supports the presupposition that natural predation on leeches is decreased in turbid water and suggests that this problem should be followed further.

Heavy siltation may eliminate other species of leeches by replacing a solid substrate with a thick layer of silt. Presumably, the efficiency of their suckers for locomotion would decrease, and sites for the deposition of cocoons would become scarce. Just such siltation in an Illinois River flood plain lake after 1956 may have led to the decline of the leech population (Paloumpis and Starrett, 1960).

J. Salinity

Not all leeches have the same tolerance to salinity changes; the erpobdellids and certain glossiphoniids (e.g., *Helobdella stagnalis* and *Helobdella lineata*) are especially sensitive to salinity, whereas some glossiphoniids (e.g., *Placobdella parasitica*), all piscicolids and perhaps the hirudinids are remarkably tolerant. For example, the average survival times (in days) of the following European and American leeches placed in water of 9.7‰ at 10°C are *Helobdella stagnalis* (1), *Helobdella lineata* (2), *Theromyzon tessulatum* (3.6), *Glossiphonia complanata* (8), *Placobdella parasitica* (23), *Erpobdella octoculata* (3.8), and *Macrobdella ditetra* (3.8) (unpublished data). Leeches can tolerate much more salinity in cold water than in warm water.

The piscicolids, *Piscicola geometra*, *Illinobdella moorei*, and *Myzobdella lugubris*, will live indefinitely under the conditions described above. In South Carolina the latter two species are commonly encountered in brackish water, to 15‰ or even higher. In Europe no leeches occur in salinity higher than 6‰, except *Piscicola geometra* which may occur up to 8‰ (Herter, 1937). The many truly marine leeches (all piscicolids) also appear to be tolerant of salinity extremes. In South Carolina the marine leech *Calliobdella carolinensis* occurs naturally in water from 32.5‰ to 4.5‰. In the laboratory this species will survive in freshwater as long as 12 days at 21.5°C and 51 days at 10°C (Sawyer and Hammond, 1973).

In certain Gulf coast rivers the freshwater species, *Helobdella stagnalis*, *Helobdella elongata*, *Placobdella parasitica*, *Illinobdella moorei*, *Erpobdella punctata*, and *Mooreobdella microstoma*, are confined to water with a total salinity less than 0.68‰ (Wurtz and Roback, 1955). Only the blue crab leech, *Myzobdella lugubris*, occurs at greater salinity (6.2‰).

In Colorado the role that total residue content of the water plays in the distribution of leeches has been investigated (Herrmann, 1970b). The maximum value of total residue (in ‰) at which each species occurs is *Helobdella stagnalis* (2.99), *Glossiphonia complanata* (0.95), *Theromyzon rude* (2.09), *Illinobdella moorei* (2.05), *Erpobdella punctata* (1.86), *Dina dubia* (1.69), *Dina parva* (1.10), *Nephelopsis obscura* (0.65), and *Haemopis* (*Percymoorensis*) *marmorata* (3.80). Similarly, the role that water conductivity (salinity) plays in the distribution of leeches in a series of saline lakes in the Southern Interior Plateau region of British Columbia has been investigated (Scudder and Mann, 1968). With the exception of *Theromyzon rude* (3200 micromhos/cm), no leeches were encountered in water with a conductivity above 1650 micromhos/cm. In laboratory tests *Nephelopsis obscura* died in two hours at 3770 micromhos/cm.

IV. Pollutional Ecology of Freshwater Leeches

A. Organic Enrichment

For two decades from about 1900 the Illinois River was subjected to increasing amounts of raw sewage and other forms of organic pollution primarily from Chicago, Peoria, and Pekin. The subsequent changes in the physical, chemical, and biological environment along a 362.1-km stretch of the river especially for the years 1913–1925, have been well documented by Richardson (1925a,b, 1928). A provocative list of index species of small bottom animals taken in the middle Illinois River 1920–1925 was presented in the approximate order of species tolerance to the varying degrees of organic pollution found in the river at that time. The category of species described as *pollutional* (more common in the pollutional zone than below it) did not include any leeches: *Tubifex tubifex*, *Limnodrilus hoffmeisteri*, *Tendipes plumosus*, *Tendipes decorus*, and *Limnodrilus claparedianus*, in that order. Some of these and other species frequently reached very large numbers in the pollutional zone. For example, the maximum concentrations to Tubificidae (chiefly *Limnodrilus hoffmeisteri*), Tendipedidae (chiefly *Tendipes plumosus*), and Sphaeriidae (chiefly *Musculium transversum*) were 71,886, 12,624 and 33,393 individuals/m^2, respectively.

The next category, *subpollutional-unusually tolerant* species, included those described as normal in the cleaner water zones but fully capable of living in pollutional or subpollutional zones. This group included *Musculium transversum, Musculium truncatum, Tendipes lobiferus, Pisidium compressum, Plumatella princeps, Hyalella azteca* (= *knickerbockeri*) and *Pelopia* (= *Tanypus*) sp., as well as six leech species: *Helobdella stagnalis*, *Mooreobdella microstoma*, *Glossiphonia complanata*, *Dina parva*, *Erpobdella punctata*, and *Helobdella elongata* (=*nepheloidea*), in that order. *Helobdella stagnalis* was locally abundant in the pollutional and subpollutional zones but with no apparent direct connection with pollution. With the possible exception of *Glossiphonia complanata*, which normally feeds on snails, all of the above are known to feed predominantly upon aquatic oligochaetes and dipteran larvae (Sawyer, 1972). The abundance of these leeches in these zones probably reflects their response to the enriched food supply and therefore constitutes an indirect, but definite, positive response to the polluted conditions. In Denmark Berg (1948) also observed that certain species of leeches were commonly abundant in polluted areas in response to the increased food supply. The only other leech listed in the Illinois River study was *Placobdella ornata* (=*rugosa*), a turtle parasite categorized as *subpollutional-less tolerant-less common* species. No leeches at all were included in the *clean water-sensitive* category.

In Chillicothe near the general region where the subpollutional leeches were found, the surface dissolved oxygen changed from 3.0–3.72 ppm in 1911–1912 to as low as 0.2–0.47 ppm (or lower) in 1920; by 1925 the oxygen concentration had increased to 0.53 ppm, indicating a partial recovery of conditions. The total numbers of leeches declined moderately from 1915 to 1920 and leveled off until 1924 after which there was a sharp increase in numbers (Table II). In middle Peoria Lake they increased from none in 1920 to 655 individuals/m^2 in 1925; in the region between Pekin and the lower end of Peoria Lake they increased from 69 individuals/m^2 in 1915 to 1807 individuals/m^2 in 1925; and in the 16.1-km stretch just above Havana they increased from under 24 individuals/m^2 in 1920 to more than 11,960 individuals/m^2 in 1925, a level described as a "veritable epidemic." This epidemic may have been partially attributable to the continuous summer flood of 1924. It is of interest to note here that Thomson (1927) described an epidemic of the fish leech *Piscicola punctata* in the nearby Rock River during the following winter, 1925–1926.

The most abundant leech between Chillicothe and Havana from 1920 to 1925 was *Helobdella stagnalis*, but *Helobdella elongata, Mooreobdella microstoma, Erpobdella punctata*, and *Glossiphonia complanata* were also well represented. In the channel of the river between Liverpool and Havana the total number of leeches increased from 1985 individuals/m^2 in 1924 to 29,107 individuals/m^2 in 1925, and outside the channel they increased from 158 individuals/m^2 in 1924 to 6689 individuals/m^2 in 1925. From one mile above Liverpool to Havana the leeches increased from less than 6% of the weight composition of the small, bottom animals in 1924 to 47% in 1925. In fact from Liverpool to Havana the total weight of the leeches rose to over 2802.1 kg/hectare (2500 lb/acre) in 1925 and they came close to exceeding either Tubificidae, Tendipedidae, or Sphaeriidae in weight. Richardson (1928, pp. 429–30) suggested that the meteoric rise in the concentration of

TABLE II

CHANGES IN THE AVERAGE WEIGHT OF THE LEECHES PER HECTARE AND THE AVERAGE NUMBER OF THE LEECHES PER SQUARE METER IN THE POLLUTED MIDDLE ILLINOIS RIVER[a,b]

	Average weight in kilograms per hectare				Average numbers per square meter[c]		
Location	1915	1920	1924	1925	1915	1924	1925
Upper Peoria Lake	3.3	0	24.4	32.9	22.7	69.4	164.0
Middle Peoria Lake		0	32.6	104.4	—	200.0	655.0
Lower Peoria Lake	3.7	0	30.4	54.8	25.0	136.0	232.0
P. + P.U.R.R. Bridge-Pekin	10.1	—	5.2	351.2	69.0	15.5	1,807.0
Pekin-Copperas Creek Dam	9.5	—	47.7	76.3	48.0	57.0	3,014.0
Copperas Creek-Liverpool	6.5	—	1.1	—	45.0	7.2	—
Liverpool-Havana	17.9	(?)	186.7	2,625.7	123.0	527.0	12,289.0
Havana-Beardstown	1.2[d]	(?)	7.8	9.2	6.0	20.3	43.1

[a]From 1915 to 1928, based on Richardson (1928). There was a partial recovery of conditions in 1925.

[b]Inside and outside the channel data combined.

[c]Chiefly *Helobdella stagnalis* in 1924 and 1925.

[d]Havana to Lagrange.

leeches led to the decline of *Musculium transversum* (Sphaeriidae), on which some of the leeches were thought to feed. Such a relationship is open to question, however.

Between 1952 and 1958 Paloumpis and Starrett (1960) investigated the benthic organisms in three Illinois River flood plain lakes in the vicinity of Havana, Illinois. Part of the area under study, Lower Quiver Lake, was investigated by Richardson between 1913–1925 when the bottom was less silty. The chemical compositions of these three lakes, which have high levels of sewage, industrial, and silt pollution, were similar. The phosphate, nitrate, alkalinity, and hardness (both as $CaCO_3$) levels were 0.0–1.2, 0.0–10.4, 96.0–220, and 157–298 ppm, respectively. The pH averaged about 8.2 (7.6–8.6) and turbidity ranged from 25 to 775 ppm. The nonleech macroinvertebrate bottom fauna consisted primarily of *Limnodrilus, Tubifex, Hexagenia limbata, Oecetis inconspicua, Tendipes, Sphaerium, Pisidium, Musculium*, and *Cincinnatia emarginata*. In contrast to these groups, the leeches constituted a relatively minor part of the bottom fauna collections. In the period 1952–1958 the dominant leech species, *Helobdella stagnalis, Helobdella elongata*, and *Erpobdella punctata*, had an average density of 80.7, 21.5, and 0.6 individuals/m^2, respectively. Each of the other common species reported, *Glossiphonia complanata, Helobdella lineata, Placobdella montifera*, and *Illinobdella moorei*, had densities of less than 1.8 individuals/m^2.

In Middle Quiver Lake in 1952 there was an unusual abundance of leeches (1453 individuals/m^2), primarily *Helobdella stagnalis*. After 1952 there was a drastic decline in the density of the leeches: in 1953, 1954, and 1955–1958 there were 598, 24, and 0 individuals/m^2, respectively. The authors unjustifiably attributed the decline in the leeches to a parallel decline in the snail, *Cincinnatia emarginata*, and the various fingernail clams. First of all it is well known that leeches usually abound particularly in the shallow, highly vegetated zone close to shore and become markedly scarcer with increasing depth (and scarcer vegetation). With this in mind, the decline in the leech population described above may in part be a result of the increasing average depths at which bottom samples were taken: in 1952, 1956, and 1958 the average depths (in meters) of bottom samples in Middle Quiver Lake were 1.25 (0.2–2.71), 2.71 (1.98–3.66), and 3.47 (2.74–4.42), respectively. Even if the decline were entirely real, it is probably more closely linked to the scarcity of dipteran and other insect larvae in Quiver Lake in 1952 than to the decline of mollusks. The predominant leech, *Helobdella stagnalis*, which constituted 84% of the leeches in Middle Quiver Lake, much prefers aquatic insects to mollusks for prey. Coincidentally, the drastic decline in the leech population was accompanied by a slight, but noticeable, increase in the dipteran insect density in this lake. The authors were probably correct in implicating silt pollution, which has been present in these lakes since the early 1930's, as a fundamental factor in the population dynamics of the bottom fauna of these lakes. A rapidly increased siltation rate after 1956 could have lead simultaneously to the decline of the mollusk and leech populations, and to the increase of the dipteran density.

Carlson (1968) investigated the summer biological conditions of the Mississippi River, above Dam 19, Keokuk, Iowa, and found no evidence of pollution. The dominant bottom fauna during the summer months were *Sphaerium transversum*, *Hexagenia*, *Tendipes plumosus*, *Coelotanypus*, *Stenochironomus, Branchiura sowerbyi, Limnodrilus hoffmeisteri, Campeloma, Lioplax subcarinata*, and various tubificids. The leeches constituted a constant, but never dominant, portion of the bottom fauna. The dominant leech species were *Helobdella stagnalis, Erpobdella punctata, Helobdella elongata*, and, to a much less extent, *Glossiphonia complanata*. The density of all leeches during the summer ranged from about 20 to 40 individuals/m^2, and the maximum density was 124 individuals/m^2 in July. This figure probably reflects the newly hatched young of *Helobdella stagnalis*.

The reports by Moore (1901) and Sawyer (1972) of the southern hirudinid *Philobdella gracilis* in the region adjacent to the lower end of the Illinois River indicate that warm water conditions suitable for truly southern species have prevailed here since at least 1900, probably independent of later organic pollution of this area.

Weston and Turner (1916) investigated the biological and chemical effects of partially purified sewage effluent on Coweeset Stream, a small otherwise unpolluted tributary of the Taunton River, Massachusetts. About 7,570,800 liters of effluent from the sewage treatment sand beds of Brockton were poured daily into the Coweeset which had an approximate daily flow of 18,927,000 liters. They established sampling stations from just above the effluent to 4.83 km below it and found that the changes in the biota were most extensive in the first 1.17 km below the source of the pollution to the point at which the steam is joined by a larger, unpolluted stream. To judge from the monthly mean figures presented below, the partially treated effluent made a tremendous impact upon the physical and chemical character of Coweeset Stream. The annual temperature of the effluent, which ranged from 5.5° to 17.8°C, had the effect of cooling the stream in the summer and warming it in the winter: the annual temperature range was 2.0°–22.1°C immediately above the effluent and was 3.0°–20.4°C in the immediate area of the effluent. The minimum concentration of dissolved oxygen was 5.64 ppm above the effluent and was 1.13 ppm in the polluted region. The concentrations of free ammonia, nitrites, and nitrates were 0.035–0.422, 0.000–0.001, and 0.120–0.580 ppm above the effluent and 2.30–12.51, 0.008–0.660, and 0.257–2.457 ppm, respectively, in the polluted region. The bottom of the stream in the polluted zone was described as an organic and mineral ooze. The stream bed, which was hard immediately above the effluent, became firm again only below the junction with the larger tributary.

The leeches, *Erpobdella punctata* and *Helobdella stagnalis*, were described as being common in the polluted portion of the stream. These two species probably fed upon certain other invertebrates which were also reported common in the polluted zone, e.g., the oligochaetes, *Tubifex tubifex* and *Nais* sp.; the dipteran larvae, *Tendipes decorus* and *Liriope* (=*Ptychoptera*) *clavipes*; and the gastropods, *Helisoma trivolvis* and *Physa heterostropha*. Unlike almost all of these species lists above, *Erpobdella punctata* and *Helobdella stagnalis*, were also commonly encountered in the unpolluted area upstream from the effluent. The only other leeches reported in the study were *Helobdella fusca, Glossiphonia complanata*, and *Placobdella hollensis*. *Helobdella fusca* feeds exclusively upon snails and was found only on the rocky bottom well above the effluent; *Glossiphonia complanata* feeds predominantly upon snails and was found only on the hard bottom regions, i.e., above the effluent and below the confluence with the larger tributary. *Placobdella hollensis*, a relative of *Placobdella ornata*, was encountered only once, attached to the submerged branch in the polluted area. Assuming that the identification of this poorly known species is correct, it is likely that this individual was carried there by a turtle and has no significance as an indicator species.

In Colorado Herrmann (1970a) surveyed the leeches along Boulder Creek 1 km upstream and 5 km downstream from the effluent of the sewage treatment plant of Boulder, Colorado. This effluent was rich in organic residue, nitrates, and phosphates. No leeches were taken in the upstream station, but the following species were taken in large numbers in the downstream station: *Mooreobdella microstoma, Dina parva, Erpobdella punctata, Haemopis marmorata*, and *Helobdella stagnalis*. In addition to leeches there were many tubificids, tendipedids, and snails. In five Danish streams described as heavily polluted Bennike (1943) found the following species: *Helobdella stagnalis*, *Erpobdella octoculata*, *Glossiphonia complanata*, and *Glossiphonia heteroclita*, in order of abundance.

Recent investigators have presented striking evidence that during 1906–1961 major biological physicochemical changes have occurred at the western end of Lake Erie near the mouths of the Maumee, Raisin, and Detroit Rivers, the principal sources of the pollution (Carr and Hiltunen, 1965; Beeton, 1961). From 1930 to 1961 the region affected by the pollution has increased from about 263 km^2 to about 1020 km^2. In the area of pollution the concentrations of calcium, magnesium, sodium (and potassium), silica, sulfates, and chlorides in 1906–1907 were 13.1, 7.6, 6.5, 5.9, 13.0, and 8.7 ppm, respectively, and in 1958 were 36.7, 10.2, 9.4, 1.6, 20.7, and 20.0 ppm, respectively. Similarly, free ammonia, nitrites, and nitrates increased from 0.013, 0.005, and 0.10 ppm in 1930 to 0.036, 0.008, and 0.225 ppm, respectively, in 1942; by 1958 free ammonia had increased to 0.092 ppm. Also, the average total phosphorus increased from 14.4 ppb in 1942 to 33 ppb in 1958. Beeton (1961) felt that changes in the nitrogen and phosphorus content might have contributed more directly to the biological changes described below than the 2.6–11.3 ppm increase in the concentration of the other major ions mentioned above. The mean annual water temperature was reported to be about 1.1°C warmer today than during the 1918–1928 period; and since 1930 very low concentrations of dissolved O_2 have been found to involve large portions of the polluted area.

Some of the important changes from 1930–1961 in the bottom fauna of western Lake Erie were nine times as many oligochaetes, six times as many gastropods, four times as many Tendipedidae, and twice as many Sphaeriidae (Carr and Hiltunen, 1965). In 1930 the bottom was dominated by *Hexagenia* but by 1961 this group was reduced to less than 1% of its former level. In 1961 the bottom fauna had become dominated by oligochaetes, tendipedids, sphaerids, gastropods, and leeches, in that order. Most of the leeches were *Helobdella stagnalis* (personal communication). In 1961 the maximum concentrations of each of these five groups in the heavily polluted regions were

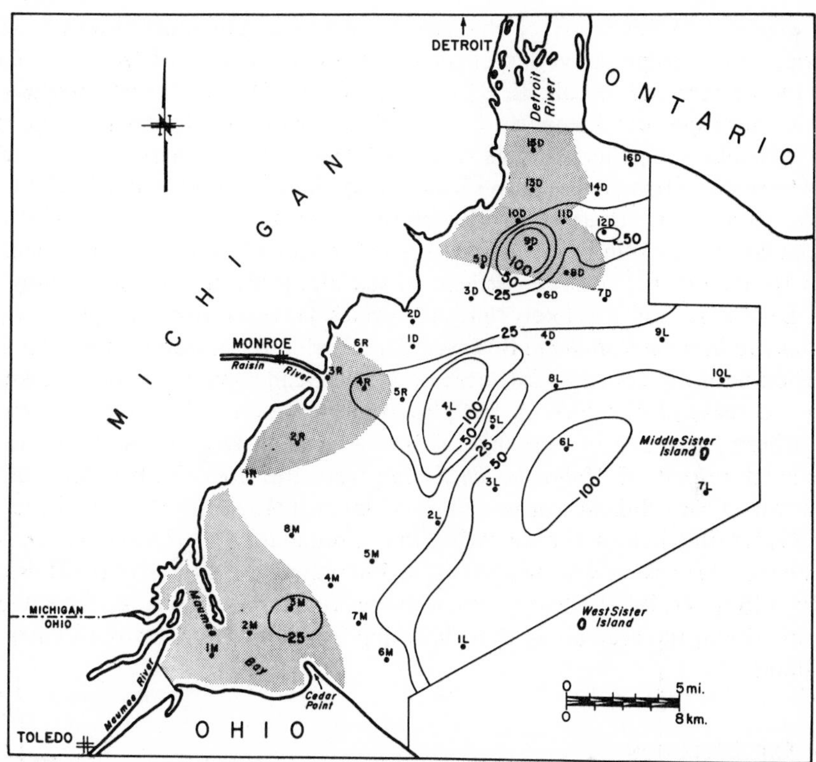

Fig. 22. Distribution of Hirudinea, chiefly *Helobdella stagnalis*, in western Lake Erie, 1961. Shaded areas are zones of heavy pollution. (Carr and Hiltunen, 1965).

39164, 1323, 2862, 1040, and 189 individuals/m^2, respectively. The authors observed that the distribution of the leeches was similar to that of the mollusks (Fig. 22). The concentration of leeches was high east of the Raisin River and near the center of the Detroit River, and the concentrations were noticeably low along the Michigan shore and in the Maumee River area. They noted a wide variation between species in their tolerance to pollution.

The dominant leech species on the bottom of shallow, unpolluted bays of Lake Superior are *Glossiphonia complanata, Erpobdella punctata, Helobdella stagnalis, Nephelopsis obscura, Haemopis grandis*, and *Dina parva*, in order of abundance (Thomas, 1966). An impressive list of macroinvertebrates associated with these leeches includes *Hexagenia, Ephemera, Stagnicola, Limnaea, Helisoma, Physa, Pisidium*, and *Sphaerium*. Unfortunately, neither the Oligochaeta nor the Tendipedinae, both reported as numerous on the bottom, were identified.

Relevant to this study of changes in the bottom fauna of western Lake Erie is the intriguing overlap of the ranges of two species of *Mooreobdella* in the western end of Lake Erie (Sawyer, 1967, 1972). *Mooreobdella fervida* is a northern species which lives over much of Canada and the northern states as far south as northern Illinois, Ohio, and Pennsylvania, where it is replaced by the closely related southern species, *Mooreobdella microstoma*. The latter extends as far north as the southwestern shores of Lake Erie, and in 1967 a population was found some kilometers up the Raisin River. There have been no recent records of *Mooreobdella fervida* along the southwestern shores of Lake Erie, and it is likely that this species has recently been replaced by *Mooreobdella microstoma* in this area. The accelerated eutrophication of this region with the concomitant increase in the mean annual temperature and the decrease of dissolved O_2 has played an important role in the apparent northern extension of the range of *Mooreobdella microstoma* and the subsequent retreat of *Mooreobdella fervida* from this area. Both species are known to eat tubificid worms and insect larvae (Moore, 1912, 1920; Miller, 1929). In this light it is noteworthy that a population of *Haemopis (Percymoorensis) terrestris* was also found in 1967 along the shores of the Raisin River (Sawyer, 1972). This species, which is often misidentified, is distributed primarily in the area along the Mississippi and Ohio Rivers from Ohio to Illinois.

B. Oil Pollution

The effects of bunker oil pollution on the planktonic and bottom fauna of the Muddy River, Massachusetts, was investigated from the autumn of 1961 until the summer of 1963 (McCauley, 1966). A surface oil film, which gradually diminished with time, excluded part of the dissolved oxygen. Sedimentation of the oil produced an oily sludge on the bottom which correlated with low biochemical oxygen demand. In the region of heaviest pollution the water contained 221.3 ppm of oil in the autumn of 1961, 0.0–2.1 ppm in the summer of 1962, and 0.8–0.9 ppm in the summer of 1963. By the summer of 1963 the oil in the sludge had increased to 1850–1931 ppm. The biochemical oxygen demand (as mg 0/mg volatile solids) in the autumn 1961, summer 1962, and summer 1963 was 0.016, 0.014–0.021, and 0.067–0.076, respectively. Similarly, the dissolved oxygen for these three periods was 0.0, 7–11.5, and 8–10 ppm, respectively. In this area of heavy pollution some organisms, such as *Gammarus, Agrion*, and *Dugesia*, were conspicuously absent, whereas *Tubifex, Tendipes*, and unidentified leeches were tolerant. There were no leeches until the summer of 1962 when a density of 215.3 individuals/m^2 was noted. From then until the summer of 1963 when the study was terminated the leech density stabilized at 10–29 individuals/m^2. Interestingly, the greatest density of leeches (527.6 individuals/m^2), as well as of *Tubifex* (26,970 individuals/m^2), occurred in the summer of 1963 in a locality just above the most heavily polluted area. The absence of leeches from the autumn 1961 to the summer 1962, in spite of presence of potential food (e.g., *Tubifex* and *Tendipes* larvae), probably reflects toxic effects of substances dissolved in the oil.

C. Heavy Metals

The effects of dissolved lead from numerous derelict lead mines on the bottom fauna of the River Rheidol in West Wales during the period 1919–1932 have been investigated (Laurie and Jones, 1938; Jones, 1940). A general improvement of the fauna in the stream correlated with a decrease in the concentration of dissolved lead. The fourteen species with were encountered in the lower Rheidol in 1919–1921, when the lead concentration was 0.2–0.5 ppm, did not include any leeches. In 1931–1932 the lead concentration had reduced to 0.02–0.1 ppm, and the bottom fauna had stabilized to a total of 103 species, of which three were leeches. *Glossiphonia complanata* was the most common species, followed closely by *Erpobdella octoculata* (= *atomaria*). Only one individual of *Haemopis sanguisuga* was encountered during this period. Laboratory experiments showed that this species is unusually tolerant of dissolved lead, surviving for 18 days in a solution of 3 ppm of lead nitrate (Jones, 1938). This compares with a survival time of 4 hours for *Lymnaea pereger*, 5.4 days for *Tendipes* sp., 6 days for *Chloeon simile*, and 7–10 days for *Tubifex tubifex*. One explanation for this apparent tolerance to lead is that *Haemopis sanguisuga* is an amphibious species which normally will climb out of contaminated water.

The toxic effects of dissolved zinc remains much longer than that of lead. Thirty-five years after the closing of lead and zinc mines along the nearby River Ystwyth, the concentration of dissolved lead was negligible but the river continued to be polluted by the effluents of the derelict zinc mines (Jones, 1958). In 1958 the concentration of zinc was 0.2–0.7 ppm, a concentration that probably was directly or indirectly responsible for the conspicous absence of leeches, oligochaetes, mollusks, and crustaceans. Along these lines, the LC_{50} of aluminum nitrate in the laboratory is 11.8–17.9 ppm against undetermined leeches (Nekipelov, 1961).

Copper sulfate is ineffective as a control for bloodsucking (hirudinid) leeches (Moore, 1923; DeJesus, 1934; Smith, 1939; Sawyer, 1973). In the laboratory copper sulfate has an LC_{50} and LC_{100} of 1.7–2.5 and 6.6 ppm, respectively, against the Asian species *Hirudo nipponia* (Kimura and Keegan, 1966) and an LC_{100} of 20 ppm against *Hirudinaria manillensis* (DeJesus, 1934). A concentration of 1.0 ppm will kill 100% of *Macrobdella decora* in 5–12$\frac{1}{2}$

hours, and a concentration of 0.2 ppm will kill them in 16–86 hours (Moore, 1923). The toxicity increases with temperature. No fatalities, however, occur in field tests, even at a concentration of 200 ppm. This survival is possibly owing to the fact that these leeches can leave noxious water for extended periods. Calcium chloride at concentrations of 1.0 ppm and 2.0 ppm is 100% fatal to *Macrobdella decora* in 4 days and 18–40 hours, respectively (Moore, 1923). The radioactive manganese (Mn^{54} from fallout) residues in the tissues of *Haemopis sanguisuga* in irrigation ecosystems in the Po Valley, Italy, exceeds that of most other species studied (4.6 picocuries/gm of ash), except tadpoles and insects (Cavalloro and Merlini, 1967). In addition this species has more stable Mn (995 μg/gm of ash) than do snakes, fish, adult frogs, and gastropods (except *Planorbis*). These observations are in agreement with Davis and Foster (1958) who found that bottom animals accumulated more radioisotopes than did fish.

D. Pesticides

Leeches have an unusual tolerance to DDT, comparable to that of certain mosquitos and houseflies. The LC_{50} of DDT is greater than 100 ppm against *Hirudo nipponia, Hirudinaria manillensis*, and *Poecilobdella* sp. (Kimura and Keegan, 1966; Kimura *et al.*, 1967). The mechanism was determined by radioactive assessment of the absorption of DDT by leeches which had been exposed for 24 hours in water containing 7.1 ppm of ring-labeled C^{14} DDT. The results showed that 16% and 17.4% of the exposed DDT was recovered from *Hirudo nipponia* and *Hirudinaria manillensis*, respectively. The chromatograms of the *Hirudo nipponia* extract demonstrated that 49% of the absorbed DDT was metabolized into nontoxic DDE (Fig. 23). However, DDE was not found in the *Hirudinaria manillensis* extract. This dehydrochlorination of DDT, which is apparently the first report for an annelid, is very similar to that observed in DDT resistant mosquitos, *Culex fatigans* and *Culex tarsalis*. Similarly, the pesticide residues in the tissues of *Helobdella stagnalis* collected from Belzoni, Mississippi, a region heavily sprayed for cotton and soybean insects, contained 0.600, 2.890, and 1.750 ppm of DDT, DDD, and DDE, respectively (Naqvi, 1972). Therefore, dehydrochlorination of DDT also appears likely in *Helobdella stagnalis* since the amount of DDD is considerably greater than DDT. The DDD has in turn been broken down further to DDE.

The uptake of Mirex by *Helobdella stagnalis* and an erpobdellid, tentatively identified as *Erpobdella* sp. (? *Mooreobdella microstoma*), was examined by exposing the leeches to a 1.0 ppm solution (Naqvi, 1972). At the end of 24 and 120 hours the concentration of Mirex in the tissues was 5.070 and 1.950 ppm for *Helobdella stagnalis* and 1.920 and 26.140 ppm for *Erpobdella*

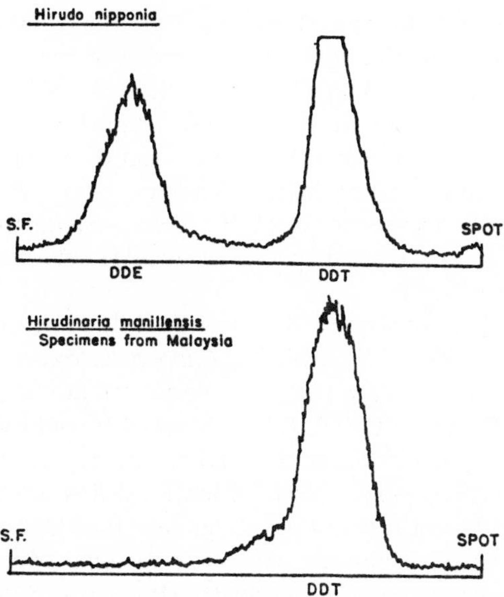

Fig. 23. Dehydrochlorination of DDT by Malayan freshwater leeches. Chromatographs of extracts of *Hirudo nipponia* (upper) and *Hirudinaria manillensis* (lower). (Kimura *et al.*, 1967).

sp., respectively. *Helobella stagnalis* appears to detoxify or metabolize Mirex since the uptake is inversely proportional to exposure time. *Erpobdella* sp., however, appears to be concentrating Mirex in the tissues since the uptake is directly proportional to exposure time. The residues of Mirex in the tissues of *Erpobdella* sp. and *Placobdella multilineata* collected from a pond on the Mississippi State University campus were 0.240 ppm and 0.400 ppm, respectively. *Erpobdella* sp. collected from a creek in Noxapater, Mississippi, an area which had been treated with Mirex two months prior, had an unusually high Mirex residue (1.756 ppm). In fact the average residue levels of annelids were greater than any other invertebrate group studied, including crustaceans, insects, and mollusks.

Using the standard WHO procedures for determining the resistance of mosquito larvae to pesticides, Kimura and Keegan (1966) have examined the toxicity of some commonly used pesticides to *Hirudo nipponia* Whitman, 1886. Dinex (LC_{50} 0.1–0.3 ppm), ICI 24223 (Imperial Chemical Industries, Ltd, England, experimental compound) (0.3–0.4 ppm), chlordane (1.0–2.3 ppm), and diazinen (1.5–7.0 ppm) are the most toxic pesticides to this species, followed by Sevin (5.5–20.0 ppm), NaPCP (Mitusi Chemical Co., Japan) (6.5–10.0 ppm), Bayer 73 (7.0–7.2 ppm), lindane (7.0–22.0 ppm), Sumithion (8.5–12.0 ppm), malathion (17.0–45.0 ppm), Baytex (20.0–28.0

ppm), and dieldrin (60.0–67.0 ppm). Keegan *et al.* (1964) have similar results with *Hirudo nipponia* and an undetermined species of *Hirudo* from Taiwan. For these species chlordane (LC_{100} 1.0 ppm) is the most effective compound, followed by lindane (5.0 ppm), dieldrin (10.0 ppm), Baytex (10.0 ppm), melathion (15.0 ppm), Sumithion (15.0 ppm), and diazinon (15.0 ppm). In concentrations suitable against insects, lindane, aldrin, and dieldrin are ineffective against the freshwater leech *Hirudinara manillensis* in the laboratory and against the terrestrial species *Haemadipsa zeylanica* in the field (Harrison and Audy, 1954).

The toxicities (LC_{100}) of Baytex, Baygon, and Dylox against the American freshwater leeches, *Erpobdella punctata, Illinobdella moorei, Piscicola salmositica, Placobdella parasitica,* and *Theromyzon* sp. are 0.5, 1.0, 1.0, 0.25, and 1.0 ppm for Baytex; 1.0, 0.5, 1.0, 0.25, and 1.0 ppm for Baygon; and 0.5, 0.125, 0.5 0.125, and 0.25 ppm for Dylox, respectively (Meyer, 1969). In each case *Placobdella parasitica* is the most sensitive species.

The LC_{100}'s of Chlorofos (Chinese), Neguvon (Baya), and Trichlorphon (Wolfen), all esters of phosphoric acids, are 2.0 ppm each in field trials against the fish parasite *Piscicola geometra* (Prost and Studnicka, 1966). Similarly, the LC_{100} of Foschlor (Polish), which contains the same active ingredient as Dylox [*0,0*-dimethyl(2,2,2-trichloro-1-hydroxyethyl)phosphonate] is 0.67–1.0 against *Piscicola geometra* (Prost and Studnicka, 1968). In field tests the LC_{100} of Masoten (2,2,2-trichloro-1-hydroxyethylphosphoric acid dimethyl ester) is about 0.25–0.50 ppm against *Piscicola geometra* (Plate, 1970). The lamprey larvicide TFM (3-trifluoro-methyl-4-nitrophenol) kills about 95% of the erpobdellids and only about 11% of the glossiphoniids. Along these lines Fujikawa and his colleagues (1945–1950) investigated the toxicity of various alkylresorcinols and diphenyl ether compounds on unidentified Japanese freshwater leeches. DeJesus (1934) investigated the well-known observation that tobacco extracts (? nicotine) are very toxic to leeches.

The U.S. Army all-purpose repellent "M-1960" consists of equal parts of *N*-butylacetanelide, 2-butyl-2-ethyl-1,3-propanediol, and benzylbenzoate along with 10% emulsifier (Audy and Harrison, 1954; see also, Walton *et al.*, 1956). Uniforms impregnated with M-1960 give almost complete protection against the freshwater leech *Hirudinaria manillensis* and the terrestrial leeches *Haemadipsa zeylanica* and *Haemadipsa picta*, even after more than six cold washes. Socks impregnated with M-1960 are of little value against the freshwater species *Hirudinaria manillensis* but are effective against the terrestrial *Haemadipsa zeylanica*. In addition, three ointments, consisting of dimethyl phthalate (DMP); ethyl β-phenyl hydraerylate; and a mixture of dimethyl phthalate, Repellent 612 (= Rutgers 612 = 2-ethylhexanediol) and dimethyl carbate, respectively, gave protection for no more than a few minutes against *Hirudinaria manillensis* and for no more than an hour against *Haemadipsa picta*. Synthetic pine oil in combination with DDT, dimethyl phthalate, or oils of *Acorus/Curcuma* is reported to have a synergistic effect against leeches (Perti and Agarwal, 1969, 1970). Ramachandran *et al.* (1971) found *N*-toluylpiperidine and *N*-benzoylpiperidine to be superior repellents to dimethyl phthalate and diethyl-*m*-toluamide against *Haemadipsa sylvestris*.

V. Summary and Conclusions

Even though leeches constitute a significant portion of the fauna of most kinds of freshwater habitats, relatively little has been specifically written on the normal or pollutional ecology of North American leeches. Most of our understanding is based on work on European species which are the same or similar to those of North America. In the United States and Canada there are about 51 nominal species of freshwater leeches belonging to the four major families of the Hirudinea. Of these only a dozen are common or occasionally associated with polluted water. By far the most important of these are *Helobdella stagnalis* and *Erpobdella punctata*. Other species occasionally associated with disturbed environments include *Glossiphonia complanata, Helobdella elongata, Helobdella lineata, Illinobdella moorei, Mooreobdella microstoma*, and *Dina parva*. Almost all of the species which are associated with polluted water feed primarily upon organisms more directly associated with pollution such as oligochaetes, insect larvae, and small crustaceans.

It is the quantitative, rather than the qualitative, composition of the leech fauna which characterizes the different types of normal or disturbed habitats. The presence and relative abundance of food organisms play an important part in the ecological distribution of the various leech species. The sucker is very important for locomotion, feeding, and reproduction. A solid substrate (e.g., rocks or pebbles) is a prerequisite for the proper functioning of the leech sucker. A muddy or sandy substrate severely restricts the distribution of leeches. Leeches as a group have their maximum concentration in submerged vegetation, more or less independent of the depth at which it is found. Leeches are scarcer at greater depths primarily because of the lack of vegetation and suitable substrates. Almost all species prefer standing water to running water. The size of ponds and lakes as a factor influencing the distribution of leeches is not so striking as the difference between standing and running water. Leeches can withstand considerable desiccation and many species are commonly found in temporary ponds and streams. Such species survive the dry periods by burrowing into the damp soil. The species which are most commonly found in running water, such as *Glossiphonia complanata* and *Piscicola geometra*, are usually the same species found in the surf zones of lakes.

Except at extremely low levels, hardness, total alkalinity, and pH have little or no direct influence on the distribution or relative abundance of leeches. However, such chemical factors can profoundly affect the occurrence of their various food organisms and, thus, indirectly determine the leech composition. Almost all species are most abundant in water with a total alkalinity ($CaCO_3$) above 60 ppm, and few species occur below 18 ppm. Similarly, all leech species are most abundant in water with a pH of 7.0 or above, and few species occur at pH 6.0 or below. Water temperature plays an important role in the reproductive biology of leeches, primarily by determining the onset of the breeding season. Heated effluents accelerate the onset of breeding. In general, leeches do not reproduce in water in which temperature does not reach 11.0°C for some time. Notable exceptions are certain piscicolid leeches of the *Piscicola-Calliobdella* group which breed at 5°–10°C. High summer temperatures (about 30.0°C or higher) limit the distribution of leeches in both natural and thermally polluted environments. In the laboratory leeches die at temperatures of 33°–35°C. Most leeches can withstand anaerobic conditions for unusually long periods and are little restricted by temporary oxygen depletion. Oxygen consumption is roughly proportional to surface area rather than weight of the leeches. The rate of oxygen consumption is higher in small individuals than in larger ones. *Piscicola*, which is normally found in highly oxygenated water of streams and lakes, has an unusually high oxygen demand.

Siltation and turbidity have profound effects on the ecology of leeches. Water turbidity reduces the natural predation on certain ectoparasitic species (e.g., *Illinobdella* and *Calliobdella*) and leads to high densities and even epidemics. Siltation changes the nature of the substrate to such an extent that leeches have difficulty moving and in depositing cocoons. Not all leeches have the same tolerance to salinity. *Helobdella* and *Erpobdella* are unusually sensitive, whereas *Placobdella* and all piscicolids have remarkable tolerance. All species can tolerate much more salinity in cold water than in warm water.

Helobdella stagnalis and *Erpobdella punctata* are commonly associated with many kinds of polluted water and can be considered "indicator species" only in terms of unusually high densities (above 500 individuals/m^2, depending upon environmental conditions). Both are common and can reach high densities under ideal natural conditions. No leech species can be classified as a *clean water—sensitive* species. Leeches can reach tremendously high densities in regions of organic enrichment. In the Illinois River in 1925 leeches reached 29,107 individuals/m^2 (24,336 individuals/yd^2) with a total combined weight over 2800 kg/hectare (2500 lb/acre). The species most commonly associated with organic enrichment are *Helobdella stagnalis* and *Erpobdella punctata*. In addition those occasionally associated with this form of pollution are *Helobdella elongata, Glossiphonia complanata, Dina parva*, and *Mooreobdella microstoma*. All of these species (with the possible exception of *Glossiphonia complanata*, which feeds primarily upon snails) feed predominantly upon organisms described as *pollutional*, such as *Tubifex, Limnodrilus*, and *Tendipes* larvae.

In streams polluted by derelict mines leeches are more tolerant of dissolved lead than zinc. *Haemopis sanguisuga* is unusually tolerant of dissolved lead, surviving for 18 days in a solution of 3.0 ppm of lead nitrate. In the laboratory the LC_{100} of copper sulfate varies from 1 to 20 ppm, depending upon the species. However, it is ineffective in the field as a control for leeches. Leeches have an unusual tolerance to DDT, the LC_{50} being greater than 100 ppm. DDT is dehydrochlorinated by at least some leeches to the nontoxic metabolite DDE. The LC_{50}'s of the major pesticides, such as Dinex, Chlordane, diazinen, lindane, Mirex, and malathion, vary in the laboratory from 0.5 to 10.0 ppm. Few, if any, of these pesticides are effective under field conditions. Leeches are relatively tolerant of bunker oil pollution.

Acknowledgments

Thanks are due Professor E. W. Knight-Jones, University of Wales, Dr. M. C. Meyer, University of Maine, for their continued support and encouragement of my interest in freshwater and marine leeches. Sincere thanks also go to Gail E. Sanford, College of Charleston, for her untiring patience during the preparation of this manuscript, and to Forrest Reaves, College of Charleston, for his help in the preparation of the illustrations. The staffs of the College of Charleston library and the British Museum (Natural History) library were more than cooperative. Many other individuals were involved in the preparation of this manuscript, but special mention should be made of John W. Tucker and Bruce A. Daniels, College of Charleston; Dr. Edwin B. Joseph, S. C. Marine Resources Division; and Dr. Syed M. Z. Naqvi, Mississippi State University.

Most importantly, I would like to acknowledge that many of the figures and ideas used in this paper were based on those of other workers on leech ecology, most notably Drs. S. A. B. Bennike, University of Kobenhavn (Figs. 8 B–C, 14, and 18); J. F. Carr and J. K. Hiltunen, Bureau of Sport Fisheries and Wildlife, Ann Arbor (Fig. 22); S. J. Herrmann, Southern Colorado State College (Fig. 19); H. L. Keegan, University of Mississippi Medical Center (Fig. 23); D. J. Klemm, University of Michigan (Fig. 20); and J. A. Sapkarev; University of Skopje, Yugoslavia (Figs. 12, 15, 16, and 17).

This work is a result of research sponsored by NOAA Office of Sea Grant, Department of Commerce, under Grant # NG-33-72 The U.S. Government is authorized to produce and distribute reprints for governmental purposes notwithstanding any copyright notation that may appear hereon.

References

Audy, J. R., and Harrison, J. L. (1954). Field tests of repellent M-1960 against leeches. *Med. J. Malaya* **8**, 240–250.

Baal, I. Van (1928). Versuche über Temperatursinn an Blutegeln. *Z. Vergl. Physiol.* **7**, 436–444.

Baker, F. C. (1922). The molluscan fauna of the Big Vermilion River, Illinois, with special reference to its modification as a result of pollution by sewage and manufacturing wastes. *Ill. Biolog. Mongr.* **7**, 1–127. Leeches: 16.

Beck, D. E. (1954). Ecological and distributional notes on some Utah Hirudinea. *Proc. Utah Acad. Sci.* **31**, 73–78.

Becker, C. D., and Katz, M. (1965). Distribution, ecology and biology of the salmonid leech, *Piscicola salmositica* (Rhynchobdellae: Piscicolidae). *J. Fish. Res. Bd. Can.* **22**(5), 1175–95.

Beeton, A. M. (1961). Environmental changes in Lake Erie. *Trans. Amer. Fish. Soc.* **90**, 153–159.

Bennike, S. and Boisen, A. (1943). Contributions to the ecology and biology of the Danish freshwater leeches (Hirudinea). *Folia Limnol. Scand.* **2**, 1–109.

Berg. K. (1938). Studies on the bottom animals of Esrom Lake. *Dan. Vidensk. Selskabs Skrifter. Naturvidensk. Math. Afdeling* (9) **8**, 1–255.

Berg, K. (1948). Biological studies on the River Susaa. *Folia Limnol. Scand.* **4**, 1–318.

Bunge, G. (1888). Über das Saverstoffbedüfnis der Schlammbewohner. *Z. Phys. Chem.* **12**.

Buu-Hoi, N. P. (1962). Repellent action of different chemical substances against earth leeches in Vietnam. *C. R. Séances Soc. Biol.* **156** (2), 277–279.

Carlson, C. A. (1968). Summer bottom fauna of the Mississippi River, above Dam 19, Keokuk, Iowa. *Ecology* **49** (1), 162–169.

Carr, J. F., and Hiltunen, J. K. (1965). Changes in the bottom fauna of western Lake Erie from 1930 to 1961. *Limnol. Oceanogr.* **10**, 551–569.

Cavalloro, R., and Merlini, M. (1967). Stable manganese and fallout radio-manganese in animals from irrigated ecosystems of the Po Valley. *Ecology* **48** (6), 924–928.

Clifford, H. F. (1969). Limnological features of a northern brown-water stream, with special reference to the life histories of the aquatic insects. *Amer. Midl. Natur.* **82** (2), 578–597.

Cristea, V. (1970). Unele Aspecte ale Biologiei reproducerii si dezvoltării la *Erpobdella testacea* Sav. (Hirudinea-Pharyngobdellae). *Stud. Cercet. Biol. Ser. Zool.* **22** (5), 447–453.

Cummings, T. R. (1969). Quality of surface waters of South Carolina: a summary of data, 1945–1968. *U.S. Geological Surv. Water Resources Div., Open-file Rep.*

Davies, R. W. (1971). A Key to the freshwater Hirudinoidea of Canada. *J. Fish. Res. Bd. Can.* **28**(4), 543–552.

Davis, J. J., and Foster, R. F. (1958). Bioaccumulation of radioisotopes through aquatic food chains. *Ecology* **39**, 530–535.

De Jesus, Z. (1934). Experiments on the control of the common water leech, *Hirudinaria manillensis. Philippine J. Sci.* **53**, 47–63.

Dixit, R. S., Saxena, B. N., and Khalsa, H. G. (1967). Evaluation of repellents against land leeches. *Labdev J. Sci. Technol., India* **5**, 140.

Elliott, J. M. (1973a). The life cycle and production of the leech *Erpobdella octoculata* (L.) (Hirudinea: Erpobdellidae) in a lake district stream. *J. Anim. Ecol.* **42**, 435–448.

Elliott, J. M. (1973b). The diel activity pattern, drifting and food of the leech *Erpobdella octoculata* (L.) (Hirudinea: Erpobdellidae) in a lake district stream. *J. Anim. Ecol.* **42**, 449–459.

Evermann, B. W., and Clark, H. W. (1920). Lake Maxinkuckee. *Indiana Dep. Conservat.* **1** (7), 304–305.

Federal Water Pollution Control Administration (1966). A report on the water quality of Charleston Harbor and the effects thereon of the proposed Cooper River rediversion. Southeast Water Laboratory, Charleston Harbor—Cooper River Project, Charleston, South Carolina.

Fjeldsa, J. (1972). Records of *Theromyzon maculosum* (Rathke 1862): Hirudinea, in N. Norway. *Norw. J. Zool.*, **20**, 19–26.

Fujikawa, F., and Nakajima, K. (1945). Toxicity tests with alkylresorcinols on leeches. *J. Pharm. Soc. Jap.*, **67**, 18.

Fujikawa, F., and Tokvoka, A. (1947). Toxicity of diphenyl ether compounds against leeches. *J. Pharm. Soc. Jap.*, **67**, 173.

Fujikawa, F., Nakajima, K., and Fujii, H. (1950). Synthesis of 2,4-dihydroxy-5-alkylbenzaldehyde. *J. Pharm. Soc. Jap.*, **70**, 22–23.

Gaufin, A. R. (1958). The effects of pollution on a midwestern stream. *Ohio. J. Sci.* **58**, 197–208.

Gouck, H. K., Taylor, J. D. Jr., and Barnhart, C. S. (1967). Screening of repellents and rearing methods for water leeches. *J. Econ. Entomol.* **60(4)**, 959–961.

Gresens, J. (1928). Versuche über die Widerstandfähigkeit einiger Süsswassertiere gegenüber Salzlösungen. *Z. Morphol. Öekol. Tiere* **12**, 706–800.

Gruffydd, L. D. (1965). Notes on a population of the leech, *Glossiphonia heteroclita*, infesting *Lymnaea pereger*. *Ann. Mag. Natur. Hist.* **8**, 151–154.

Hall, F. G. (1922). The vital limits of exsiccation of certain animals. *Biolog. Bull.* **42**, 31–51.

Harrison, J. L., Audy, J. R., and Traub, R. (1954). Further tests of repellents and poisons against leeches. *Med. J. Malaya* **9**, 61–71.

Hatto, J. (1968). Observations on the biology of *Glossiphonia heteroclita* (L.). *Hydrobiologia* **31**, 363–384.

Herrmann, S. J. (1970a). Systematics, distribution and ecology of Colorado Hirudinea. *Amer. Midl. Natur.* **83(1)**, 1–37.

Herrmann, S. J. (1970b). Total residue tolerances of Colorado Hirudinea. *Southwest. Natur.* **15(2)**, 269–272.

Herter, K (1929). Temperaturversuche mit Egeln. *Z. Vergl. Physiol.* **10**, 248–271.

Herter, K. (1937). Die Ökologie der Hirudineen. *In* "Klassen und Ordnungen des Tierreichs" (H. G. Bronns, ed.), Band 4, Abt. 3, Buch 4, Teil 2, Lief. 3, pp. 321–496.

Hilsenhoff, W. L. (1963). Predation by the leech, *Helobdella stagnalis*, on *Tendipes plumosus* (Diptera: Tendipedidae) larvae. *Ann. Entomol. Soc. Amer.* **56**, 252.

Hilsenhoff, W. L. (1964). Predation by the leech, *Helobdella nepheloidea*, on larvae of *Tendipes plumosus* (Diptera: Tendipedidae). *Ann. Entomol. Soc. Amer.* **57**, 139.

Hoffman, G. L. (1967). "Parasites of North American Freshwater Fishes," pp. 288–298. Univ. of California Press, Berkeley.

Hynes, H. B. N. (1960). "The Biology of Polluted Waters." Liverpool Univ. Press, Liverpool. 202 pp.

Jones, J. R. E. (1938). The relative sensitivity of aquatic species to lead in solution. *J. Anim. Ecol.* **7**, 287–289.

Jones, J. R. E. (1940). The fauna of the river Melindwr, a lead-polluted tributary of the river Rheidol in North Cardiganshire, Wales. *J. Anim. Ecol.* **9**, 188–201.

Jones, J. R. E. (1958). A further study of the zinc-polluted River Ystwyth. *J. Anim. Ecol.* **27**, 1–14.

Kalbe, L. (1966). Zur Ökologie und Saprobiewertung der Hirudineen im Havelgebiet. *Int. Rev. Ges. Hydrobiol.* **51** (2), 243–277. (Twenty-one species and subspecies of leeches discussed).

Keegan, H. L., and Weaver, R. E. (1964). Studies of Taiwan leeches. II. Field tests of effectiveness of insect repellents against aquatic leeches at Cha'o Chow, Pingtung, Taiwan. *Bull. Inst. Zool. Acad. Sinica* **3** (2), 83–92.

Keegan, H. L., Poore, C. M., Weaver, R. E., and Suzuki, H. (1964). Studies of Taiwan leeches. I. Insecticide susceptibility-resistance tests. *Bull. Inst. Zool. Acad. Sinica* **3**, 39–43.

Keegan, H. L., Weaver, R. E., Fleshman, P., and Zarem, M. (1964). Studies of Taiwan leeches. III. Further tests of repellents against aquatic blood-sucking leeches. *Bull. Inst. Zool. Acad. Sinica* **3** (2), 93–106.

Kenk, R. (1949). Animal life of temporary and permanent ponds in southern Michigan. *Misc. Publ. Univ. Mich. Mus. Zool.* **71**, 1–66.

Kimura, T., and Keegan, H. L. (1966). Toxicity of some insecticides and molluscicides for the Asian blood-sucking leech, *Hirudo nipponia* Whitman. *Amer. J. Trop. Med. Hyg.* **15** (1), 113–115.

Kimura, T., Keegan, H. L., and Haberkorn, T. (1967). Dehydrochlorination of DDT by Asian blood-sucking leeches. *Amer. J. Trop. Med. Hyg.* **16** (5), 688–690.

Klein, L. (1962). "River Pollution. II. Causes and Effects." Butterworths, London and Washington, D.C.

Klemm, D. J. (1972a). The leeches (Annelida: Hirudinea) of Michigan. *Mich. Acad.* **4** (4), 405–444. (A thorough account with sound ecological conclusions, but with an uncritical systematic treatment.)

Klemm, D. J. (1972b). Freshwater leeches (Annelida: Hirudinea) of North America. Biota of Freshwater Ecosystems. Environmental Protection Agency Identification Manual No. **8**, pp. 1–53.

Kolkwitz, R., and Marsson, M. (1909). Ökologie der teirischen Saprobien. *Inter. Rev. Ges. Hydrobiol.* **2**, 126–152.

Kopenski, M. (1972). Leeches (Hirudinea) of Marquette County, Michigan. *Mich. Acad.* **4** (3), 377–383.

Kulajew, S. I. (1929). Die Öekologie der Hirudinea des Stammes *Herpobdella* (Blainv. 1818) im Zusammenhang mit ihrem Verhalten gegen das Vertrocknen. *Bolchewo, Russia. Station Biologique. Bulletin.* (*Zap. Biol. Stan. Bolsheve*) **3**, 59–73.

Kussat, R. H. (1969). A comparison of aquatic communities in the Bow River above and below sources of domestic and industrial wastes from the city of Calgary. *Can. Fish Cultur.* **40**, 3–31.

Laurie, R. D., and Jones, J. R. E. (1938). The faunistic recovery of a lead polluted river in north Cardiganshire, Wales. *J. Anim. Ecol.* **7**, 272–286.

McAnnaly, R. D., and Moore, D. V. (1966). Predation of the leech *Helobdella punctatolineata* upon *Australorbis blabratus* under laboratory conditions. *J. Parasitol.* **52** (1), 196–197.

McCauley, R. N. (1966). The biological effects of oil pollution in a river. *Limnol. Oceanogr.* **11** (4), 475–486.

MacKenthun, K. M. (1969). The Practice of Water Pollution Biology. U. S. Dep. of Interior, Fed. Water Pollut. Contr. Administration. Div. of Tech. Support, pp. 1–281.

Madsen, B. L. (1963). (Ecological investigations on some streams in East Jutland. 2. Planarians and leeches.) *Flora Fauna* **69**(**4**), 113–125. (In Danish; English summary).

Mann, K. H. (1953). The life history of *Erpobdella octoculata* (L.). *J. Anim. Ecol.* **22**, 197–207.

Mann, K. H. (1955a). Some factors influencing the distribution of freshwater leeches in Britain. *Proc. Int. Ass. Theoret. Appl. Limnol.* **12**, 582–587.

Mann, K. H. (1955b). The ecology of the British freshwater leeches. *J. Anim. Ecol.* **24**, 98–119.

Mann, K. H. (1956). A study of the oxygen consumption of five species of leech. *J. Exp. Biol.* **33**, 615–626.

Mann, K. H. (1957a). The breeding growth and age structure of a population of the leech *Helobdella stagnalis* (L.). *J. Anim. Ecol.* **26**, 171–177.

Mann, K. H. (1957b). A study of a population of the leech *Glossiphonia complanata* (L.). *J. Anim. Ecol.* **26**, 99–111.

Mann, K. H. (1958). Seasonal variation in the respiratory acclimatization of the leech *Erpobdella testacea* (Sav.). *J. Exp. Biol.* **35**, 314–323.

Mann, K. H. (1961a). The oxygen requirements of leeches considered in relation to their habitats. *Proc. Int. Ass. Theoret. Appl. Limnol.* **14**, 1009–1013.

Mann, K. H. (1961b). The life history of the leech *Erpobdella testacea* and its adaptive significance. *Oikos* **12**, 164–169.

Mann, K. H. (1961c). "*Leeches (Hirudinea): Their Structure, Physiology, Ecology and Embryology.*" Pergamon, Oxford.

Mason, W. T., Jr., Anderson, J. B., Kreis, R. D., and Johnson, W. C. (1970). Artificial substrate sampling, macroinvertebrates in a polluted reach of the Klamath River, Oregon. *J. Water Pollut. Contr. Fed.* **42** (8, Part 2), R315–R328.

Meyer, F. P. (1969). A potential control for leeches. *Prog. Fish-Cultur.* **31** (3), 160–163.

Meyer, M. C. (1940). A revision of the leeches (Piscicolidae) living on freshwater fishes of North America. *Trans. Amer. Microsc. Soc.* **59** (3), 354–376.

Meyer, M. C. (1946). Further notes on the leeches (Piscicolidae) living on freshwater fishes of North America. *Trans. Amer. Microsc. Soc.* **65**, 237–249.

Miller, J. A. (1929). The leeches of Ohio. *Ohio State Univ. Contribut. Franz Theodore Stone Lab.* **2**, 1–38.

Miller, J. A. (1937). A study of the leeches of Michigan with keys to orders, suborders and species. *Ohio J. Sci.* **37**, 85–90.

Moore, J. E. (1964). Notes on the leeches (Hirudinea) of Alberta. *Natur. Hist. Papers Nat. Mus. Can.* **27**, 1–15.

Moore, J. E. (1966). Further notes on Alberta leeches (Hirudinea). *Natur. Hist. Papers Nat. Mus. Can.* **32**, 1–11.

Moore, J. P. (1901). The Hirudinea of Illinois. *Bull. Ill. State Lab. Natur. Hist.* **5**, 479–547.

Moore, J. P. (1906). Hirudinea and Oligochaeta collected in the Great Lakes region. *Bull. Bur. Fish.* **25**(598), 153–172.

Moore, J. P. (1912). Leeches of Minnesota. *Geol. Natur. Hist. Surv. Minnesota, Zool.* **5** (pt III, Classification), 64–150.

Moore, J. P. (1918). The Leeches (Hirudinea) *In* "Fresh Water Biology" (H. B. Ward and G. C. Whipple, eds.), pp. 646–660, Wiley., New York.

Moore, J. P. (1920). The leeches. Lake Maxinkuckee, a physical and biological survey. *Ind. Dep. Conservat.* **2**, 87–95.

Moore, J. P. (1922). The fresh-water leeches (Hirudinea) of southern Canada. *Can. Field-Natur.* **36**, 6–11, 37–39.

Moore, J. P. (1924). The leeches (Hirudinea) of Lake Nipigon. *Univ. Toronto Stud. Biol.* **23**, 17–30.

Moore, J. P. (1959). Hirudinea. *In* "Freshwater Biology" (W. T. Edmondson, ed.), 2nd. ed. pp. 542–557. Wiley, New York.

Mozley, A. (1932). A biological study of a temporary pond in western Canada. *Amer. Natur.* **66**, 235–249.

Muttkowski, R. A. (1918). The fauna of Lake Mendota: a qualitative and quantitative survey with special reference to the insects. *Trans. Wis. Acad. Sci.* **19** (1), 374–482. Hirudinea, 391–392.

Naqvi, S. M. Z. (1972). Personal communication. The data on Mirex uptake and residues kindly supplied by Dr. Naqvi, Mississippi State University, are based on a manuscript co-authored with Dr. Armando A. DeLaCruz, "Mirex incorporation in the environment and residues in non-target organisms". It is being submitted to *Environ. Pollut.*

Nekipelov, M. I. (1961). On the toxicity of nitrates for water organisms. *Zool. Zh. Moscow* **46**, 932–936. (In Russian; English summary). (An incomplete and unconfirmed reference describing the reaction of *Hirudo medicinalis officinalis* to semilethal nitrate concentrations; a similar unconfirmed paper by the same author, translated, "The toxic effect of aluminum compounds on marine organisms," deals with the minimum lethal concentrations of aluminum nitrate on leeches).

Ökland, J. (1964). The eutrophic lake Borrevann, Norway. An ecological study on shore and bottom fauna with special reference to gastropods, including a hydrographic survey. *Folia Limnol. S cand.* **13**, 1–337.

Paloumpis, A. A., and Starrett, W. C. (1960). An ecological study of benthic organisms in three Illinois River flood plain lakes. *Ameri. Midl. Natur.* **64** (2), 406–435.

Patrick, R., Cairns, J., Jr., and Roback, S. S. (1967). An ecosystematic study of the fauna and flora of the Savannah River. *Proc. Acad. Natur. Sci. Philadelphia*, **11**(5), 109–407.

Pawlowski, L. K. (1936). Zur Okologie der Hirudineen fauna der Wigryseen. *Arch. Hydrobiol. Rybactwa* **10**, 1–47.

Pawlowski, L. K. (1955). Observations biologiques sur les sangsues. *Bull. Soc. Sci. Lett. Lodz (III)* **6**, 1–21.

Pennak, R. W. (1953). "Freshwater Invertebrates of the United States." Ronald Press, New York. Hirudinea, 302–320.

Pennak, R. W. (1958). Some problems of fresh-water invertebrate distribution in the Western States. *In* "Zoogeography" (C. L. Hubbs, ed., Amer Ass. for the Advan. Sci. Publ. No. 51.)

Perti, S. L., and Agarwal. P. N. (1969). Use of synthetic pine oil in pest control. *Pesticides* **2** (7), 45.

Perti, S. L., and Agarwal, P. N. (1970). Synergism in insecticides. *Labdev J. Sci. Technol. Part B* **8** (2), 67–71.

Pickavance, J. R. (1971). Pollution of a stream in Newfoundland: effects on invertebrate fauna. *Biol. Conserv.* **3** (4), 264–268.

Plate, G. (1970). Masoten für die Bekampfung von Ektoparasiten bei Fischen. *Arch. Fishereiwiss.* **21** (3), 258–267.

Prost, M., and Studnicka, M. (1966). (Investigations on the use of organic esters of phosphoric acid in the control of external parasites of farmed fish. I. Control of the invasion of *Piscicola geometra* (L.)).—*Med. Wet.* **22(6)**, 321–330. (In Polish; English, Russian, German, and French summaries.)

Prost, M. and Studnicka, M. (1968) (Investigations on the use of organic esters of phosphoric) in the control of external parasites of farmed fish. IV. Therapeutic efficacy of the preparation "Foschlor." *Med. Wet.* **24** (2), 97–101. (In Polish; Russian, French and German summaries.)

Ramachandran, P. K., Koshy, T., Sastry, K. G. K., Singh, S. P. Srinivasan, M. N., and Ganguly, S. K. (1971). Studies on leech repellents. *J. Econ. Entomol.* **64** (5), 1293–1294.

Ribbands, C. R. (1946). Experiments with leech repellents. *Ann. Trop. Med. Parasitol.* **40**, 314–319.

Richardson, R. E. (1925a). Illinois River bottom fauna in 1923. *Bull. Ill. Natur. Hist. Surv.* **15**, 391–423.

Richardson, R. E. (1925b). Changes in the small bottom fauna of Peoria Lake, 1920 to 1922. *Bull. Ill. Natur. Hist. Surv.* **15**, 327–388.

Richardson, R. E. (1928). The bottom fauna of the Middle Illinois River, 1913–1925. Its distribution, abundance, valuation and index value in the study of stream pollution. *Bull. Ill. Natur. Hist. Surv.* **17**, 387–475.

Ricker, W. E. (1952). The benthos of Cultus Lake. *J. Fish. Res. Bd. Can.* **9**, 204–212.

Rosca, D. I., and Oros, I. (1962). Research on the penetration of Phosphorus-32-labeled Phosphate into the body of leeches (*Hirudo medicinalis*). *Acad. Repub. Pop. Rom. Filiala Cluj Stud. Cerc. Biol.* **13** (2), 347–353.

Sandner, H. (1951). Badania and Fauna Pijawek. *Acta Zool. Oecol. Univ. Lodziensis* **4**, 1–50.

Sapkarev, J. A. (1963). Die Fauna Hirudinea Mazedoniens. I. Systematik und Ökologie der Hirudinea des Prespa-Sees. *Bull. Sci.* **8** (1/2), 7–8.

Sapkarev, J. A. (1968). The taxonomy and ecology of leeches (Hirudinea) of Lake Mendota, Wisconsin. *Trans. Wisconsin Acad. Sci.* **56**, 225–253.

Sarah, H. H. (1971). Leeches found on two species of *Helisoma* from Fleming's Creek, Michigan. *Ohio J. Sci.* **71** (1), 15–20.

Sawyer, R. T. (1967). The leeches of Louisiana, with notes on some North American species. *Proc. Louisiana Acad. Sci.* **30**, 32–38.

Sawyer, R. T. (1968). Notes on the natural history of the leeches (Hirudinea) on the George Reserve, Michigan. *Ohio J. Sci.* **68**(**4**), 226–228.

Sawyer, R. T. (1970). Observations on the natural history and behavior of *Erpobdella punctata* (Annelida: Hirudinea). *Amer. Midl. Natur.* **83** (1), 65–80.

Sawyer, R. T. (1971). The phylogenetic development of brooding behaviour in the Hirudinea. *Hydrobiologia* **37** (2), 197–204.

Sawyer, R. T. (1972). "North American Freshwater Leeches, Exclusive of the Piscicolidae, with a Key to All Species." Univ. of Illinois Press, Urbana.

Sawyer, R. T. (1973). Bloodsucking freshwater leeches: observations on control. *J. Econ. Entomol.* **66**(2), 537.

Sawyer, R. T., and Hammond, D. L. (1973). Distribution, ecology and behavior of the marine leech, *Calliobdella carolinensis*, parasitic on the Atlantic menhaden in epidemic proportions. *Biol. Bull.* **145**(2), 373–388.

Saxena, B. N., Khalsa, H. G., and Pillai, K. R. M. (1969). Evaluation of repellents against land leeches. II. *Def. Sci. J.* **19** (2), 93–96.

Scudder, G. G. E., and Mann, K. H. (1968). The leeches of some lakes in the southern interior plateau region of British Columbia. *Syesis* **1** (1/2), 203–209.

Smith, A. J. (1967). The effect of the lamprey larvicide, 3-trifluoro-methyl-4-nitrophenol, on selected aquatic invertebrates. *Trans. Amer. Fish. Soc.* **96** (4), 410–413.

Smith, M. W. (1939). Copper sulfate and rotenone as fish poisons. *Trans. Amer. Fish. Soc.* **69**, 141–157.

Soos, A. (1965–1969). Identification key to the leech (Hirudinoidea) genera of the world, with a catalogue of the species. I–IV. *Acta Zool. Acad. Sci. Hung.* **11**, 417–463; **12**, 145–160, 371–407, **13**, 417–432; **15**, 151–201, 397–454.

Sterba, G. (1963). Über die Bedeutung des Anasthetikums MS-222 Sandoz (ethyl-m-amino-benzoate) für aquatile Wirbeltiere und Wirbellose. Vol. D. *Bull. Inst. Oceanogr.* (*Monaco*) 1D 127–133.

Thomas, M. L. H. (1966). Benthos of four Lake Superior bays. *Can. Field-Natur.* **80** (4), 200–212.

Thompson, D. H. (1927). An epidemic of leeches on fishes in Rock River. *Bull. Ill. State Natur. Hist. Surv.* **17**, 195–201.

Thut, R. N. (1969). A study of the profundal bottom fauna of Lake Washington. *Ecol. Monogr.* **39**, 79–100.

Traub, R., Wisseman, C. L., Jr., and Audy, J. R. (1952). Preliminary observations on a repellent for terrestrial leeches. *Nature (London)* **169**, 667–668.

Tucker, D. S. (1958). The distribution of some freshwater invertebrates in ponds in relation to annual fluctuations in the chemical composition of the water. *J. Anim. Ecol.* **27**, 105–123.

Walton, B. C., Traub, R., and Newson, H. D. (1956). Efficacy of the clothing impregnants M-2065 and M-2066 against terrestrial leeches in North Borneo. *Amer. J. Trop. Med. Hyg.* **5**(**1**), 190–196.

Webster, E. J. (1967). An autoradiographic study of invertebrate uptake of DDT-Cl^{36}. *Ohio J Sci.* **67** (5), 300–307.

Weston, R. S., and Turner, C. E. (1917). Studies on the digestion of a sewage-filter effluent by a small and otherwise unpolluted stream. *Contrib. Sanit. Res. Lab. Sewage Exp. Sta. Mass. Inst. Technol.* **10**, 1–96.

Wilhm, J. L., and Dorris, T. C. (1966). Species diversity of benthic macroinvertebrates in a stream receiving domestic and oil refinery effluents. *Amer. Mid. Natur.* **76**, 427–449.

Wright, S. (1955). Limnological survey of western Lake Erie. *U.S. Fish Wildl. Serv. Spec. Sci. Rep. Fish.* **139**, 1–341.

Wurtz, C. B., and Bridges, C. H. (1961). Preliminary results from macro-invertebrate bioassays. *Proc. Penn. Acad. Sci.* **35**, 51–56.

Wurtz, C. B., and Roback, S. S. (1955). The invertebrate fauna of some gulf coast rivers. *Proc. Acad. Natur. Sci. Philadelphia* **107**, 167–206.

CHAPTER 5

Aquatic Earthworms (Annelida: Oligochaeta)

R. O. BRINKHURST and D. G. COOK

I. Introduction

Aquatic oligochaetes are familiar objects to anyone with the most cursory acquaintance with organically polluted water. Sludge worms may come to dominate the bottom fauna so completely as to form a patchy red carpet, and hence the whole group has become so associated with gross organic pollution that their discovery in "clean" environments causes undue alarm. Segmented worms, mostly oligochaetes but sometimes polychaetes in fresh water (*Manyunkia speciosa*, for instance, is common in the mouth of the Detroit River as shown by Hiltunen, 1965), occupy sandy to muddy substrates in streams, rivers, ponds, and lakes. A reasonable number of oligochaetes are found in estuaries and on seashores, and some are even found beyond the continental shelf (Brinkhurst and Jamieson, 1971*; Cook, 1970,

*This contains an account of the taxonomy of world aquatic oligochaetes.

TABLE I

THE TUBIFICIDAE OF NORTH AMERICA, EXCLUDING THOSE ENDEMIC TO LAKE TAHOE AND OFFSHORE SPECIES.[a]

Genus	Species			
	Widely distributed or cosmopolitan	Eastern	Western	Pan-American
Tubifex	*tubifex*	*newaensis*		
		ignotus		
		kessleri		
		americanus		
		pseudogaster		
		longipenis		
Psammoryctides		*curvisetosus*	*californianus*	
Limnodrilus	*hoffmeisteri*	*maumeensis*	*silvani*	*cervix*
	udekemianus	*angustipenis*		
	claparedeianus			
	profundicola			
Isochaeta		*hamata*		
Peloscolex		*ferox*	*oregonensis*	*variegatus*
		aculeatus		*gabriellae*
		benedeni		*apectinatus*
		intermedius		*nerthoides*
		carolinensis		*multisetosus*
		freyi		
		superiorensis		
		dukei		
Potamothrix		*hammoniensis*		
		bavaricus		
		moldaviensis		
		vejdovskyi		
Ilyodrilus		*templetoni*	*perrierii*	
			fragilis	
			frantzi	
Rhyacodrilus		*coccineus*		*montana*
				sodalis
Branchiura	*sowerbyi*			
Monopylephorus	*rubroniveus*	*lacteus*		
	irroratus			
	parvus			
Bothrioneurum	*vejdovskyanum*			
Aulodrilus	*limnobius*	*americanus*		
	pigueti			
	pluriseta			
Telmatodrilus				*vejdovskyi*
Clitellio		*arenarius*		

TABLE I (*continued*)

Genus	Species			
	Widely distributed or cosmopolitan	Eastern	Western	Pan-American
Limnodriloides		*arenicolus*		
		medioporus		
Phallodrilus		*coeleprostatus*		
		parviatriatus		
		obscurus		
Spiridion		*?insigne*		
Smithsonidrilus		*marinus*		
Adelodrilus		*anissoetosus*		

[a]From Brinkhurst and Jamieson, 1971.

1971). A number of different families are involved and with practice they can be told apart quite easily. The tiny Aeolosomatidae which locomote via cilia on the prostomium and the rare Opistocystidae need not concern us too much here. These fragile worms are mostly less than 2 mm long, though chains of individuals may reach 10 mm, but they are not usually found in field collections unless specifically searched for. The Naididae are also small and fairly fragile. Some of these worms have eyes, and they are often found in chains of asexually reproducing individuals—both clues to their identity being valuable but not diagnostic. Many species can swim in a clumsy fashion, and they are often found amongst weeds, but a few, such as *Dero* species, live in sediments like the sludge worms or Tubificidae.

Tubificids are generally larger (both longer and wider) than naidids, being measured in centimeters rather than millimeters, and they are somewhat more robust. They are generally a good red color, though this may be obscured by the warty body wall in *Peloscolex* or the rich supply of coelomic cells in *Bothrioneurum* and the frequently estuarine *Monopylephorus*. The tubificids often coil up into a tight grub-screw shape when touched, but they cannot swim and none have eyes. Their setae are an order of magnitude larger than those of the Naididae in general, but there is some overlapping in all of these characteristics. While this familial distinction may sound hard to make, it is, in fact, quite simple in practice.

The Enchytraeidae are often more white to pink than red in color, and their setae are distinctive, being simple-pointed (apart from one European genus) and either straight or gracefully sigmoid. While the worms may coil like tubificids, they are often stiffer, and most are found in wet soils rather than or as well as aquatic sediments. The setae of the Lumbriculidae may be simple-pointed or slightly bifid with small upper teeth, but they are distinguishable from all but the "earthworm" sized worms [mostly *Sparganophilus tamesis* (= *eiseni*), fam. Glossoscolecidae—Brinkhurst and Jamieson, 1971] by having only two setae per bundle. Most are thicker and longer than sludge worms (except for *Stylodrilus heringianus*) but not so heavily built as the earthworms." The exceedingly long and thin *Haplotaxis gordioides* (Haplotaxidae) is instantly recognizable both in general appearance and setal form (Brinkhurst and Jamieson, 1971).

While there are only a dozen North American lumbriculids, there are about 60 tubificids (Table I). These are divided up into a number of groups (brackish plus marine, cosmopolitan, pan-American, eastern, and western) and there are about 20 significant species belonging to the genera *Tubifex*, *Limnodrilus*, *Peloscolex*, *Potamothrix* (= *Euilyodrilus*), *Rhyacadrilus*, *Branchiura*, *Aulodrilus*, and *Bothioneurum*. Many of these can be identified by examining the setae and male genitalia on temporary mounts made quite simply by placing the worm on a slide beneath a cover slip using Ammans lactophenol as a mounting medium (1 part carbolic acid, 1 part lactic acid, 2 parts glycerol, 1 part water) after which they should be left for a couple of days to clear. Simplified regional keys are available in Brinkhurst *et al.* (1968) and Brinkhurst (1968).

Comprehensive (though somewhat outdated) keys to the North American fauna can be found in Brinkhurst (1964, 1965c) and Brinkhurst and Cook (1966). Corrections to nomenclature were included in Brinkhurst (1966), but the current usage and that followed herein is detailed on a world basis in Brinkhurst and Jamieson (1971). The keys in the latter volume are comprehensive, and they identify genera then species for each family. More usable keys exclude all but local species and key these out directly regardless of genera, as it is much simpler to obtain an identification using hard-part characters such as setae.

II. Basic Ecology

In general, the tubificids (plus most lumbriculids and some naidids) are adapted to a burrowing life in soft sediments. Deriving most if not all of their nutrition from bacteria, they ingest large volumes of sediment continuously in order to extract the small fraction of nutrient material contained therein. Their ability to withstand considerable oxygen depletion in their environment is therefore an essential adaptation to their niche (or profession) within the community. Recent studies (Brinkhurst, 1972) suggest that competition is avoided by selective digestion of the bacteria within the sediment, which leads to a degree of collaboration as the feces of one species of worm become the preferred food for another species. This is probably the ecological basis of the very close clumping of individuals which makes quantitative ecological studies difficult owing to the sampling problems it raises. All worms selectively ingest fine organic particles in preference to coarser inorganic material, but whether each species selectively ingests specific bacteria according to their ability to digest them has yet to be established. In normal soft sediments worms may be abundant but there will be many species in most places (up to 15 tubificids in the River Thames at Reading, United Kingdom, for example), and many other types of organisms will be found in those parts of the same locality where there are fewer bacteria but more oxygen.

In stony streams and lake margins the lumbriculids and naidids tend to replace the tubificids, which are found in such localities when sufficient organic matter is introduced to maintain a thick bacterial slime on the substrate.

In general, the quantity and quality of organic matter reaching the sediment appears to play a more dominant role in determining which tubificid species will be found in any given locality than do all of the commonly measured physical and chemical parameters of the water body or the sediment. Specific organic inputs might be expected to affect the spectrum of bacterial species present, which in turn determines the worm species and their relative abundance.

III. Worms as Pollution Indicators

There have been few physiological studies of tubificids in relation to their degree of tolerance of various pollutants (and even fewer on worms of other families) so that most of their pollution biology is actually inferred from field studies. The heavy silting, high organic content of silt and associated deoxygenation below organic effluents (from sewage works, packing plants, food processing plants, etc.) tend to increase the share of tubificid habitats on the river or lake bed. Here the number of worm species may decline but the total number of worms may drastically increase in contrast to normal localities, and at the same time the numbers and kinds of other benthic species may decline. Oligochaetes and some other soft-bodied forms (flatworms, molluscs, leeches, etc.) are more tolerant of pesticides than arthropods, but less tolerant of heavy metal ions, and anything that reduces bacterial activity (acids) will reduce worm populations accordingly. Hence, the abundance of worms relative to other organisms as well as the relative abundance of various species of worms may be of interest. The absolute number of worms may be significant in cases of self-evident organic pollution, i.e., where the sediments literally stink. In such instances, a few very resistant species, i.e., *Limnodrilus hoffmeisteri*, may be so abundant as to be thought of as "indicators" per se, but one should be aware that all such "indicators" have been around a long time, certainly far longer than man with his special knack of modifying environments to his own detriment. Hence "indicator" species *must* be present in "clean" environments, though their microhabitats therein may be "naturally" polluted by leaves or other rotting organic material dropping out of the system.

A few specific instances of field and laboratory studies should suffice to show how worms can be used to detect the sources and nature of pollution as well as the degree of damage done.

In the River Derwent, England, the entire river below Derby sometimes failed to pass muster by the limits imposed on sewage effluents at the time (20 ppm BOD, 30 ppm suspended solids) and it was dominated faunistically by *T. tubifex* and *L. hoffmeisteri* (Table II). After the installation of a sewage disposal works, twelve tubificid species were found at one time or another at those same stations and worms were much scarcer in both absolute and relative terms (Brinkhurst, 1965a).

TABLE II

THE TUBIFICIDAE RECORDED FROM THE RIVER DERWENT IN 1958 AND IN 1959–1962[a]

	Before improvement (stations)				After improvement (stations)			
	4	6	7	13 (2 surveys)	4	6	7	13 (7 surveys)
T. tubifex	+	+	+	+	+	+	+	+
T. templetoni	–	–	–	–	–	–	+	–
T. ignota	–	–	–	–	–	+	–	–
L. hoffmeisteri	+	+	+	+	+	+	+	+
L. udekemianus	–	–	–	–	–	+	–	+
L. claparedeanus	–	–	–	–	–	+	+	+
P. hammoniensis	–	–	–	–	+	+	+	+
P. moldaviensis	–	–	–	–	+	–	–	–
R. coccineius	–	–	–	–	+	+	+	–
B. vejdovskyanum	–	–	–	–	–	+	+	+
P. barbatus	–	–	–	–	+	+	+	+
A. pluriseta	–	–	–	–	–	–	–	+
Number of species	2	2	2	2	6	9	8	8
Maximum in any one survey	2	2	2	2	5	7	5	5

[a]From Brinkhurst and Jamieson, 1971.

In the St. Lawrence Great Lakes, it is possible to recognize three species associations of tubificids. One consists of *T. tubifex*, *Peloscolex multisetosus*, and several *Limnodrilus* species and is characteristic of organically polluted bays and harbours (Green Bay, Hamilton Bay, Toronto Harbour, the upper Bay of Quinte, and others). A second consists of an "eutrophic" assemblage that includes *Aulodrilus* and *Potamothrix* species together with *Limnodrilus* and *Peloscolex ferox*; while the third (clean water) group includes all sorts of species not often encountered in the other areas. In this third assemblage, however, we still find *L. hoffmeisteri*, and *T. tubifex* is often found in the samples—a further cautionary note to those using the indicator species

TABLE III

DATA ON THE DISTRIBUTION AND ABUNDANCE OF THE THREE MOST ABUNDANT TUBIFICIDS FOR EACH OF THE FOUR ZONES IDENTIFIED IN FIG. 1, AND FOR HANLAN'S POINT, TORONTO HARBOUR[a,b]

	Zone				
	1	2	3	4	Hanlan's Point
No. stations	8	11	22	2	1
Average no./m² (thousands)					
T	115	55	18	3	12
L	58	33	21	8	16
P	24	16	12	12	2
	197	104	51	23	30
Average no./m_2 (thousands) calculated from dry weight plus gut content					
Total	194	106	59	19	—
%					
T	58	53	36	14	40
L	30	31	41	33	53
P	12	16	23	52	7
Average dry weight (gm/m_2) calculated for worms minus gut content					
T	22.9	11.0	3.7	0.7	2.4
L	17.5	9.8	6.3	2.4	4.9
P	2.3	1.6	1.1	1.2	0.2
	42.7	22.4	11.1	4.3	7.5
Average dry weight (gm/m²) of worms plus gut contents (all species)	98.5	48.0	26.5	9.0	—
Ratio dry weight plus gut content/dry weight without gut content	2.3:1	2.1:1	2.4:1	2.1:1	—

[a]From Brinkhurst, 1972.
[b]T = *Tubifex tubifex*; L = *Limnodrilus hoffmeisteri*; P = *Peloscolex multisetosus*.

concept being evident here. Even in Toronto Harbour, where most of the area is dominated by sludge worms, the patterns of relative abundance of identified species strongly reflect the influence of the major polluting input, the Don River (Table III, Figs. 1–8). Details of this story are documented by Brinkhurst (1969, 1970, 1972) and the system was looked at from a com-

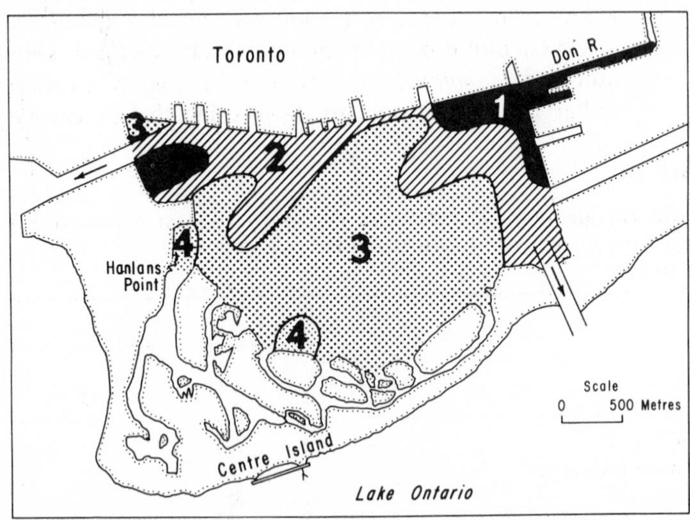

Fig. 1 Abundance (no./m²) of tubificids in Toronto Harbour. Zone 1 (solid), > 125,000; zone 2, 75,000–125,000; zone 3, 25,000–75,000; zone 4, >25,000.

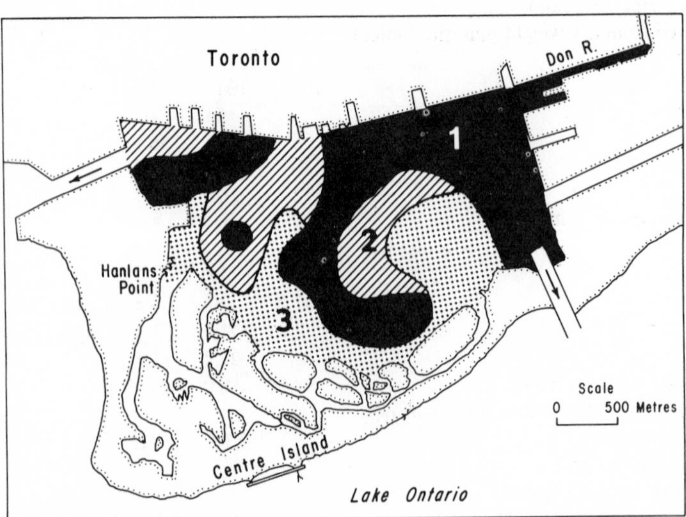

Fig. 2. Abundance of *Tubifex tubifex* in Toronto Harbour as percentage of all tubificids. Zone 1 (solid), >50%; zone 2, 25–50%; zone 3, >25%.

munity standpoint as well as in diversity terms in the Bay of Quinte, a smaller model of the whole Great Lakes system (Johnson and Brinkhurst, 1971). References to the work of other contributors to this story (e.g., J. Hiltunen; D. M. Veal and D. S. Osmond, Table IV) will be found in these summary statements and in a paper by Howmiller and Beeton (1970) on the oligo-

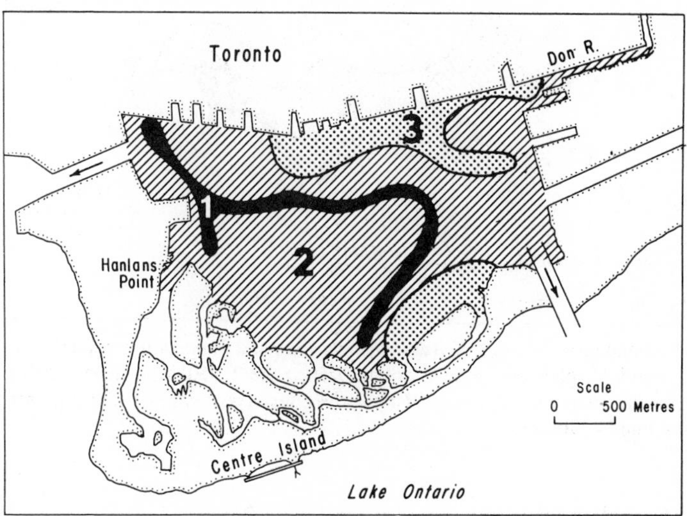

Fig. 3 Abundance of *Limnodrilus hoffmeisteri* in Toronto Harbour as percentage of all tubificids. Zone 1 (solid), > 50%; zone 2, 25–50%; zone 3, < 25%.

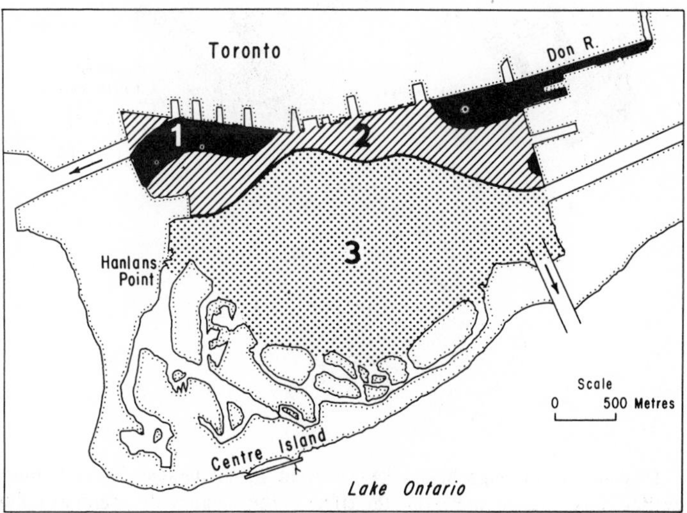

Fig. 4. Abundance (no./m^2) of *Limnodrilus udekemianus* in Toronto Harbour. Zone 1 (solid), > 5000 (mean, 9500); zone 2, 1000–5000 (mean, 2500); zone 3, 0–500 (mean, 360).

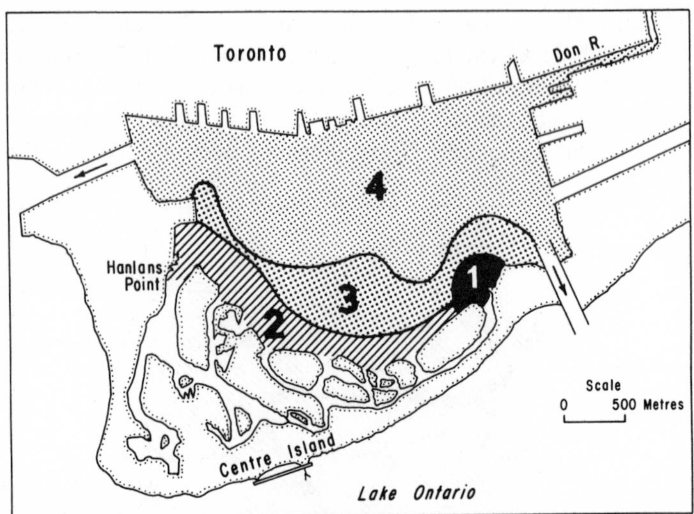

Fig. 5. Abundance (no./m²) of *Aulodrilus pluriseta* and *Potamothrix vejdovskyi* in Toronto Harbour. *Aulodrilus pluriseta:* zone 1 (solid), 9250; zone 2, 750–3750 (mean, 2000); zone 3, 250–750 (mean, 350); zone 4, absent. *Palamothrix vejdovskyi*: zones 1 and 2 present (mean, 800); zones 3 and 4, absent.

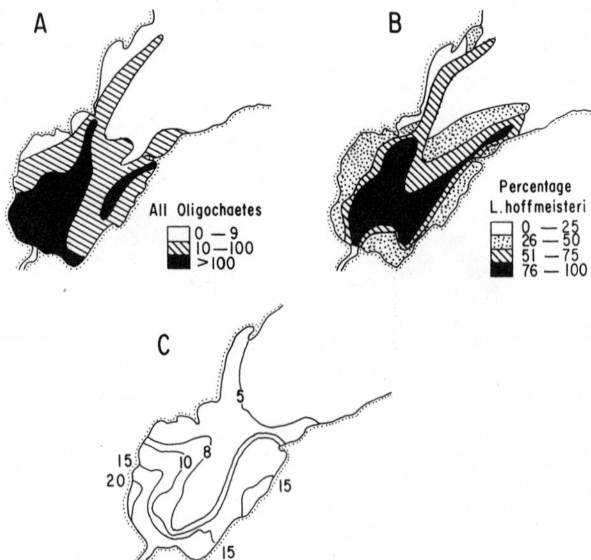

Fig. 6. Distribution of oligochaetes (A), percent *L. hoffmeisteri* (B) and chlorosity (C) in Saginaw Bay, Lake Huron, indicating the effect of the Saginaw River and its pollutants on the bay.

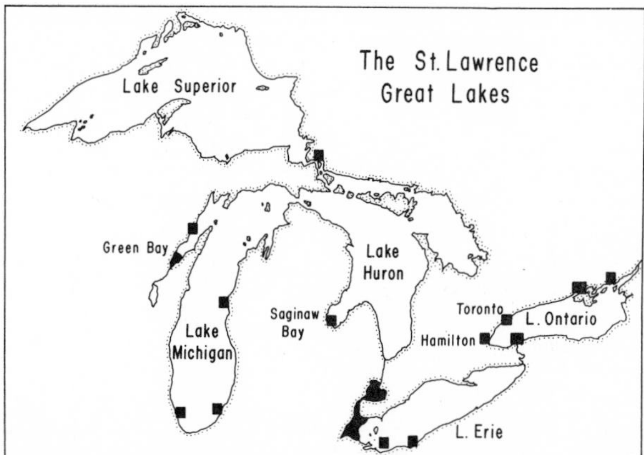

Fig. 7. Distribution of "polluted water" assemblage of tubificids in the St. Lawrence Great Lakes (dominated by *L. hoffmeisteri* and *T. tubifex*, includes *P. multisetosus*, *L. cervix*, *L. maumeensis*, and *B. sowerbyi*, perhaps *I. templetoni*). Note that relative abundance is important here, as shown in Table III. These illustrations (Figs. 7–8) indicate the *commonest* species found in the area indicated.

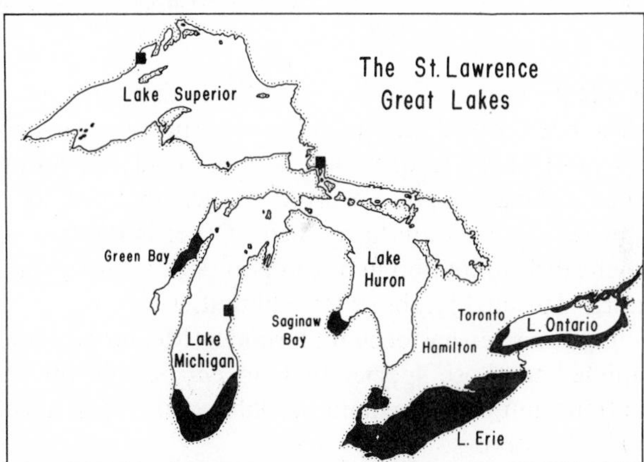

Fig. 8. Distribution (in black) of "eutrophic lake" assemblage of oligochaetes in the St. Lawrence Great Lakes (includes *Aulodrilus*, *Potamothrix* (= *Euilyodrilus*), *Peloscolex ferox*, *P. freyi* plus *L. hoffmeisteri*). White areas contain both *L. hoffmeisteri* and *T. tubifex* (hence relative abundance is important) plus *T. kessleri americanus*, *Rhyacodrilus*, and *Aulodrilus americanus*.

chaetes of Green Bay, Lake Michigan. A number of figures and tables are added to this paper as an appendix to indicate the sort of data available.

In 1965b, Brinkhurst reviewed the literature on the biology of tubificids in relation to various types of effluents. Unfortunately we still do not have comparative physiological data on the tolerance to various pollutants of a

TABLE IV

THE INSHORE FAUNA OF LAKE ERIE[a,b]

Genus	Species	Western basin	Central basin	Eastern basin
Limnodrilus				
	hoffmeisteri	98	64	44
	cervix	76	19	4
	maumeensis	33	2	0
Tubifex				
	tubifex	7	38	24
Peloscolex				
	ferox	5	64	8
Potamothrix				
	moldaviensis	5	43	32
Stylodrilus				
	heringianus	0	38	24

[a]From Veal and Osmond, 1969.

[b]Values are percentages of sampling locations at which each species was found.

spectrum of identified tubificid species. Even the work on respiratory physiology has been done on the most tolerant species or on species mixtures, and most of the information is of an academic nature in that it does not indicate the levels of dissolved oxygen that would bring about changes in species composition in the field. Other laboratory studies have focused on niche discrimination in mixed populations of worms in polluted sites particularly Toronto Harbour (Brinkhurst, 1972), but no comparable study of clean water sites has been undertaken. Meaningful tolerance tests should be applied to those species that do *not* occupy the most grossly polluted situations, but all of the data available on these is, alas, taxonomic or is based on field surveys.

Hence the oligochaetes can be used in field surveys (because the taxonomy is now in an advanced state, rendering the species easy to identify with a little practice), and many such surveys have been published so that interpretation of species distributions can be made, but there is, as yet, no good correlation between observed distributions and those environmental parameters usually measured.

Two special instances may be noted. The exotic species *Branchiura sowerbyi* has been introduced to Europe and North America, probably via the tropical greenhouses in botanic gardens (in one of which it was initially discovered). It reaches large numbers and attains sexual maturity in heated effluents, but there is little critical data to support this contention. Brackish water species (such as *Paranais litoralis*) may be found in inland waters affected by the saline discharges from salt deposits utilized by chemical industries, such as those along the St. Clair River and the River Werra in Germany (Wachs, 1963).

References

Brinkhurst, R. O. (1964). Studies on the North American aquatic Oligochaeta. I. Naididae and Opistocystidae. *Proc. Acad. Natur. Sci. Philadelphia* **116**, 195–230.
(The nomenclature on the second family is badly out of date. The majority of the Naididae are as reported here.)

Brinkhurst, R. O. (1965a). Observations on the recovery of a British river from gross organic pollution. *Hydrobiologia* **25**, 9–51. (A case history study)

Brinkhurst, R. O. (1965b). The biology of the Tubificidae with special reference to pollution. *In* "Biological Problems in Water Pollution," Third Seminar–1962, Cincinnati, U.S.P.H.S.

Brinkhurst, R. O. (1965c). Studies on the North American aquatic Oligochaeta. II. Tubificidae. *Proc. Acad. Natur. Sci. Philadelphia* **117**, 117–172.
(Nomenclature has changed and several species have been added to the list—see Brinkhurst and Cook, 1966; Brinkhurst and Jamieson, 1971.)

Brinkhurst, R. O. (1966). Taxonomical studies on the Tubificidae (Annelida, Oligochaeta). *Int. Rev. Ges. Hydrobiol. Suppl.* **51**, 727–742.
(Generic definitions are reviewed and nomenclature updated.)

Brinkhurst, R. O. (1968). Keys to water quality indicative organisms (southeastern United States). U. S. Dep. of the Interior, Fed. Water Pollut. Contr. Administration.
(A simplified key—nomenclature reasonably up to date. Useful as far north as lower Canada.)

Brinkhurst, R. O. (1969). Changes in the benthos of Lakes Erie and Ontario. In *Proc. Conf. Changes Biota of Lakes Erie and Ontario, April 16–17, 1968. Bull Buffalo Soc. Natur. Sci.* **25**, 45–65.

Brinkhurst, R. O. (1970). Distribution and abundance of tubificid (Oligochaeta) species in Toronto Harbour, Lake Ontario. *J. Fish. Res. Bd. Can.* **27**, 1961–1969.

Brinkhurst, R. O. (1972). The role of sludge worms in eutrophication. U. S. Environmental Protection Agency, Washington, D.C. Ecol. Res. Ser. EPA-R3-72-004, August 1972. (A comprehensive review of studies on three polluted water species.)

Brinkhurst, R. O., and Cook, D. G. (1966). Studies on the North American aquatic Oligochaeta. III. Lumbriculidae and additional notes and records of other families. *Proc. Acad. Natur. Sci. Philadelphia* **118**, 1–33. (Updates Parts I and II which should be used in conjunction with this paper).

Brinkhurst, R. O., and Jamieson, B. G. M. (1971). "Aquatic Oligochaeta of the World." Univ. of Toronto Press, Toronto.
(The definitive source for descriptions and nomenclature. Keys for specialists, not regional studies.)

Brinkhurst, R. O., Hamilton, A. L., and Herrington, H. B. (1968). Components of the bottom fauna of the St. Lawrence, Great Lakes. Great Lakes Inst. Univ. of Toronto. No. PR 33. 49 p. + illustrations. (Simple keys to tubificids.)

Cook, D. G. (1970). Bathyal and abyssal Tubificidae (Annelida, Oligochaeta) from the Gay Head-Bermuda transect, with descriptions of new genera and species. *Deep Sea Res.* **17**, 973–981.

Cook, D. G. (1971). The Tubificidae (Annelida, Oligochaeta) of Cape Cod Bay. II: Ecology and systematics, with the description of *Phallodrilus parviatriatus* Nov. Sp. *Biol. Bull.* **141**, 203–221.

Hiltunen, J. K. (1965). Distribution and abundance of the Polychaete, *Manayunkia speciosa* Leidy, in western Lake Erie. *Ohio J. Sci.* **65**, 183–185.

Howmiller, R. P., and Beeton, A. M. (1970). The Oligochaete fauna of Green Bay, Lake Michigan. *Proc. Conf. Great Lakes Res., 13th, 1970, Int. Ass. Great Lakes Res.* pp. 15–46.

Johnson, M. G., and Brinkhurst, R. O. (1971). Associations and species diversity in benthic macroinvertebrates of Bay of Quinte and Lake Ontario. *J. Fish. Res. Bd. Can.* **28**, 1683–1697.

(A more statistical approach to field surveys—too elaborate for forensic biology.)

Wachs, B. (1963). Zur Kenntnis der Oligochaeten der Werra. *Arch. Hydrobiol.* **59**, 508–514. (An inland saltwater situation.)

CHAPTER 6
Bryozoans (Ectoprocta)

JOHN H. BUSHNELL

I. Introduction

The freshwater Ectoprocta compose only a very small percentage of the total species of the phylum, between 1 and 1.5%. There are approximately 50 freshwater forms. The total number varies only slightly among systematists, any variation in total number being a reflection of disagreement with respect to the validity or classification of a few controversial species. At least minor taxonomic disagreement is found among researchers of essentially all animal taxa, and the questionable validity of a few ectoproct species will in no real way affect the discussion of the pollution ecology of the group as a whole.

A. North American (Nearctic) Ectoprocta

Essentially 19 species of ectoprocts are known for North America (Bushnell, 1973), and all of these have been reported for the United States. At least one half of this number can be considered common in the United States, certain ones more so in one region of the country than another. At first glance the number of ectoproct species in the United States, or in the world, may seem almost trivial—even smaller than the number of freshwater sponges and somewhat greater than the number of hydrozoans. This initial interpretation of ectoproct importance is deceiving, however, as species of Ectoprocta are truly ubiquitous in terms of habitats. They are found in the vast majority of lakes, ponds, streams, marshes, and even roadside ditches, when careful examinations of these habitats are made. For example, I have found species of ectoprocts in 35 of the last 37 habitats (a combination of the aquatic habitats given above) sampled in Colorado from the plains to the subalpine zone. A similar frequency of occurrence is common in the midwestern and New England states. Not only are they usually found in the entire range of aquatic habitats, but they are very frequently one of the dominant organisms of an aufwuchs community, sometimes even of an epibenthic community. They are thus of substantial importance in the aquatic ecosystem and of potential importance as pollution indicator organisms. The ubiquity and frequent, sometimes even nuisance, abundance of the ectoprocts is not as yet fully recognized by a majority of aquatic scientists.

There are six families of freshwater Ectoprocta, and species of five of these families are found in North America. All these North American (Nearctic) species are also found in the United States proper (see list under Nearctic, Table I). These families belong to two classes, Gymnolaemata and the Phylactolaemata. The Gymnolaemata include the vast assemblage of marine ectoproct species, but two families, Paludicellidae and Hislopiidae, have some strictly freshwater species. Only the former family has species (2) occurring in the United States. The class Phylactolaemata is composed of species found only in fresh water. There are four families in the Phylactolaemata—Fredericellidae, Plumatellidae, Lophopodidae, and Cristatellidae. Only one biologist, Lacourt (1968), believes an additional family of Phylactolaemata should be recognized, taking its family name from the genus of the monotypic species *Stephanella hina* (Oka, 1908). The family Plumatellidae has the greatest number of species of any freshwater family. The species in this family belong to seven genera. Only one species, *Hyalinella orbisperma* (Kellicott, 1882; Bushnell, 1965b), can still be considered limited to the United States (Michigan). However, *Pectinatella* (*Cristatella*) *magnifica* (Leidy, 1851) was peculiar to eastern North America until its introduction into Europe just prior to 1900. There are, however, 19 species endemic at present to a single faunal region of the world. The highest endemicity occurs in the Neotropical (6), Ethiopian (6), and Oriental (4) faunal regions. No species of freshwater Ectoprocta, except possibly *Lophopodella carteri* (Hyatt, 1866) in Ohio (Masters, 1940), are known to have been other than accidentally introduced into North America. Only four species of Ectoprocta are totally cosmopolitan, i.e., recorded from all major faunal regions of the world, but several other species are common and sometimes very common faunal elements in a variety of aquatic habitats in two, three, or more faunal regions of the world.

B. Interrelationships of Suprageneric Taxa

The phylactolaematous species are separated from the freshwater gymnolaematous species principally on the basis of the presence of an epistome, or flap of tissue arching over the mouth, a horseshoe-shaped lophophore (instead of a circular lophophore and associated ring of tentacles as in the gymnolaemates), and the production of internal asexual dormant structures, the statoblasts. By contrast to this last characteristic the gymnolaematous species give rise to truncated external branches, which cease growing shortly after budding from the parent colony, and then, like the statoblasts of phylactolaemates become darkened and heavily sclerotized. They are now called hibernacula. Again, like statoblasts, the hibernacula remain dormant over dry or cool periods, or for a period during the summer, and then germinate to form the initial individual zooids of new colonies when favorable conditions for growth prevail. The criteria for discriminating the four families within the fresh water class Phylactolaemata are well chosen. The family Fredericellidae is characterized by linear branching colonies, fewer tentacles than other families, a nearly circular lophophore (with only a slight, and difficult to observe, horseshoe shape), and the production of only one kind of (nonadherent) statoblast (piptoblast) lacking peripheral float cells. The family Plumatellidae has linear branching colonies, although some forms have sequential clusters of contiguous zooids which are often hyaline and gelatinous. All species in the family form statoblasts with peripheral float cells (floatoblasts) and many also form adherent nonfloating statoblasts (sessoblasts). These have not been found in collections of certain species and

TABLE I[a]

DISTRIBUTION OF FRESHWATER ECTOPROCTA WITHIN MAJOR FAUNAL REGIONS

Ectoprocta	Cosmopolitan	Nearctic	Neotropical	Palearctic[b]		Ethiopian	Oriental	Australian	Endemic to one region
				European and Mediterranean Subregions	Siberian and Manchurian Subregions				
Phylactolaemata:									
Fredericellidae									
Fredericella sultana	×	×	×	×	×	×	×	×	
Fredericella australiensis		×	×	×	×			×	
Plumatellidae									
Plumatella repens	×	×	×	×	×	×	×	×	
Plumatella emarginata	×	×	×	×	×	×	×	×	
Plumatella fruticosa		×	×	×	×				
Plumatella casmiana		×	×	×	×	×	×		
Plumatella longigemmis			×			×	×		
Plumatella marcusi			×						×
Plumatella vorstmani					×		×		
Plumatella siolii			×						×
Plumatella javanica		×	×	×		×	×	×	
Plumatella ruandensis						×			×
Plumatella marlieri						×			×
Plumatella fungosa		×		×	×				
Hyalinella punctata	×	×	×	×	×	×	×	×	
Hyalinella carvalhoi			×						×
Hyalinella vaihiriae		×						×	
Hyalinella lendenfeldi			×					×	
Hyalinella minuta					×		×		
Hyalinella orbisperma		×							×
Hyalinella africana						×			×
Hyalinella indica						×	×		
Afrindella philippinensis						×	×		
Afrindella tanganyikae					×	×	×		
Stephanella hina					×				×
Stolella indica		×				×	×		
Stolella evelinae		×	×						
Stolella agilis			×						×
Gelatinella toanensis					×		×		
Internectella bulgarica				×					×

TABLE I (*continued*)

Ectoprocta	Cosmopolitan	Nearctic	Neotropical	Palearctic[b]: European and Mediterranean Subregions	Palearctic[b]: Siberian and Manchurian Subregions	Ethiopian	Oriental	Australian	Endemic to one region
Lophopodidae									
Lophopus crystallinus		×		×	×				
Lophopodella carteri		×			×	×	×	×	
Lophopodella capensis						×			×
Lophopodella thomasi						×			×
Lophopodella stuhlmanni						×	×		
Lophopodella pectinatelliformis							×		×
Pectinatella magnifica		×		×					
Pectinatella gelatinosa					×		×		
Cristatellidae									
Cristatella mucedo		×		×	×				
Gymnolaemata:									
Paludicellidae									
Paludicella articulata		×	×	×	×		×	×	
Paludicella pentagonalis			×				×		
Pottsiella erecta		×	×						
Hislopiidae									
Hislopia lacustris			×				×		
Hislopia corderoi			×						×
Hislopia malayensis							×		×
Hislopia moniliformis							×		×
Hislopia cambodgiensis							×		×
Echinella placoides					×				×
Arachnoidia ray-lankesteri						×			×
	4	19	20	14	19	19	24	10	19
Total, endemic, by region		1	6	1	2	6	4	0	

[a]Permission to use Table I was granted by Academic Press, London. It was taken from the paper by J. H. Bushnell, titled The freshwater Ectoprocta: A zoogeographical discussion. *In* Larwood, G. P. (ed.) (1973), "Living and Fossil Bryozoa: Recent Advances in Research." London (Academic Press).

[b]Total species in palearctic is 24.

genera, e.g., *Hyalinella*. None of the species in the Plumatellidae have floatoblasts with obvious terminal extensions or peripheral ornamentation, viz., hooks or spines. One species, *Plumatella casmiana* Oka, 1907, produces at least 3 clearly different architectural varieties of floatoblasts. The tentacle numbers of polypides in the family Plumatellidae range roughly between 25 and 65.

The family Lophopodidae and Cristatellidae differ fundamentally from the two preceding families in their growth form. They lack the clear linear plantlike branching discernible with care even in more gelatinous forms of the Plumatellidae. The Lophopodidae form globular colonies with few to many zooids. Groups of zooids within a globular colony bear a more or less common coelom during their life. Except for newly budded zooids, all zooids in the preceding two families eventually demonstrate essentially discrete individual coelomic regions, whether separated by septa or not. The species of Lophopodidae are always hyaline to yellowish in color, gelatinous, and unsclerotized, and the polypides have, generally, 55 to 85 tentacles. The colonies are slowly motile, particularly when small in size. Only large floatoblasts are produced, and they have either attenuated termini or a partial or complete single line of peripheral hooks. The monotypic family Cristatellidae is characterized by an elongated, gelatinous, caterpillarlike colony with a unique flat base ("sole"). Tentacles number 80 to nearly 100 and the colony is slowly motile. Only large floatoblasts are produced, and they have a double line of hooks around the complete periphery. For general references for clarification of terminology and further reading on morphology or systematics consult Pennak (1953), Edmondson (1959), Hyman (1959), and Ryland (1970).

C. Marine and Brackish Water Ectoprocta

While a detailed discussion of these forms and their ecology is not appropriate in this paper, a few remarks are permissible. A variety of species have been found in brackish waters, but one species, *Victorella pavida*, is a notably euryhaline form, even being reported from freshwater habitats (Brattstrom, 1954). Species are found in a number of estuaries where pollution is evident, e.g., Chesapeake Bay and the Thames River. Dr. K. S. Rao (personal communication) states that the species is common at the mouths of several rivers in India stationed just below where large cities are releasing abundant raw sewage and other pollution products. The species responds by forming "summer hibernacula" when exposed to pollution (Carrada and Sacchi, 1964; Ryland 1970). One aspect of pollution, adaptability to sedimentation or silt deposition, has been of particular importance to some researchers on marine and brackish gymnolaemates. Relatively few species of ectoprocts can tolerate much clay or silt deposition. One of the interstitial species of *Monobryozoon* is, however, found in mud. Other forms survive siltation either by having their feeding zooids attached to a specialized stalk or by the movement of unique colonial structures that raise the entire colony and are able to maintain the feeding zooids at a level above silt accumulation. The latter are lunulitiform colonies (see Ryland, 1970) confined to warm water. In addition to their capacity for vertical movement, these colonies (like various other marine forms) have vibracular setae that help sweep the colony surface clear of particles and settling larvae. As in the case of so many sedentary benthic organisms, most marine ectoprocts are highly vulnerable to fine particle deposition, their fine ciliary feeding mechanism ultimately being rendered ineffective. Where rivers increase their fine sediment load through increased land erosion and irrigation diversion, ectoprocts in estuarine and in adjacent and more seaward locations are jeopardized. Detailed studies on the specific effects of pesticides and various industrial pollutants on marine ectoprocts have not been conducted. There are numbers of scattered reports in the literature, however. Nearly 30 years ago, Richards and Cutkomp (1946), investigated the DDT sensitivity of the chitinous cuticle of various animals including *Bugula* sp., finding this animal one of the most sensitive. Miller (1946), in the same year, reported that concentrations of copper above 0.3 mg/liter kill *Bugula neritina* larvae and completely inhibit colony growth. Long-term experimentation of multiple pollutional effects on marine ectoprocts and a review of the pollution related literature for this large group would be a valuable endeavor.

II. Phylactolaemata: Fredericellidae and Plumatellidae

A. Systematics

1. Fredericellidae: One Genus, Fredericella; Two Species

The two species in this family are found in North America. *Fredericella sultana* (Blumenbach, 1779) is cosmopolitan in North America and in the world. The rarer world form, *Fredericella australiensis* Goddard 1909, has been found only in Wyoming and the central plateau of Mexico in North America. While the species status of *F. australiensis* has been questioned in the past, there is general agreement among recent workers that the two species are valid.

2. Plumatellidae: Seven World Genera, 28 Species; 3 North American Genera, 11 Species

The species status, particularly that of the Nearctic forms, is considered sound. One bryozoologist (Lacourt, 1968) considers species of *Stolella* and *Gelatinella* should be placed in the genus *Plumatella*. More research is needed for general agreement. Lacourt (1968) also considered *Hyalinella orbisperma*, on the basis of the inadequate original description, a dubious species. He was unaware of the new species description and new generic designation and habitat description given by Bushnell (1965b). The most common and widely recorded species in the United States are *Plumatella repens* (Linnaeus, 1758), *Plumatella emarginata* Allman 1844, *Plumatella fruticosa* Allman 1844, *Plumatella Casmiana* Oka 1907, and *Hyalinella punctata* (Hancock, 1850). A sixth species, *Plumatella fungosa* (Pallas, 1768) is somewhat less common. The remaining species are rare, and often localized in distribution.

B. General Ecology

1. Life cycles

In temperate regions overwintered statoblasts germinate in the spring at temperatures of approximately 10° to 24°C, depending on the species. *Fredericella sultana* colonies tolerate lower temperatures, and statoblasts germinate at lower temperatures than other species. The lowest temperature at which statoblasts of this species will germinate is not known precisely. Colonies develop rapidly by budding as temperatures rise in the late spring and early summer. When colonies mature in late spring, they produce statoblasts. This early production of large numbers of statoblasts is particularly well documented for two common species of *Plumatella*, *P. repens*, and *P. casmiana*. Each zooid of these two species may produce a large number of statoblasts, certain zooids of certain species sometimes have over 20 in each zooid, e.g., *P. repens* (Bushnell, 1973). Considerably more floatoblasts and sessoblasts are produced by various species of the genus *Plumatella* than by species of other genera in either the Fredericellidae or the Plumatellidae. Toward midsummer colonies of several hundred or several thousand individuals can be seen to undergo progressive degeneration. The older central portions of the colony deteriorate first. Small isolated colonies representing the younger remnant branch tips of the original large colonial masses remain. Statoblasts which were produced by these spring colonies are released with colony deterioration, and in late July and August they germinate (in the case of floatoblasts, often on new substrates). Thus, a second generation of colonies have, by late August or September, increased in size and finally produce large numbers of statoblasts. These colonies deteriorate, often with the onset of colder autumn temperatures. Their statoblasts will overwinter to produce the late spring colonies in the following year. A third generation of colonies has been reported by Bushnell (1966) for *P. repens*. These colonies were derived from statoblasts produced in August by second generation colonies. The germination of third generation colonies takes place in September and October. Some of these colonies, while generally small, produce a few statoblasts which then overwinter. *Plumatella fungosa* and *P. casmiana* produce a large number of statoblasts, many within each zooid, while *P. fruticosa* and several other species produce fewer statoblasts. More work is needed to determine whether all species within these two families regularly have two statoblast-produced generations in the seasonal cycle. Personal observations suggest at least two generations for *P. emarginata* and *P. casmiana*. Less is known about this aspect of the life cycle for the species of the less common genera of Plumatellidae. Species of *Hyalinella* produce, in general, fewer statoblasts than the more common species of *Plumatella*, but the zooids of the narrowly restricted *H. orbisperma* produce several statoblasts, more than usually seen in the cosmopolitan *H. punctata*. Zooids of *F. sultana* usually produce one to three statoblasts (piptoblasts), rarely more, and *F. austrailiensis* about the same. However, in many temperate locations the former species probably produces statoblasts throughout the year. While no long-term field studies have been done on these two families in the tropics and subtropics, it is assumed that there is a more or less continual sequence of generations throughout the year.

With respect to the sexual aspect of the life cycle of freshwater ectoprocts much more needs to be known. Mature ovaries and testes with sperm have been observed from time to time. Larvae have been collected in the plankton at various times, usually in late spring or early summer. Occasionally larvae have been seen in the late summer or early autumn, e.g., *F. sultana*. Production of larvae appears to be of short duration, one to two weeks usually, occasionally longer, and at the time of release only some of the zooids in colonies in the vicinity where larvae are collected appear to have mature gonads. In cold north temperate or arctic regions larvae have not been found. The reasons for the erratic occurrences of larvae and the possible complex of environmental stimuli contributing to sexuality are not known. By contrast sexuality in many marine forms is a regular predictable phenomenon, colonies in one vicinity producing larvae often for several weeks (see, for example, Mawatari, 1951). It appears, therefore, that among freshwater forms a majority of colonies are derived from asexual structures rather than larvae.

2. *Habitat and Environmental Factors*

Fredericella sultana is a widely adaptable species found frequently in rivers, numbers of oligotrophic and eutrophic lakes, and reservoirs. The colonies are often found in colder streams and high mountain lakes. They are not so luxuriant in these habitats, but this is doubtless because of the nutrient impoverishment—not low temperatures. The species is also found in strongly marl lakes (Bushnell, 1966). Dense growths of the species are found on logs and plants and in upright tufts rooted in the sand or mud in medium-to-large lakes and slow streams. In ponds or small lakes where highly eutrophic conditions prevail, marked by dense plant growths and phytoplankton blooms, *F. sultana* disappears or fails to grow. This is doubtless related to higher daytime temperatures. In habitats of these kinds in both western and eastern North America the author has found small colonies only early in the season or again late in the season when lower temperatures prevail. The species grows best at 20°C or less. It is the only species known to survive at near zero temperatures under the ice all winter (Bushnell, 1966; Lacourt, 1968). While cosmopolitan from approximately the Arctic Circle to the sub-antarctic regions it is scarce in tropical and subtropical regions. It is not found in waters of extremely high pH, but has been found at a pH of 5.3 (Tanner, 1932).

While *F. sultana* is one of the most commonly available species of freshwater ectoprocts, the other species found in North America, *F. australiensis*, is rare, having been found only in shallow arid land waters of Mexico and Wyoming. It has been found on vegetation near the bottom, and luxuriant colonies are found at higher temperatures than *F. sultana*.

Among the North American Plumatellidae, *P. repens* is common in large lakes and small lakes, streams, ponds of all sizes, marshes, and roadside ditches. It is found on the stems and leaves of a large variety of rooted aquatic plants, although, like many species, generally shunning *Chara*. It is a common species in extremely eutrophic situations—less common or absent in markedly oligotrophic waters, in habitats with water colored yellow or brown by organic decay acids, and/or with a neutral or lower pH. The species stops budding at 9°C and will not survive below 5° or 6°C, but it is as tolerant (or more so) of high temperature as other freshwater species. This must in part explain its commonness compared to other species in small ponds, shallow pools, and ditches were insolation causes a sharp rise in temperature. I have found the species in such habitats with summer daytime temperatures during a hot spell repeatedly reaching 37°C. Colony growth at this temperature was impaired, but feeding behavior appeared normal.

Plumatella casmiana is found in medium to large streams, ponds, small to large natural lakes, and shallow reservoirs. While it is found in the Great Lakes region, it is becoming apparent that it is probably more common in the western United States than in other North American regions. It is the most common species in several gravel pit ponds and irrigation reservoirs along the eastern foothills of the Rocky Mountains. It is also the only species thus far found in shallow irrigation reservoirs in the central Great Plains of eastern Colorado and western Kansas. These habitats are notably high in total residues (e.g., 1740 mg/liter in Nee Noshe Reservoir, Arkansas River drainage), and various salts, e.g., sulfates, chlorides, and bicarbonates. It has not been found in montane and alpine lakes in the west, i.e., lakes surrounded by coniferous forests or tundra. The species is found on water plants, rocks, sticks, logs, and floating planks, sometimes forming dense honeycomb formations (Bushnell and Wood, 1971) on wood substrates in shallow water *Typha* stands.

Plumatella emarginata is found in streams and in medium to large lakes, less commonly in small lentic bodies of water with lush growths of vegetation. It is more common in streams than other species of Plumatellidae, especially the otherwise common *P. repens* (Bushnell, 1966). *Plumatella emarginata* is less tolerant of high temperatures than *P. repens* and certain other freshwater Ectoprocta. This is certainly a partial explanation for its usual occurrence in streams and medium to large lakes. However, *P. emarginata* does not tolerate extremely low temperatures either. Both laboratory and field research indicates that it is more stenothermal than several other common species of the Plumatellidae.

Plumatella fruticosa is largely limited to northerly holarctic streams or lakes, or to mountain lakes in more southerly latitudes. It is found at lower altitudes primarily in states bordering Canada. In Michigan the species is more often encountered in the northern half of the state (Bushnell, 1965a). Essentially the same distributional pattern is evident for Europe. In Colorado this organism is absent from collections made on the plains or foothills, 3000–6000 foot altitude. However in the montane zone, at 9500–11,000 feet, it is the dominant ectoproct, found in tiny glacial ponds or lakes on both the eastern and western slopes of the Rocky Mountains. Most of the collection sites in the United States are within coniferous evergreen forests, and with few exceptions in water yellow to brown in coloration and near neutral or below in pH. It is not found in extremely warm or eutrophic freshwater habitats. The species grows on rocks, submerged branches of shoreline shrubs, floating or submerged logs and branches, or on water plants such as *Sparganium* and *Nuphar*. It is fairly frequently association with sponges, colonial peritrichs, and green *Stentor*.

Plumatella fungosa is found on substantial substrates like concrete walls, bridge supports, rocks, mollusk shells, and heavy branches. It is seen mostly in the eastern United States from Maine to the Illinois River, but recently Bushnell (1968) described colonies from a reservoir in the western mountains of Colorado. It is widely distributed in Europe, from below sea level in the Netherlands to medium mountain altitudes. It is found in shallow to deep waters of slowly flowing streams, ponds, and lakes and in brackish water.

Hyalinella punctata is found usually in mesotrophic or eutrophic ponds or lakes. The distribution is largely confined to eastern North America. It grows best at a temperature of 18° to 26°C and is sometimes found in habitats with a pH of over 9.0. It attaches to branches, bark, leaves, stones, and the leaves and stems of several aquatic plants. Lacourt (1968) says it has been found in brackish water.

The remaining species of Plumatellidae known for North America are very restricted in occurrence. *Hyalinella orbisperma* is known only from Michigan, either in small lakes or ponds or in the backwater pools of streams. In all cases the habitats are rich in rooted aquatic plants and algae. *Stolella evelinae* has only been reported from a small eutrophic lake in Michigan (Bushness, 1965b) and *Stolella indica* from Michigan and Pennsylvania. *Plumatella javanica* was recently identified by Lacourt (1968) from earlier Lake Erie collections (probably specimens from Louisiana are also this species), and *H. vaihiriae* is known only from a river in Utah.

3. Food Chain Relationships

Members of the Fredericellidae and Plumatellidae, like species in other families, feed on acellular algae (Tenney and Woolcott, 1966), protozoa, bacteria, organic detritus, and small metazoans brought from the ambient water to the mouth by rows of tentacular cilia. The size of the organisms eaten by different species is partially determined by the size of the mouth opening of their respective polypides. Much more research is necessary to determine which food organisms, of appropriate size for ingestion, are preferred or rejected by the different species.

Ectoprocts serve as incidental food and probably at certain times of their great abundance as significant fare for a number of organisms. Insect larvae are of primary importance as predators. Trichoptera larvae have been observed frequently eating live polypides and statoblasts of *P. repens* (Bushnell, 1966). Sometimes entire substrates have been cleaned of colonies within a few days by insect larvae and snails. *Lymnaea* and *Helisoma trivolvis* are gastropods which have been observed by me as predators on polypides and statoblasts. Elmidae and Tendipedidae larvae have also been observed as predators. Other reports (Marcus, 1926; Rusche, 1938) assert that turbellarians, e.g., *Dendrocoelum* and *Polycelis* eat live colonies and that tendipedids eat live colonies and statoblasts. Leidy (1851) observed predation on *Plumatella emarginata* by larvae of *Hydrophilus*. All the cases cited have related to predation on species of Plumatellidae. No authentic reports are available for predation on the two species of *Fredericella*. Osburn (1921) found the remains of several species of marine Ectoprocta in the stomachs of two species of eider ducks and *Plumatella* statoblasts in the stomachs of several species of freshwater fish. Dendy (1963) presents very striking evidence of predation by fish, most notably bluegills. Dense growths of *Plumatella* were observed in experimental farm ponds. In ponds where the ectoprocts were protected from fish there were always numerous branches free from the substrate and colony surface. When not protected, the fish grazed on the colonies, removing all free branches. These colonies appeared more like cylinders, with all these ragged loose branches removed. Colonies loosely attached to pieces of plants in aquaria and in ponds were all taken by bluegills. K. S. Rao (personal communication) states that fisheries researchers in India recently observed fish feeding on freshwater ectoprocts.

C. Pollution Ecology

Perhaps less is known about the pollution ecology of the freshwater Ectoprocta than about any of the other invertebrate taxa discussed in this volume. There are a modest number of scattered reports or personal communications relating to the presence, absence, or deformities of Ectoprocta in polluted waters, but nothing in the way of good measurements of actual levels of pollution where these observations were made. Some conclusions can be inferred from field observations and limited experimentation combined with knowledge about the "normal" ecology of the more common species and my personal experience. The only laboratory data in this chapter on tolerances of colonies and statoblasts to selected pollutants are those derived from experimentation begun while this chapter was in preparation. This quantitative information is valuable only as it suggests in a preliminary way the sensitivity of ectoproct species (in comparison to other taxa) to certain pollutional stresses. Most of this laboratory study has been done on a single North American species, *Plumatella casmiana*. More elaborate studies will be conducted in the near future.

Any study of the effects of pollution on the Ectoprocta must give particular attention to the pollution tolerances of the statoblasts and the hibernacula. Even if colonies should be killed by pollution, any resistance and continuing viability of their sclerotized asexual structures can obviously insure the eventual propagation of the species through germination in the same habitat when pollution is alleviated or in new habitats to which they may be transported by stream flow, wind, birds, etc.

1. *Field and Laboratory Observations*

There are no published records on either species of *Fredericella* from sites known to be classic areas of high pollution. This is true for freshwater ectoproct species in general. *Fredericella australiensis* has been collected from a small alkali pond in Wyoming (Rogick, 1945) and from the gray silty turbid waters of a lake and very shallow irrigation reservoir in the Chihuahuan desert of north central Mexico (Bushnell, 1971). All of these locations and certain other world reports indicate that the species appears to tolerate alkali waters and, therefore, has the ability to osmoregulate in situations where one or another of the concentrated salts in solution are not directly toxic. At both desert sites the water was kept continually turbid by only a moderate wind which kept the flocculent silt zone of the bottom agitated and continually carried more silt and clay sediment from the open, arid surrounding hills. The bottom at the very shallow location could not be seen clearly even when the water depth was no more than 6 inches, and large and small newly formed colonies were numerous on every *Ludwigia palustris* plant. This ability to tolerate siltation is shown to some degree by *F. sultana* collected from turbid shallow marl lakes and from highly turbid small silt-laden streams. Absence of light does not impair ectoproct growth as several species have become a nuisance by clogging the inlet pipes to city water supplies. Just recently I was consulted by the city of Sacramento, California, concerning dense ectoproct and sponge growths in the covered cement storage tanks which feed directly into the city water supply lines.

Fredericella sultana is sometimes found growing on decaying leaves contiguous to a black odoriferous lake bottom with much organic decay. The open growth habit and loose attachment to the substrate permits the species to grow up and away from the bottom. This loose open colonial design may be a partial explanation for *Fredericella*'s tolerance to turbidity and siltation as there is no dense colony surface on which silt can collect.

Fredericella sultana statoblasts kept for several weeks in putrescent debris in small closed jars have shown more than 50% viability when allowed to germinate in clear stream water. No observations however, have been made on their viability when stored in raw sewage or sludge.

The generally lower temperatures preferred by *F. sultana* for optimal growth suggest that the species would be more sensitive than certain others to thermal pollution. Several colonies of *F. sultana* and *P. repens* were exposed to rather rapid increases in temperature, e.g., 7°–10°C, in 30 minutes beginning at 14° and 20°C. The behavior of *F. sultana*, on the basis of very limited experimentation, supports the above contention. The polypides revealed an early and sporadic movement of the tentacles and finally all polypides retracted into their zooecia. While a number of *P. repens* polypides withdrew during the same period, several would again protrude partly or entirely. Sometime after the upper temperature level was reached nearly all the *P. repens* polypides protruded while all *F. sultana* polypides remained retracted for nearly two hours. Many finally protruded but others never did, and polypide degeneration was observed for several of these latter within 48 hours.

To my knowledge the only well-defined observation of the presence of *F. sultana* in a multipolluted habitat is a recently received personal communication from Dr. K. S. Rao. He states that colonies of *F. sultana* were found "sparingly" in a decidedly polluted location in a small spring fed tributary stream in the city of Indore, India. This small 6- to 8-foot wide stream had a temperature of 29° to 30°C at observation times during February and March, 1972. The stream dries up during May. The colonies were found about

50 yards downstream from where two small open ditches were carrying the nearly black waste effluent from two factories. One of the factories was a steel compressor factory and the other a groundnut oil extraction factory. Effluent entering the stream had an iridescent oily surface. Fecal contamination in the river was great and people upstream from the factories regularly washed clothes in the river. At the collection sites the water was extremely malodorous. The slowly flowing water was so turbid that a Secchi disc could scarcely be seen held six inches below the surface. This is an unusually polluted habitat, having several kinds of pollution, high turbidity, and, for *F. sultana* particularly, high temperature.

The colonies of *F. sultana* in the highly polluted habitat described above were small and sparse, but associated with them were abundant robust growths of *P. repens* and three species of sponge. These colonies are described as "abnormally healthy" with very wide zooecia and an unusually thick zooecial wall. *Plumatella repens*, in effect, seemed to be thriving. The unusual thickness of the sclerotized chitinous zooecia may be a response to various environmental disturbances. I have noted this for species of *Fredericella* and *Plumatella casmiana* collected where turbidity and silt deposition was pronounced, for *P. emarginata* (thick black zooecia) in turbid, mildly polluted stream waters and in a black-bottom putrescent forest ditch, and for *P. repens* in stagnant ditches and occasional collections from colored brown waters.

Rogick and Brown (1942) collected *P. repens* from a stream in Puerto Rico polluted by livestock. *Plumatella emarginata* has been collected from about 300 feet below where waste effluent from an industrial chemical plant was entering the Kalamazoo River, Michigan (Bushnell, 1966). The oligochaetes *Dero* and *Tubifex* were abundant in the area.

K. S. Rao (personal communication) has seen *P. emarginata* in lake Dhakuria in Calcutta, where sewage pollution is high. He has also found this species and *P. repens* and *H. punctata* in Museum Pond, a small pond inside the Indian Museum "heavily polluted by animal feces, washermen, construction, etc." This is an unusual location for *H. punctata* which is otherwise known only from clear waters of slow streams and lakes. Species of *Hyalinella*, as well as *Gelatinella* and *Stephanella*, have a growth form which keeps all their zooids close to the surface of their substrate, while some of the common species of *Plumatella* and species of *Fredericella* may be tightly adherent or they may project numerous branches free of the substrate. This latter ability allows these more versatile species an opportunity to escape silt accumulation on the substrate and to avoid the intense intraspecies and interspecies competition for food prevailing when numerous individuals are tightly clustered together close to a substrate during periods of nutrient impoverishment. This gives some species of *Plumatella* and *Fredericella* an advantage under certain kinds of environmental stress.

Richardson (1928) failed to find Ectoprocta in collections from a pollutional zone of the middle Illinois River. He did find *P. fruticosa* in the next, or subpollutional, zone, along with tendipedids, leeches, and sphaeriids. Lacourt (1968) cites records for *P. fungosa* in western Europe near harbors and in polluted waters and for the dark color of colonies living in close association with the alga *Phormidium fragile. Plumatella casmiana* has been found surviving in very turbid silty waters of shallow, alkali, depression reservoirs on the prairie in eastern Colorado.

There is one study of particular interest that concerns the effects of fertilizers and other organic enrichment on ectoproct growth (Dendy, 1963). While *P. casmiana* and *P. magnifica* were found in some of the farm ponds studied, *P. repens* was the predominant ectoproct. Ponds were fertilized with N–P–K (nitrogen–phosphorus pentoxide–potash) with ratios of 0–8–2 and 8–8–2. Each month 100 pounds of fertilizer was added to the enriched ponds. Ectoproct colonies in unfertilized ponds were small in size and number while much richer growths of "Bryozoa" were observed in fertilized ponds and on settling plates in ponds fertilized with both ratios of fertilizer. On the shore of one pond kept unfertilized for many years a "piggery" was built. The runoff was highly "fertilized" and the formerly "clear reservoir pond" soon had an abundant plankton bloom and "vigorous growths of Bryozoa clogged the drain-pipes repeatedly." This study rather clearly demonstrates that certain *Plumatella* species not only do not suffer phsyiological stress by some addition of fertilizer chemicals or natural organic pollution, but in fact thrive on the enrichment.

2. *Laboratory Data on the Sensitivity of Colonies and Statoblasts to Selected Pollutants*

During the past few months a series of experiments have been conducted in our laboratory to determine the effects of selected potential pollutants on both colonies collected in the field and statoblasts produced by colonies in the field. The viability of statoblasts following exposure to various pollutional stress is obviously of critical importance in predicting the ultimate success of ectoproct species. Even though colonies may be quite sensitive to low levels of various pollutants, often with very low survival times, the ability of their statoblasts to survive these lethal levels will allow ectoproct species to propagate new colonies when environmental conditions improve. Statoblasts released by colonies into streams may be carried for great distances, particularly floatoblasts. Large numbers of them, depending on the stream, doubtless pass through zones where they are exposed to a wide variety of moderate to heavy pollution. Time spent in these pollutional zones will depend on how often they are temporarily caught in debris along the way. Statoblasts with marginal hooks have a greater chance of being held for long periods. Their sensitivity to pollution will dictate how successful they are in forming new colonies when they finally reach clear water, either in the stream, in backwaters, reservoirs, lakes, or marshes. Statoblasts of various species are known to survive freezing and drying for months, even years (Brown, 1933; Rogick, 1940; Bushnell, 1973), and passage through the digestive tracts of cold-blooded vertebrates, (and even with low viability, through ducks). Quantitative information on the resistance to pollution has been, however, unavailable.

Nearly all of the laboratory data acquired in the present study relate to *P. casmiana*, one of the common world species. This organism was the most easily obtainable in quantity during the summer and autumn in several ponds and small lakes in the vicinity of Boulder, Colorado, along the eastern foothills of the Rocky Mountains. Some incidental information is available for *P. repens*. Colonies and statoblasts maintained under experimental conditions in the laboratory were kept in 50 or 100 ml of culture medium at room temperature or at 5°C in an incubator. Culture media were changed either every 24 hours, or in some cases, every 48 hours. Most statoblasts retained for future germination study were removed directly from colonies collected in the field and allowed to dry in the refrigerator at 5°C. Some statoblasts were taken from colonies which had been exposed to lethal pollutant levels in cultures kept at room temperature. When these were removed from dead colonies, they were allowed to dry and remain at 5°C until used for germination experiments. Some statoblasts were kept in fine mesh nylon bags in sewage or clear anoxic water and maintained at 5°C. Their germinability, after different periods of time in cold storage, was later determined in culture dishes at room temperature (23°–24°C).

a. Sensitivity of P. casmiana to Three Heavy Metals and Arsenic. Four chemicals were used in these tests: mercury, copper, cadmium, and arsenic. The following salts were used to obtain a series of concentrations of each of the above: mercuric chloride, copper sulfate, cadmium sulfate, and sodium arsenite. In each case where concentrations are given the amount is mg/liter of the heavy metal or arsenic (not of the total compound), e.g., mg/liter of arsenic in a sodium arsenite experimental medium. The results of these laboratory tests allow us to assign the ectoprocts (one common species, at least) to an approximate position on the spectra of sensitivities of aquatic organisms for these four toxic chemicals.

A parallel series (2 replicates) of culture media were prepared using mercuric chloride. Concentrations used were 0.2 mg/liter (one replicate only), 0.5 mg/liter, 1.0 mg/liter, 2.0 mg/liter, 3.0 mg/liter, 5.0 mg/liter, and the control. A total of 100 (plus or minus 7) zooids of *P. casmiana* were exposed to each concentration. At 5 mg/liter all animals were dead after 2 hours. Eight percent survived 24 hours of 2 mg/liter (all dead at 40 hours). Sixty percent of the zooids kept in 1.0 mg/liter mercury solutions died within 72 hours, while the remaining zooids died during an additional 6 days.

With the concentration of 0.5 mg/liter of mercury there was no appreciable mortality (11%) during the first 7 days. All additional zooids died, however, within the next 48 hours. The controls and the 0.2 mg/liter mercuric chloride cultures were checked daily for 20 days, with no significant deleterious effect evident for zooids in 0.2 mg/liter mercury. Both groups revealed a mortality of slightly over 50% of their zooids when observations were discontinued. The incipient lethal level (concentrations the animal could presumably tolerate indefinitely) is thus 0.2 mg/liter of mercury or slightly higher, but less than 0.5 mg/liter. An estimated LC_{50} (lethal concentration for 50% of the zooids) for 48 hours is 1.1–1.3 mg/liter. Concentrations above 1.0 mg/liter are rapidly lethal, while in concentrations of 0.5 to 1.0 mg/liter zooids survive several days. At these latter concentrations, however, only a few polypides were observed out of the zooecia and feeding, seven at the 72-hour observation. The polypides of most of them are permanently retracted after 48 hours. Living zooids can be identified after retraction as they exhibit muscular contractions upon being touched with a probe. ReVelle and ReVelle (1973) have given what are presumably selected lower lethal levels of mercury: 0.03 mg/liter for *Scenedesmus* (an alga), 2.0 mg/liter for bacteria, and 3.0 mg/liter for snails. *Plumatella casmiana* appears to be in the more sensitive end of the range. It was of interest in following several of the different chemical sensitivity experiments on *P. casmiana* that one or more oribatid mites, *Hydrozetes*, (common aufwuchs community associates of the ectoproct) were often present on the small wood substrate pieces placed in the experimental dishes. In 2.0 mg/liter mercury cultures, 3 mites were recorded alive after 5 days (observation then discontinued) and the mites were alive in the two highest mercury concentrations for 2 days dying after the second day in 5 mg/liter solutions. The arthropod was obviously more resistant than the ectoproct.

Replicate experiments using 5 concentrations of copper (0.5 mg/liter, 1 mg/liter, 3 mg/liter, 5 mg/liter, and 10 mg/liter) demonstrated very clearly that concentrations of 1 mg/liter and higher were rapidly lethal. Only approximately 7% of polypides were alive after 24 hours in 1 mg/liter solutions. These were all withdrawn into their zooecia and gave only a brief vital response to the probe. At the 48-hour observation, none were alive. All polypides kept in the higher concentrations succumbed in the first 24 hours, presumably almost instantly in the highest. Mites were still alive in the highest concentration after 2 days. Nearly 50% of the ectoproct zooids survived for 7 days in 0.5 mg/liter copper solutions, but all were dead after 9 days. Slightly fewer than 50% of the total zooids in the control culture were alive after 20 days. It is obvious that the incipient lethal level is somewhat below 0.5 mg/liter. This is remarkably close agreement with the copper sensitivity of the marine ectoproct *Bugula neritina* (Miller, 1946), where it was found that copper concentrations of greater than 0.3 mg/liter "kill larvae and completely inhibit growth of attached forms." The ectoproct tolerance is likewise closely similar to that of the bluegill for concentrations of copper in copper sulfate (Warren, 1971), where 90% of the fish survive the concentration of 0.5 to 0.6 mg/liter for 96 hours (essentially the same as the ectoproct at 0.5 mg/liter) and 15% of the fish survived nearly 1.0 mg/liter for 96 hours (higher tolerance than for the ectoproct). The TL_m (median tolerance limit) for 96 hours for the fish was estimated to be 0.74 mg/liter. The TL_m, for 96 hours (or for 48 hours) for the ectoproct would presumably be somewhere between 0.5 and 1.0 mg/liter. This ectoproct species falls into a rather sensitive range for copper, as copper sulfate is often used to kill algal growths, often at a concentration of 1 mg/liter. It is apparent that treatment of ponds or impoundments for very many hours without dilution could also kill certain ectoprocts. Certain industrial effluents or mining effluents, giving streams or other waters more than a very brief copper level of 1 mg/liter or more, would be deleterious to colonies. Possible synergistic effects of copper in combination with modest levels of other chemical pollutants would probably lower the potential lethal level.

Colonies of *P. casmiana* placed in increasing concentrations of cadmium sulfate (0.1, 0.5, 1.0, 3.0, 6.0, 10.0, and 14.0 mg/liter) were not markedly affected below 3.0 mg/liter. Slightly less than 25% of the zooids were alive in the control and in 0.1 mg/liter and 0.5 mg/liter after 17 days, when observations were discontinued. All zooids in 1.0 mg/liter had died by the 16th day, so the incipient lethal level appears to be between 0.5 and 1.0 mg/liter. All zooids were dead on the eleventh, tenth, ninth, and sixth day in 3.0, 6.0, 10.0, and 14.0 mg/liter, respectively. It was surprising that nearly 25% lived for 5 days in the highest concentrations used, as the general lethal range for aquatic organisms ranges from 0.01 to 10.0 mg/liter (ReVelle and ReVelle, 1973). All concentrations of cadmium of 1 mg/liter and above are obviously increasingly lethal, but none were immediately lethal. Polypides were not protruded after the second day in the two highest concentrations.

Several colonies representing a total of 130 (plus or minus 10) zooids in each concentration were reared in cultures of sodium arsenite containing 1.0, 3.0, 5.0, 10.0, 20.0, and 45.0 mg/liter arsenic. An initial 6-hour check of all colonies was made and those in 20 and 45 mg/liter were dead, so their survival was presumably almost instantaneous. Zooids in 10 mg/liter survived less than 96 hours, those in 5 mg/liter less than 7 days, and those in 3 mg/liter less than 12 days. The control and the 1 mg/liter cultures were observed for 22 days, at which time approximately one-third of the zooids were alive in each. Both had several polypides protruded with tentacles extended in a normal feeding posture. The TL_m concentrations for 96 hours is presumably an amount between 5 and 10 mg/liter and the incipient lethal level an amount between 1 and 3 mg/liter. One feature of death in arsenic solutions, not noted for experimental animals in other inorganic or pesticide cultures, was the position of the polypides. Particularly in solutions above 5 mg/liter the polypides often died in a protruded state with the tentacles much contracted and often folded over the mouth. In other cases the polypides retracted prior to death, often remaining retracted one or more days before response with a probe could no longer be elicited. Slightly more than 50% of polypides were protruded in a concentration of 10 mg/liter after 48 hours, but not at 72 hours. At the lower concentrations for which a lethal effect was eventually apparent many polypides remained normally protruded for a long period, but for only 96 hours in 5 mg/liter.

ReVelle and ReVelle (1973) have stated that aquatic organisms are sensitive to arsenic (in sodium arsenite) at levels of 1–45 mg/liter. It is obvious that *P. casmiana*, as in the case of copper and mercury particularly, is quite sensitive with an incipient lethal level of little more than 1 mg/liter arsenic. Incidental observations on three or four mites present at the three highest concentrations revealed the mite to be more tolerant than the ectoproct. Mites were still alive after 8 days at 10 mg/liter, but survived only 3 days at 20 mg/liter, and less than 6 hours at 45 mg/liter.

Some natural wells and springs contain up to 4 mg/liter of arsenic, but undisturbed fresh waters generally only have 1–10 μg/liter. Certain contaminated waters in England have reached concentrations of 12 mg/liter of arsenic. A number of industrial wastes contain arsenic, as do many insecticides and herbicides. Trace amounts come from rock weathering and fertilizers, but considerable amounts may enter fresh waters from time to time through mining operations. Arsenic is cumulative in organisms, so that small repeated sublethal levels introduced into fresh waters may ultimately affect the mortality rate, at least of more sensitive organisms.

A series of preliminary experiments to test the viability of statoblasts of *P. casmiana* were conducted using copper sulfate, mercuric chloride, cadmium sulfate, and sodium arsenite. *Plumatella casmiana* is an unusual ectoproct in that it produces three distinctly different kinds of floatoblasts, as well as sessoblasts (Wiebach, 1963; Vigano, 1968; Wood, 1971). One of the floatoblasts is a highly fragile, minimally sclerotized, form called a leptoblast, and the other two are heavily sclerotized (chitinous) forms called pycnoblasts and intermediate floatoblasts. The last is more elongate with less peripheral float coverage than the pycnoblast. Wood found that colonies produced from intermediate floatoblast germination formed both leptoblasts and more intermediate floatoblasts; colonies resulting from leptoblast germination formed both pycnoblasts and more leptoblasts; colonies formed from pycnoblast germination produced only intermediate floatoblasts. It was only intermediate floatoblasts that overwintered to produce new colonies the following spring. All data from viability experiments in the present discussion and in later discussions in this chapter (unless otherwise indicated) are for the overwintering intermediate floatoblasts of *P. casmiana*.

Thirty floatoblasts taken from the colonies which had been reared in the heavy metal and arsenic cultures discussed above were permitted to remain in these concentrations for 3 to 4 weeks, regardless of whether the colonies which produced them died immediately or not. The petri dish tops were then removed and the culture media with the statoblasts were then permitted to air dry at 5°C in an incubator. After several weeks in a dry state 30 statoblasts from each of the now dry culture dishes were removed and placed in petri dishes in clear Millipore filtered (0.8 μm) lake water at room temperature. A parallel series of experiments were run, the only difference being that separate statoblasts (not from the dead colonies) were kept in the same series of metal and arsenic concentrations for a comparable period, then dried in the refrigerator, and later placed in clean cultures.

Of those statoblasts exposed to 1.0, 5.0, and 10 mg/liter copper sulfate solutions 20–25%, 3–6%, and 3%, respectively, had germinated after 20 days. About 30% had germinated in the control in the same period. More germination time will be allowed in future experimentation to determine whether a higher percentage of germinations occur in general and whether the maximum number of germinations, in the higher concentrations, is merely delayed, i.e., if given sufficient time, will more closely approach the percentage of the control. We have found this latter to be the case in similar sewage exposure experiments.

There was, during 20 days observation, one germination for statoblasts pretreated with 3 and 5 mg/liter of mercuric chloride. Two statoblasts pretreated in 6 mg/liter cadmium sulfate and 4 statoblasts pretreated in 14 mg/liter cadmium sulfate germinated, but there were no germinations for statoblasts kept in 45 mg/liter cadmium sulfate solutions for 4 weeks. Either statoblast germination was severely delayed or this very strong solution is fully lethal. As many as 4 previously untreated statoblasts germinated when kept in 14 mg/liter cadmium sulfate cultures at room temperature.

The total of 60 statoblasts (in each concentration), pretreated for up to 4 weeks and dried in solutions of 10, 20, and 45 mg/liter sodium arsenite, exhibited an unexpected germination pattern. Only 3.2% (2) from 10 mg/liter culture solutions germinated, while 11.6% (7) and 31.6% (19) in 20 and 45 mg/liter, respectively, germinated after 20 days. The last percentage is nearly as high as the control.

In a few cases during these experiments and during the following experiments with pesticide and sewage waters the two valves of statoblasts would separate, but no well formed polypide was seen. Germinations were recorded only when the polypide was protruded with tentacles fully visible. While further research is needed, it is evident that colonies of *P. casmiana* are generally in the more sensitive range with respect to their tolerance of the 4 chemicals used (with the possible exception of cadmium), all well-known pollutants. However, preliminary evidence indicates that statoblasts may remain viable after prolonged exposure to high concentrations of these chemicals, concentrations which are immediately lethal to colonies and most other organisms. Their tolerance to several chemicals in combination or the absolute upper tolerance level to the chemicals used is not yet known.

Some extremely tentative data resulting from a single experiment just completed is perhaps germane to the present discussion. Molybdenum mining is common in several locations in the western United States. Research on biological effects of high molybdenum concentrations is now being conducted in Colorado at the University of Colorado. Molybdenum in streams below the mines and in Dillon Reservoir on the western slopes of the Rocky Mountains has been recorded at 300 and 400 ppb and higher. The study of this heavy metal is of particular interest, because, unlike other heavy metals, it is a micronutrient in trace amounts. Most other heavy metals have been studied because of their presumed toxicity to all organisms. The data acquired recently are only for germinability of *P. casmiana* floatoblasts. Numbers of floatoblasts were placed in fresh Millipore-filtered stream water. These cultures (one replicate) with a total of 30–35 statoblasts each consisted of a control and three cultures to which sodium molybdate had been added. The molybdenum content of these solutions was 10, 100, and 300 mg/liter. Statoblasts were randomly selected, as they were in all experiments on germination, from colonies collected at the same time and on the same substrate, then dried and refrigerated for three months. At the end of 14 days the following percentages of germination were reported: control, 30%; 10 mg/liter Mo, 40%; 100 mg/liter Mo, 40% and; 300 mg/liter Mo, 60%. The highest molybdenum concentration had twice the number of germinations as the control and significantly more than any of the experimental controls used in studies reported in this paper (i.e., germination in all controls after 14 days). It is obvious that considerable further experimentation is needed, and there is no verifiable explanation at present for this effect on germination of statoblasts. If additional data prove consistent with these early observations, then one factor seems of possible importance. The only other inorganic compound used which elicited greater germination with increasing concentrations (though percentage is not as high) was sodium arsenite. The molybdenum and the arsenic were both sodium compounds. Can sodium be a stimulus to germination? Higher levels of sodium arsenite must be fully lethal to statoblasts—perhaps, also, very high levels of sodium molybdate.

But at sublethal levels sodium may stimulate germination, while the statoblast wall protects the developing animal from the larger toxic molecule. After germination zooids are killed rapidly in the arsenic, but survive in the highest molybdenum concentration, dying after several days presumably from starvation. Further experiments using the same compounds and also additional sodium compounds such as sodium carbonate and sodium acetate will be conducted in the future.

b. Tolerance of P. casmiana to Pesticides and Herbicides. The tolerance of *P. casmiana* colonies to a series of concentrations of 2 organochlorine pesticides (dieldrin, toxaphene), one organophosphorus pesticide (malathion) and three herbicides (2,4-D butyl ester; MCPA; DNBP) was studied. The results of these just completed studies must be considered crude estimates, but they offer at the very least a general range of tolerances. The actual upper limits of tolerance and tissue accumulation rates must await further study. Studies on the pesticides are notably difficult. As in the case of other toxic chemicals, variations of the pH, hardness, and temperature of the water may all affect the toxicity. Organochlorines are known to persist and accumulate in freshwater animals and plants, adsorbed to particulate matter in the water and in the bottom soil and debris. Organophosphorus, by contrast, breaks down rather rapidly and there appears to be little or no indication of persistence or accumulation in organisms (Muirhead-Thomson, 1971). The degree of solubility in water is another variable which makes precise assessment of the actual levels of the toxic ingredients difficult to determine. The possible relative toxicity of the solvent, often acetone, must be taken into consideration when evaluating the effects of the pesticides.

The most significant information obtained for *P. casmiana* (some observations on *P. repens*) pertain to pesticide levels which do not seem to impair feeding, produce abnormal behavior, increase mortality, or retard budding. Pesticide levels in which normal colony activity is maintained for a period of one month to nearly ten weeks (depending on the experiment) are presumed to be concentrations below the incipient lethal level for this ectoproct. Concentrations which elicited unusual behavior, e.g., frequent erratic tentacle movement, repeated regurgitation of food by polypides, etc., and reduced longevity, were considered toxic (above the incipient level), even though a few zooids continued to respond with feeble contractions after contact with a probe. TL_m values were not determined or even roughly estimated as a possible range. The best that we can provide at present are approximate pesticide values, below which behavior and longevity remain normal for an extended period of observation. Above these values behavior and longevity are both affected, but surprisingly, in some instances, small numbers of zooids remain alive in a wide range of increasing pesticide concentrations. Between 100 and 200 zooids were used in each concentration of each pesticide. The total number includes at least 2 replicates at each concentration (9 replicates were used in the case of malathion). Standards using pure chemicals (except malathion) were prepared in acetone and kept in volumetric flasks. Dilutions with fresh pond water were used to prepare new culture concentrations each 48 hours. A commercial malathion solution, containing 50% malathion, was used to prepare the malathion culture solutions which were also replaced every 48 hours. In acetone controls using 0.1 mg/liter acetone no abnormality was noted for colonies. Only the strongest pesticide culture solutions exceeded this acetone value. Comments on high acetone concentrations appear in the subsequent discussion.

Plumatella casmiana is more sensitive to the two organochlorine pesticides than to malathion or the three herbicides. The greater tolerance to herbicides was expected, and is consistent with most information available for other aquatic invertebrates and fish. Experiments with both dieldrin and toxaphene were followed for 32 days. Zooid survival for this period was reduced, even in the lowest concentrations of both of these pesticides. While approximately 40% of the control were alive after 32 days, only 13, 20, 10, and 9% of zooids were alive after the same number of days in 0.0005, 0.003, 0.01, and 0.02 mg/liter, respectively, of toxaphene. There was no survival past 25 days in 0.1 mg/liter (but 25% survived 12 days) and no survival after 6 days in 1.0 mg/liter (but 23% survived 3 days). At 5 mg/liter there was no survival after 6 hours. However, 50 statoblasts were maintained in 5 mg/liter cultures for 30 days and then dried. When returned to pond water at room temperature 15 of these statoblasts germinated in 14 days, essentially the same germination number as in the untreated control. The time for 50% mortality in concentrations from 0.0005 to 0.1 mg/liter varied erratically from 7 to 12 days, while there was 50% survival after 25 days in the pure water control. Budding (2 buds) was noted only in the lowest concentration (0.0005 mg/liter) of toxaphene, and also in the control. There is no immediate explanation for the long term survival of the few zooids in concentrations of 0.0005 to 0.02 mg/liter toxaphene, but it is evident that even at the lowest concentrations used, growth and survival were decidedly affected. Appropriate acetone controls for this series of pesticide concentrations did not significantly reduce longevity when compared with pure water controls. While it is difficult without more laborious methods to ascertain the ages of all zooids in each of the colonies used in experimentation, it was observed (in all chemical testing) that older zooids died earliest, in general.

In dieldrin experiments 50% of the control animals survived for 20 days and 28% survived for 32 days. In dieldrin concentrations of 0.001, 0.003, and 0.01 mg/liter, 14, 12, and 13% of animals survived for 32 days. In all these three concentrations 50% of the zooids were dead in 13 days (plus or minus one day). Budding (growth) was observed in the 0.01 mg/liter cultures, but no budding occurred in higher concentrations. In 0.1 mg/liter dieldrin 50% survived 6 days and the last zooid died in 18 days. In 1.0 mg/liter solutions 50% survived 4 days and all were dead in 12 days. In the highest concentrations no polypides were protruded after 48 hours. Tentacles were often observed to move erratically in all dieldrin concentrations. As in toxaphene, several zooids survived for the full 32-day observation period in concentrations up to 0.01 mg/liter. Muirhead-Thomson (1971) states that the TL_m value for 96 hours for naiades of one species of stonefly was reported as 0.039 mg/liter. While the *P. casmiana* data are insufficient for such exactitude, they suggest a similar order of tolerance. Holden (1972) states that "a percentage" of fish were dead after 48 hours in 0.1 mg/liter dieldrin (nearly 50 zooids were dead in this concentration at 48 hours) and the minimal concentration found lethal to rainbow trout after several weeks exposure in the laboratory was between 0.001 and 0.01 mg/liter. The ectoproct appears slightly more sensitive to these same concentrations, but obviously tissue accumulations would have to be compared. Holden reports the highest concentrations in rivers of northern England as 540 ng/liter dieldrin (used in textile industry), but generally organochlorine levels are below 50 ng/liter. ReVelle and ReVelle (1973) give figures of 11,000 ppb for dieldrin in water below chlorinated hydrocarbon plants, an obvious toxic zone for sensitive organisms. Toxic effects from these pesticides are not considered a serious threat for most rivers, but certainly higher levels maintained for several days will affect survival. Holden, however, cautions that a "flush" of pesticides of short duration may cause death several days later "not easily related to the discharge." Muirhead-Thomson (1971) likewise notes that a loss of insecticides in "well-aerated static water" caused by "adsorption or codistillation" may result in data not immediately applicable to effects from these pesticides observed in flowing water. Two factors discussed by ReVelle and ReVelle (1973) are germane to the present discussion, one being the stimulation of bacterial growth by breakdown products of degradable pesticides. Also, productivity of phytoplankton cultures is reduced 70 to 94% by exposure of 4 hours to 1.0 mg/liter dieldrin and toxaphene. This information complicates the assessment of pesticide data for organisms one step up in the food chain which feed on bacteria and phytoplankton.

In cultures prepared from a 50% commercial malathion spray solution there was no zooid survival longer than 96 hours in concentrations above 3.3 mg/liter. Even zooids which survived 24 hours were retracted and it was difficult to elicit contraction responses by the use of a probe. In concentrations of 3.3 or 0.3 mg/liter 50% of the zooids (of a total of more than 120) survived for 14 and 13 days, respectively, compared to 11 days for the control. Over a 68-day observation period budding floatoblast germination were observed in these malathion cultures and in the control. Irregular trembling and erratic movement of the tentacles were observed from time to time in the two pesticide concentrations, but never in the control cultures. In spite of this aberrant behavior in the pesticide, 2 and 3 zooids, respectively, in 3.3 and 0.3 mg/liter were alive after 68 days. All control animals were dead after 58 days. It is unusual that animals in any reasonable concentration of pesticide should live longer than in the control. However, the actual concentrations of malathion are suspect. The malathion cultures were the only pesticide or herbicide cultures prepared from a commercial source. All other stock solutions were carefully prepared by a chemist from pure chemicals dissolved in acetone and kept in volumetric flasks for the regular preparation of new cultures. Also, the sensitivity of the ectoproct to all other herbicides, pesticides, and other chemicals used was consistently within the range of sensitivities known for various other aquatic organisms. Only in malathion cultures was the survival much higher than expected. For example, Muirhead-Thomson (1971) includes malathion in the "high" to "very high" categories with respect to its toxicity to fish, and the 96 hour TL_m value for two species of stonefly naiades is known to be 0.05 ppm (mg/liter). This is well below the level of the concentration in which several ectoproct colonies survived. Even though prudence is required in evaluating the results of commercial malathion solutions, it can be said that the ectoproct zooids survived very well in 2 regularly replenished culture concentrations of malathion (and/or derivative products plus solvents) from which the odor of malathion was readily detectable at all times. Admitting some degradation, adsorption, etc., of the malathion, it seems doubtful that the ectoproct can be considered an extremely sensitive animal.

The sensitivities of *P. casmiana* to the 3 herbicides, 2,4-D butyl ester, MCPA, and DNBP were generally comparable. A total of 100 to 135 zooids (including replicates) were exposed to a series of concentrations. In concentrations of 0.0005 to 0.1 mg/liter there was decidedly no increase in mortality when compared to the controls. In a concentration of 1.0 mg/liter of 2,4-D butyl ester there was no significant mortality, in fact zooids in this concentration did slightly better than the control, 50% surviving for 16 days (50% surviving for 15 days in the control). One replicate sample was followed for 30 days with 11% of zooids still alive (13% in the control). In 1.0 mg/liter MCPA 50% of zooids survived for 11 days (50% for 26 days in the control), but 12% were alive after 30 days (11% in the control). Thus the data for one month do not reflect any adverse effects from the MCPA.

In 1.0 mg/liter DNBP 50% of zooids were dead after 18 days (50% after 26 days in the control). All zooids were dead in this concentration after 26 days, while just over 38% of the control zooids were still alive after 32 days. By contrast in 0.1 mg/liter DNBP zooid survival (38%) after 32 days was essentially identical to the control. Appropriate acetone controls for all lower herbicide concentrations up to and including 1.0 mg/liter had no effect on zooid mortality. However, at concentrations of 3 to 10 mg/liter survival time, compared to pure water controls, was reduced by approximately 20 to 50%. Survival was also much increased if the herbicide culture solutions were allowed to stand for 6 to 8 hours before the ectoproct was added, permitting much of the acetone to dissipate. For example, when acetone was permitted to evaporate from a 10 mg/liter solution of 2,4-D butyl ester 50% of zooids survived for more than 9 days, but when zooids were placed immediately in fresh concentrations of 5 mg/liter 2,4-D butyl ester (more than 80% of zooids survived 96 hours) all were dead after five days. Likewise, a few zooids survived 14 days in preevaporated 10 mg/liter cultures of MCPA. Several zooids survived for up to 14 days in preevaporated cultures containing 3 mg/liter DNBP, but there was no survival after 24 hours in this concentration when zooids were placed directly in the freshly prepared herbicide medium. There was no zooid survival even in preevaporated cultures of 5 mg/liter DNBP and above, or in preevaporated cultures of 15 mg/liter and above 2,4-D butyl ester and MCPA. Floatoblasts germinated after storage for one week in preevaporated 10 mg/liter 2,4-D butyl ester, but considerable extra work must be done to determine statoblast viability following exposure to higher concentrations of both herbicides and pesticides.

Only a few small colonies of *P. repens* were available, and they were placed in MCPA solutions. They died rapidly, within 48 hours in concentrations above 0.1 mg/liter (i.e., in 1.0 mg/liter). Erratic tentacle movements and reduced longevity were observed for zooids in the 0.1 mg/liter MCPA cultures. The number of zooids used was inadequate for any substantial conclusions. The observations merely suggest that *P. repens* may be more sensitive to this herbicide than *P. casmiana*.

It is apparent that *P. casmiana* is markedly less sensitive to the herbicides than to the organochlorine pesticides, toxaphene and dieldrin (comparisons with malathion are tenuous for reasons previously discussed). Mortality was not affected after at least one month in herbicide concentrations of 1.0 mg/liter (with the possible exception of DNBP), and many zooids lived for several days in concentrations of 10 mg/liter MCPA and 2,4-D butyl ester. This was expected, for as Holden (1972) and others have stated insecticides are more toxic to invertebrates and to fish than are herbicides. For example, the 48-hour TL_m for bluegills in the herbicide diquat is 80–210 ppm and fish mortality is negligible below 10 mg/liter of 2,4-D butyl ester, with no great accumulation in the fish. In the case of *P. casmiana*, mortality was negligible at 1.0 mg/liter or below. Holden (1972) indicates that the flora is markedly affected at concentrations approaching 10 mg/liter. In the treatment of ponds and small lakes, where concentrations have reached nearly this amount, the ectoprocts could have been adversely affected, but such a level would be much higher than that derived from runoff following terrestrial use of herbicides. Holden (1972) provides information from a study on 3 species of stoneflies, where the 48 hour LC_{50} values following exposures to organochlorine and organophosphorus pesticides was 0.001–0.20 mg/liter. Herbicides, however, were less toxic with LC_{50} values of 0.15–80.0 mg/liter. The sensitivity of *P. casmiana* falls in the lower end of both these ranges with respect to organochlorine pesticides and herbicides.

c. Tolerance of Statoblasts to Sewage Pollution. Several small-mesh nylon bags were made and 50 statoblasts were placed in each. These were then submerged in raw influent sewage water (taken from the intake of the Boulder and Denver, Colorado, sewage plants), primary (malodorous) sedimentation basin sludge, semisolid sludge, and clean anoxic pond water. Oxygen was driven from the latter by heating Millipore-filtered pond water in a reagent bottle in a water bath. Wet nylon bags with statoblasts were weighted with stones and lowered into pure anoxic water and into closed containers with all the above grades of sewage. They were then kept at 5°C in an incubator for 2, 7, and 14 days. Similar packages of statoblasts were kept only in influent raw sewage at the treatment plant. The mean pH of this water when brought into the laboratory was 6.9. A D.O. meter failed to register any oxygen concentration in the sludge and in the anoxic water. The moving influent sewage water had a D.O. reading of 3–4 ppm.

The purpose of these experiments was to determine simply whether some statoblasts did remain viable after selected periods of time in city sewage composed, obviously, of both home and industrial wastes. Later experimentation will determine more precise viability rates. After 2, 7, and 14 days in sewage statoblasts were washed and placed in clean Millipore-filtered stream water. Several statoblasts kept for 2 days and 7 days in polluted influent water (from Denver and Boulder sewage plants) germinated. Statoblasts emersed (in the field) in the intake canal of the Boulder sewage treatment plant also germinated. There were no germinations of statoblasts stored in Boulder influent water for two weeks, but 3 statoblasts (of 50) placed directly in Denver influent water at room temperature germinated after 30 days. Germinations in influent water after 30 days were always less than half that of the controls. The 30-day percentage germination of treated statoblasts never exceeded 20%, while germination in the controls often exceeded 40%. Thirty floatoblasts of *P. repens* were also kept for 48 hours in the influent sewage water from the Boulder treatment plant. Five of these had germinated after 23 days.

Several (10%) *P. casmiana* statoblasts kept for 2 days in semiliquid sludge taken from the Boulder preliminary sedimentation basin germinated, but 20% germinated after storage for 2 days in semisolid sludge in the next treatment basin. No germinations were observed after 30 days for statoblasts kept in Boulder treatment plant sludge for one week or more or in Denver treatment plant primary and secondary sludge, or effluent water. Five germinations in 18 days were recorded for statoblasts kept in clear anoxic water for one week, but only 2 germinations after 30 days following a 2-week storage in anoxic water. One additional observation, not related to city sewage, was made: *P. casmiana* statoblasts were stored in a refrigerator for over 3 months in a covered finger bowl containing much decaying plant material. The contents of this markedly putrescent finger bowl were then uncovered and allowed to dry. Then statoblasts were removed and placed in clear stream water at room temperature. First germinations occurred in less than 48 hours, and in less than 3 weeks over 60% had germinated. Germinations were both earlier and greater in number than the control (statoblasts stored and then dried in clear water) during the three weeks of observation.

The time for the first and succeeding germinations was progressively delayed for all statoblasts exposed to various grades of city sewage. Statoblasts in the control germinated earliest followed at several day intervals by those in influent raw sewage, then those in preliminary sedimentation basins, and finally those in semisolid sludge. Observation periods were always 30 days or less, so it is very possible that a greater number of germinations would have occurred for polluted waters if observations had been extended. Nevertheless, it is apparent that some statoblasts of *P. casmiana* remain viable after several days exposure to raw sewage (*P. repens* statoblasts viable after at least 2 days in raw sewage) and sludge, to anoxia, and to natural decay (and presumably concomitant anoxic conditions). This information strongly suggests that statoblasts can survive passage through at least certain mixed pollutional zones and ultimately colonize more favorable habitats. Statoblasts which germinated after exposure to pollution revealed seemingly normal polypides. They lived no longer than a few days, but no attempt was made to supply them with adequate food. Proper culture of such young colonies to determine possible postexposure effects on growth must await further experimentation.

The upper temperature limit, important from a thermal pollution standpoint, tolerated by statoblasts of ectoproct species is not known. In our laboratory, however, there was no germination after 25 days for *P. casmiana* statoblasts exposed to temperatures of 40°–42°C for 72 hours in both a wet state (in a water bath) and dried in an oven.

III. Phylactolaemata: Lophopodidae and Cristatellidae

A. Systematics

1. Lophopodidae: 3 Genera, Lophopus, Lophopodella, *and* Pectinatella*; 8 Species*

Three species in this family occur in North America, one species of each of the three genera: *Lophopus crystallinus* (Pallas, 1768); *Lophopodella carteri* (Hyatt, 1866); *Pectinatella magnifica* (Leidy, 1851). The genera in this family are taxonomically sound, as are the North American species. Certain species of *Lophopodella*, not found in the Western Hemisphere, have been collected from only one or two locations. One biologist (Lacourt, 1968) has proposed that species of *Pectinatella* be placed in a separate family, *Pectinatellidae.* Presently there is no general agreement on this proposal, so the present author is retaining all three in the Lophopodidae.

2. Cristatellidae: One Genus; One Species

There is only one monotypic species in this family, *Cristatella mucedo* Cuvier 1798, and the validity of this species cannot be challenged. *Cristatella mucedo* is a holarctic form, with an essentially circumpolar pattern of occurrence records.

B. General Ecology

1. Life Cycles

The life cycle of members of these two families is similar to the life cycle of fredericellid and plumatellid species. By contrast, however, species of the Lophopodidae and Cristatellidae appear to have only one generation in temperate climates, i.e., statoblasts germinate in late spring or early summer to form summer colonies which produce statoblasts that remain dormant over winter. There is thus no evidence that the statoblasts produced during one summer regularly germinate during that same summer to produce a second generation of colonies. Gametes, however, are produced during the summer by mature colonies, and the resulting larvae settle to produce additional summer colonies. No observations of larvae have been reported for the rarer species of Lophopodidae. The author has collected several larvae during the evening in August from around *Cristatella mucedo* colonies. Seemingly the most abundant collections of larvae for any freshwater ectoproct were those of *P. magnifica* by Hubschman (1970), while investigating larval substrate preferences. He was able to collect hundreds of larvae, which were generally released between 9 PM and midnight. Perhaps, more larval collections should be attempted at night.

2. Habitat and Environmental Factors

Of the three species of Lophopodidae found in the United States, *P. magnifica* is the most common. Like other members of the family it is found east of the Great Plains from the Canadian border area to Florida. This species attains its greatest size in eutrophic waters of lakes, ponds, and slowly flowing larger streams, where microscopic organisms are sufficiently abundant to sustain the closely packed zooids. Wood and stones are the usual substrates. Under optimal conditions the bulbous or spindle-shaped colonies reach a diameter of 20 cm or more. The species thrives at temperatures generally between 20° and 28°C. At temperatures much below 20°C disintegration is usually rapid. Other organisms are frequently found living in the gelatinous base or center of the colonies, e.g., protozoans, turbellarians, gastropods, and crustaceans (Brooks, 1929). Dendy has found tendipedids abundant in the gelatinous colonial masses, and one species in particular, *Tendipes pactinatellae*, most abundant and seemingly confined to colonies of this ectoproct (Dendy and Sublette, 1959; Dendy, 1963).

Lophopus crystallinus is known only from the Schuylkill River near Philadelphia, the Illinois River, and Lake Erie (Lacourt, 1968). It has been found from scattered lentic and lotic (including estuaries) habitats in central Europe, Great Britain, Ireland, southwestern Russia, and Iran. There have been no reports of this species in recent years. The usual substrate is an aquatic plant.

Lophopodella carteri is confined to the eastern United States in North America, from Michigan and New Jersey south to Virginia and Kentucky. While absent from Europe, it is found in particular abundance in the southern and more easterly portions of Asia. The species is found on rooted aquatic plants and in algal mats, and on stones and floating wood substrates. It is found primarily in the eutrophic waters of isolated pools or backwaters of streams and in small lakes and ponds. Like *P. magnifica*, it grows best at warmer temperatures. Other species of Lophopodidae are confined to a warmer habitat in the southeastern Asia region or Africa.

The monotypic and holarctic *Cristatella mucedo* is found largely in small or large lakes and ponds in the northern United States and in Canada. It is found only in montane lakes in the Rocky Mountains of Colorado. I have recently identified this species from concrete basins (fed by a reservoir) near the Nevada border, which are the sources of water for the city of Sacramento, California. Luxuriant colonies have been found in both eutrophic (up to a pH of 9.8) and oligotrophic waters. The species is markedly eurythermal, with colonies growing at temperatures of well below or well above 20°C in the United States. Lacourt (1968) states that the species "thrives" at temperatures from 9° to 30°C, being found in some cold lakes of Europe at timberline. Aquatic plants are frequent substrates, but stones and wood are also used.

3. *Food Chain Relationships*

Food of the species of these two families is the same as that for the families discussed previously, but more research is needed to determine actual food preferences. Marcus (1934) states that predators on *L. crystallinus* include snails, trichoptera larvae, and oligochaetes. Brooks (1929) reports that young *P. magnifica* colonies are eaten by flatworms, and Marcus (1926) observed *Hydrozetes lacustris* (oribatid mite) eating *C. mucedo* colonies and statoblasts. The author doubts that there is much actual predation by this mite. The oribatid is observed often as a codominant on wood substrates with ectoprocts, particularly species of *Plumatella*. The mites frequently enter zoaria after the zooids have died, presumably eating bacteria and decay products, but no observations have revealed predatory attacks on colonies of ectoprocts. Osburn (1921) found statoblasts of *Pectinatella* in the stomachs of several species of young fish. It is quite possible that young gelatinous colonies may be tempting morsels for fish. The author has found parts of colonies and statoblasts in the stomachs of yellow perch, and these fish were observed nibbling at colonies of this ectoproct. *Lophopodella carteri* may be immune, at least to fish predation. Rogick (1957) says that when this species is crushed in the vicinity of fish, the fish die rapidly. She reports that similar observations were made using the congeneric Asian specis *P. gelatinosa.*

C. Pollution Ecology

Information on the disturbed ecology of species in these families is scanty. *Pectinatella magnifica* is known only from uncontaminated water. Richardson (1928) found *Plumatella fruticosa* from a subpollutional zone in the Illinois River, but it was only farther downstream in clear water that *P. magnifica* was present. The species is not known to occur in turbid water. Likewise, *L. crystallinus* is not common in turbid or polluted waters. Older reports record the species from rivers and from Lake Erie in the United States and from rivers in Europe. Many of these same habitats are known to have suffered mild to severe pollution in the past two decades. A search for the species in these now polluted locations would be of value. *Lophopodella carteri* is not generally found in polluted habitats, but it apparently tolerates natural decay and some sewage pollution. Rao (1973; personal communication) reports that it was found abundant in a portion of a river system in India where the water was muddy and polluted by influent sewage from villages. In one area of the river, where leaves were malodorous and black from organic decomposition, colonies were found. They were like "black lumps," but still alive and seemingly normal in appearance and behavior. Similar scattered observations are consistent with the above. However, Rogick (1957) reports that studies by the Pennsylvania Fish Commission found "very abnormal growths" of this ectoproct after fertilizer was added to fish ponds. With respect to thermal effects, Toriumi (1963, 1967) found that the terminal spine region of the statoblast was narrower, and the tentacle numbers of polypides were reduced, when colonies were kept at 30°C. The latter could reduce feeding efficiency where colonies are exposed to thermal pollution.

IV. Gymnolaemata: Paludicellidae (and Hislopiidae)

A. Systematics

1. Paludicellidae: 2 Genera; 3 Species in North America

The species status of *Paludicella articulata* (Ehrenberg, 1831) and *Pottsiella erecta* (Potts, 1844) are clearly defined. Both of these species are known from the United States, and a third species *Paludicella pentagonalis* has been reported from Guatemala (Rogick and Brown, 1942). The species status of these three ectoprocts has not been questioned. *Paludicella articulata* is the most widely distributed species.

2. Hislopiidae

No species in this family have been reported to date from North America. Most of the species have a narrow distribution.

B. General Ecology

1. Life Cycle

New colonies of these freshwater gymnolaematous ectoprocts develop in late spring or early summer, following germination of the uniqe external sclerotized hibernacula. Whether hibernacula produced by these summer colonies germinate in the same season is not known. Colonies form their hibernacula in late summer and autumn and these structures then overwinter. There are no substantial data relating to the sexual stage of the life cycle.

2. Habitat and Environmental Factors

P. articulata is found commonly in the eastern half of the United States (exclusive of the Gulf states) and adjacent Canada and Greenland. It is found in lakes, streams, and sometimes marshes. Substrates include stones, leaves, shells, logs, and sometimes aquatic plants. It is often found closely associated with bryophytes, and also with both Chlorophyta and Cyanophyta (Bushnell, 1966). The colonies are sometimes nearly covered with adherent diatoms. The species will survive colder water down to 5°C, but is not found in water much over 24°C. It is generally absent from highly eutrophic waters or water with an exceedingly high pH.

Pottsiella erecta has been found on rocks and mussel shells in streams and lakes in Pennsylvania, Texas, Virginia, and Tennessee (Sinclair and Isom, 1963). Recently the author confirmed the identity of this species from collections made by E. Everitt in Louisiana. *Paludicella pentagonalis* (not known in the United States) has been collected only from a lake in Guatemala.

3. Food Chain Relationships

Colonies of *P. articulata* feed on very small organisms. The diminutive size of the polypides precludes their feeding on the largest organisms available to certain of the more robust phylactolaemates. Essentially nothing is known about perdators of paludicellid or hislopiid ectoprocts.

V. Indicator Species and Endangered Species

There are no species of freshwater ectoprocts that can be considered indicators of pollution. While some have been found in variously polluted waters, they are not unique to these waters, nor do the colonies consistently display pronounced luxuriance under polluted conditions. All species have been found most often in clean or only mildly disturbed habitats. The author has made limited searches for ectoprocts in rivers immediately below where considerable industrial effluent was entering these streams, and, except for the cases already cited, ectoprocts were not found. These searches have been made mostly in eastern Ohio, the major rivers in the immediate vicinity of Pittsburgh, Pennsylvania, and the Grand River where it passes through Grand Rapids, Michigan. No ectoprocts were found in association with the commonly encountered pollution indicator organisms, but they were found in rural districts much farther downstream. From our experimentation and from the published literature it is evident that certain species will tolerate intermediate levels of pollution, some zooids showing greater than expected tolerance to certain pollutants, e.g., *P. casmiana*, and possibly even *P. repens*. However, observations and data available suggest, generally, that freshwater ectoprocts are comparatively sensitive to environmental disturbance. It would be more correct to say that most species are pollution indicators more by their usual absence, than by their presence, in markedly polluted waters. By contrast, statoblasts, ostensibly evolved to sustain prolonged desiccation, cold and freezing, and to promote dissemination, may inadvertently be structures preadapted to withstand man-caused stresses in an age of pollution. While upper tolerance levels are not known, some statoblasts remain viable after sustained exposure to rather strong concentrations of heavy metals, arsenic, pesticides, raw sewage, and presumably potential bacterial attack.

Is it possible to identify endangered species among ectoprocts? This is admittedly difficult, and, of necessity, must be conjectural. *Lophopus crystallinus*, with occurrence records only from two larger rivers and Lake Erie in the United States, and no new published occurrences for essentially 30 years, would seem to be a candidate. It has a wide occurrence pattern in parts of Europe, but recent records are scanty. The species has rarely been found as abundantly as many other species. There are certain old reports of the occurrence of the species in high mountain lakes (it survives low temperatures), e.g., in the Pyrenees. The ability to live in cold mountain lakes is a possible refuge habitat for pollutionally sensitive species, but the rarity of occurrence in these habitats is not encouraging. Certainly all rare species with very circumscribed distributions must be considered endangered as pollution increases. Ryland (1970) states that the Broads area of East Anglia in the British Isles has received the most study. "Pollution and the increasing turbidity of the broadland waterways during the summer months" have reportedly resulted in a "disturbing decline" of *C. mucedo*. *F. sultana*, *P. repens*, and even *L. crystallinus* have remained widely distributed in this region. However, the wide holarctic distribution in the mountains, and in northern boreal habitats of the United States, Canada, Europe, and Asia precludes the placing of *C. mucedo* on the endangered species list. Much of the world in which it is found is still largely free of severe pollution.

Those ectoproct species which grow rapidly and whose productivity of relatively pollution tolerant statoblasts is high would appear to have some long term mathematical advantages for continued dissemination. Species of parasites, other invertebrates, and birds (producing only one or two eggs), whose growth rate and egg production are low, are among those animals most endangered when their environment is disturbed. *Lophopus crystallinus*, certain species of *Hyalinella* , and some other ectoproct species have a much lower statoblast production rate than, for example, *P. casmiana* or *P. repens*. Colonies of *P. repens* may double in number of zooids in only three days, perhaps even less (Bushnell, 1966), and the zooids may live for up to 53 days, each producing numerous statoblasts.

Acknowledgements

Appreciation is extended to Lon Ulrich, chemist with the Colorado Department of Agriculture, for preparation of the standard pesticide solutions and to K. S. Rao and C. Kodadek for assistance with the pollutional experiments using *Plumatella casmiona.*

References

Brattstrom, H. (1954). Notes on *Victorella pavida. Lunds Univ. Arsskr. Afd.* **50**, 1–29.

Brooks, C. M. (1929). Notes on the statoblasts and polypides of *Pectinatella magnifica. Pro. Acad. Natur. Sci. Philadelphia* **81**, 427–441.

Brown, C. J. D. (1933). A limnological study of certain fresh-water Polyzoa with special reference to their statoblasts. *Trans. Amer. Microsc. Soc.* **52**, 271–316.

Bushnell. J. H. (1965a). On the taxonomy and distribution of freshwater Ectoprocta in Michigan. Part I. *Trans. Amer. Microsc. Soc.* **84**, 231–244.

Bushnell, J. H. (1965b). On the taxonomy and distribution of freshwater Ectoprocta in Michigan. Part III. *Trans. Amer. Microsc. Soc.* **84**, 529–548.

The two above references and Part II (not cited in text, 1965, pp. 339–358. in *Trans. Amer. Microsc. Soc.*) provide detailed systematic and morphological (including variations) information on a majority of U.S. ectoprocts.

Bushnell, J. H. (1966). Environmental relations of Michigan Ectoprocta and dynamics of natural populations of *Plumatella repens. Ecol. Monogr.* **36**, 95–123.

This paper provides extensive ecological information on U.S. ectoprocts.

Bushnell, J. H. (1968). Aspects of architecture, ecology, and zoogeography of freshwater Ectoprocta (Bryozoa). *Atti Soc. Ital. Sci. Natur. Museo Civico Storia Natur. Milano* **108**, 129–151.

Bushnell, J. H. (1971). Porifera and Ectoprocta in Mexico: architecture and environment of *Carterius latitentus* (Spongillidae) and *Fredericella australiensis* (Fredericellidae) *Southwest. Natur.* **15**(3), 331–346.

Bushnell, J. H. (1973). The freshwater Ectoprocta: a zoogeographical discussion. *In* "Living and Fossil Bryozoa Recent Advances in Research" (G. P. Larwood, ed.), pp. 503–521. Academic Press, London.

Bushnell, J. H., and Wood, T. S. (1971). Honeycomb colonies of *Plumatella casmiana* Oka (Ectoprocta: Phylactolaemata). *Trans. Amer. Microsc. Soc.* **90** (2), 229–231.

Carrada, C. C., and Sacchi, C. F. (1964). (Ecology of *Victorella*). *Vie Milieu* **15**, 389–426.

Dendy, J. S. (1963). Observations on Bryozoan ecology in farm ponds. *Limn. Oceanogr.* **8**, 478–482.

Dendy, J. S., and Sublette, J. E. (1959). The Chironomidae (=Tendipedidae: Diptera) of Alabama with descriptions of six new species. *Ann. Entomol. Soc. Amer.* **52**, 506–519.

Edmondson, W. T. (ed.) (1959). "Freshwater Biology," 2nd ed., pp. 495–507. Wiley, New York.

The ectoproct chapter in this book provides general information and an up-dated key for most U.S. species.

Holden, A. V. (1972). The effects of pesticides on life in fresh waters. *Proc. Roy. Soc. London B* **180**, 383–394.

Hubschman, J. H. (1970). Substrate discrimination in *Pectinatella magnifica* Leidy (Bryozoa). *J. Exp. Biol.* **52**, 603–607.

Hyman, L. H. (1959). "The Invertebrates, Smaller Coelomate Groups." McGraw-Hill, New York.

An excellent detailed discussion of morphology, terms, etc., on ectoprocts.

Kellicott, D. S. (1882). Polyzoa. Observations on species detected near Buffalo, New York. *Proc. Amer. Soc. Microsc.* **4**, 217–229.

Lacourt, A. W. (1968). A monograph of the freshwater Bryozoa—Phylactolaemata. *Zool. Verh. Uitgegeven Rijksmuseum Natuur. Histo. Leiden* No. 93, 1–159.

A valuable monograph on freshwater Phylactolaemata only. Morphology and world distribution are thoroughly discussed. It does contain several, as yet, controversial taxonomic changes.

Leidy, J. (1851). On *Plumatella diffusa* n. sp. *Proc. Acad. Natur. Sci.* **5**, 261–262.

Marcus, E. (1926). Beobachtungen und Versuche an lebenden Süsswasser bryozoen. *Zool. Jahrb. Abt. Syst. Ökol. Geogr. Tiere* **52**, 279–350.

Marcus, E. (1934). Über *Lophopus crystallinus* (Pall.). *Zool. Jahrb. Abt. Anat.* **58**, 501–606.

Masters, C. O. (1940). Notes on subtropical plants and animals in Ohio. *Ohio J. Sci.* **40**, 147–148.

Mawatari, S. (1951). The natural history of the common fouling Bryozoan, *Bugula neritina* (Linnaeus). *Misc. Rep. Res. Inst. Natur. Resources, Tokyo* Nos. 19–21, 47–54.

Miller, M. A. (1946). Toxic effects of copper on attachment and growth of *Bugula neritina. Biol. Bull.* **90**, 122–140.

Muirhead-Thomson, R. C. (1971). "Pesticides and Freshwater Fauna." Academic Press, New York.

Osburn, R. C. (1921). Bryozoa as food for other animals. *Science* **53**, 451–453.

Pennak, R. W. (1953). "Fresh-water Invertebrates of the United States. Ronald Press, New York.

Provides a thorough discussion of the anatomy, ecology, etc., of the U.S. ectoprocts. Some of the variety designations in the key have more recently been confirmed as valid species.

Rao, K. S. (1973). Studies on fresh-water Bryozoa. III. The Bryozoa of the Narmada system. *In* "Living and Fossil Bryozoa: Recent Advances in Research" (G. P. Larwood, ed.) Academic Press, London.

ReVelle, C., and ReVelle, P. (1973). "Environment: The Scientific Dimension." Houghton-Mifflin, Boston, Massachusetts (in press).

Richards, A. G., and Cutkomp, L. K. (1946). Correlation between the possession of a chitinous cuticle and sensitivity to DDT. *Biol. Bull.* **90**, 97–108.

Richardson, R. E. (1928). The bottom fauna of the Middle Illinois River, 1913–1925. *Bull. Illinois Natur. Hist. Surv.* **17**, 387–475.

Rogick, M. D. (1940). Studies on freshwater Bryozoa. XI. The viability of dried statoblasts of several species *Growth* **4**, 315–322.

Rogick, M. D. (1945). Studies on fresh-water Bryozoa. XVI. *Fredericella australiensis* var. *Browni* n. var. *Biol. Bull.* **89**, 215–228.

Rogick, M. D. (1957). Studies on fresh-water Bryozoa, XVIII. *Lophopodella carteri* in Kentucky. *Trans. Kentucky Acad. Sci.* **18**, 85–87.

Rogick, M. D., and Brown, C. J. D. (1942). Studies on fresh-water Bryozoa. XII. A collection from various sources. *Ann. N. Y. Acad. Sci.* **43**, 123–143.

Rusche, E. (1938). Hydrobiologische Untersuchungen an niederrheinischen Gewassern. X. Nahrungsaufnahme und Nahrungsauswertung bei *Plumatella fungosa* (Pall.). *Arch. Hydrobiol. Stuttgart* **33**, 271–293.

Ryland, J. S. (1970). "Bryozoans," Hutchinson, London.

While largely concerned with marine forms, this book provides detailed morphological, environmental, etc., information on all groups of ectoprocts. A good reference for general background.

Sinclair, R. M., and Isom, B. G. (1963). The occurrence of certain Bryozoans in Tennessee waters. Tennessee Stream Pollut. Contr. B., Publ. No. 10, 1–8.

Tanner, V. M. (1932). Ecological and distributional notes on the fresh-water sponges and Bryozoa of Utah. *Utah Acad. Sci.* **9**, 113–115.

Tenney, W. R., and Woolcott, W. S. (1966). The occurrence and ecology of freshwater Bryozoans in the Headwaters of the Tennessee, Savannah, and Saluda River systems. *Trans. Amer. Microsc. Soc.* **85** (2), 241–245.

Toriumi, M. (1963). Analysis of intraspecific variation in *Lophopodella carteri* (Hyatt) from the taxonomical view-point. V. Intraspecific groups distinguished by shape of the spinoblast. *Bull. Mar. Biol. Station Asamushi, Tohoku Univ.* **11**, 161–166.

Toriumi, M. (1967). Analysis of intraspecific variation in *Lophopodella carteri* (Hyatt) from the taxonomical view-point. IX. Additional observations on the variation of tentacle number. *Bull. Mar. Biol. Station Asamushi, Tohoku Univ.* **13**, 13–20.

Vigano, A. (1968). Note su *Plumatella casmiana* Oka (Bryozoa). *Rivista Idrobiol.* **7**, 421–468.

Warren, C. E. (1971). "Biology and Water Pollution Control." *Saunders, Philadelphia, Pennsylvania.*

Wieback, F. (1963). Studien uber *Plumatella casmiana* Oka (Bryozoa). *Vie Milieu* **14**, 579–596.

Wood, T. S. (1971). Colony development in species of *Plumatella* and *Fredericella* (Ectoprocta). Ph.D. Dissertation. Univ. of Colorado.

CHAPTER 7

Crayfishes (Decapoda: Astacidae)

HORTON H. HOBBS, Jr. and EDWARD T. HALL, Jr.

I. Introduction

In this discussion of the tolerance of crayfishes to pollutants of various sorts, we have attempted to summarize all available data relating to those species that occur on the North American Continent. Many, if not most, of these data were derived from studies not primarily oriented toward the effects of a polluting agent on crayfishes; rather, they originated from an interest in such facets of the biology of these animals as comparative physiology, environmental adaptations, and, secondarily, from studies in natural history. To these data, we have added personal observations and results of experiments that have not been reported elsewhere.

Inasmuch as available data were not obtained with the same, or even similar, aims in view, they are not always entirely comparable, and perhaps we have been remiss in failing to make extrapolations in certain instances that might render them more nearly so. Obviously, many of the details of the experiments utilized by several investigators cited have been omitted, but references are provided that will lead the reader to the sources from which our presentations were extracted.

There is a real danger implicit in inferences that might be drawn from a summary article such as this. One is inclined to assume that since one species of crayfish is reported to be able to tolerate a certain concentration of some substance for an extended period of time, all crayfishes probably have an equivalent potential. As suggested in several of the studies cited below, the approximately 300 species of crayfishes occurring in North America most certainly do not possess the same limits of tolerance to alterations in their environment! The tolerance limit cited is most often based on crayfish in the intermolt stage, and the concentration cited might well have been lethal to the same animals had they been undergoing a molt. Furthermore, the sex products (eggs and spermatozoa) in crayfishes are exposed to the external environment at the time of fertilization, and both they, the newly hatched, and young crayfish may be more vulnerable than are adults to a variety of polluting agents. Thus all of the tolerance limits cited should be viewed with the realization that we do not know the tolerance limits of all stages in the life cycle of a single species of crayfish to a single chemical element or compound. Furthermore, little is known of their tolerance to variations in physical factors of the environment.

An attribute of the crayfish that is not shared with a number of aquatic animals is an ability to live out of the water for long periods of time. So long as the gills are covered by a film of moisture, most crayfishes seem to exhibit no ill effects in a terrestrial environment. Their ability to move to the air–water interface where the gills on one side may be exposed to the air causes them to appear to be much more tolerant to lowered aquatic oxygen tensions than they are. This enables them to exist in a medium in which the oxygen content is distinctly below that which would be required were they to remain submerged. Should the oxygen concentration in the water of a crayfish burrow drop below the requirement of the inhabitant, the animal can climb above the surface of the water, and there, unmolested, expose its gills to air that is virtually saturated with water. In open bodies of water, however, a crayfish finding its way to the surface, particularly during the daylight hours, would expose itself to predators, a danger equivalent to the anoxic condition existing below the surface!

It is highly improbable that pollution had anything to do with the evolution of the burrowing habit in crayfishes. However, those species which construct excavations in the banks of streams or ponds are better able to tolerate contamination of their habitat than are those that characteristically live in open water beneath rocks or in debris, as was demonstrated in a grossly polluted stream in Gordon County, Georgia. In the Conasauga River near Resaca, not one crayfish could be found on the floor of the stream bed, but a number of specimens of an undescribed species, the young of which are frequently found in "open water," were found in burrows in the banks of the stream. Those species that had been collected previously in "open water" in the same area were no longer there.

The nomenclature of the crayfishes cited below is adapted from Hobbs (1972). References to original descriptions and ranges of each species may be found there.

II. Channelization and Siltation

A. Channelization

Regardless of the purity and potential productivity of a stream, unless proper cover exists to which crayfishes can retreat during the daylight hours (particularly during molting periods), few streams with a moderate to swift current that have been channelized can support anything like the crayfish fauna that was there prior to the dredging operations. Not only do such modifications of the stream bed remove the immediate population from the stream but they also seriously deter the invasion of a subsequent one. In the absence of undercut banks and accumulations of debris, which serve as hiding places, and organic deposits on which the crayfishes largely depend for their energy source, a depauperate crayfish fauna is almost inevitable. Even in channelized sluggish streams, we have consistently found crayfishes to be exceedingly rare, and, in the absence of aquatic vegetation other than algae, frequently none have been found.

"Soil" removed in channelizing and dredging operations is usually placed on one or both banks of the stream, and with subsequent erosion, silts and sands are returned to the new stream bed, often causing problems such as are discussed immediately below.

B. Siltation and Related Problems

Pollution resulting from silt deposits has been observed to destroy or greatly diminish crayfish populations in many localities in the eastern part of the United States. Moreover, there is every reason to believe that wherever streams are carrying heavy silt loads the welfare of crayfishes in them may be seriously threatened. Although we are aware of only one published record of the effect of siltation on a crayfish population, scores of similar instances could be cited by us.

In discussing their observations on the effects of strip mining on small stream fishes in east central Kentucky, Branson and Batch (1972) stated that, "During the highest turbidity in Leatherwood Creek, silt loads were measured at over 3,000 ppm. The bottom of the stream in some places was covered to a depth of 2 to 6 inches of clay . . . and the bottom fauna and flora was virtually eliminated (mayflies and crayfish, for example, were reduced by 90%)."

Among the several localities in Virginia in which the senior author has observed the effects of silt deposits on crayfish populations are two that were rather impressive. In one of them, much of the water from a small stream was deflected from the stream bed to wash gravel, and the effluent from the operation, laden with rock dust, was returned to the bed farther down stream. Upstream where clear water flowed freely around and under the rock litter of the bed, representatives of two species of crayfishes, *Cambarus b. bartonii* (Fabricius) and *Cambarus longulus* Girard, were present in at least moderate numbers. Downstream from the junction of the stream and the effluent from the rock-crushing operation, however, the rock litter in the stream bed was virtually cemented to the bottom. This alteration effectively obliterated any cover that the crayfishes might have utilized for retreat, and few benthic organisms of any kind, including crayfishes, could be found.

Similar conditions were observed to have resulted from a mining operation on a small piedmont stream in the James River drainage system. The silt carried by the small river had settled on the stream bed where it had become compacted, and the rich bottom fauna that characterized the area immediately upstream from the discharge of the effluent from the mine was greatly reduced for several hundred yards below the pollution source. Only a few juvenile crayfishes were found in that part of the rocky stream bed that was overlaid with the silt deposit.

Deep deposits of soft silt in a sluggish stream or lentic habitat seem to be equally detrimental to crayfish populations as those that become compacted and literally pave the stream bed.

Impoundments in which silt is continuously or frequently deposited on the pond or lake bed support, in the absence of vegetation, depauperate crayfish populations. The combination of obliterating benthic retreats and the predatory activity of aquatic and semiaquatic vertebrates on the crayfish inhabitants assures the failure of the establishment of reproducing populations in such areas.

For years, erosion products from farm lands have contributed negatively to crayfish populations in many small, sluggish streams, and the upsurge of urbanization within the past few decades has been conspicuous in destroying numerous crayfish habitats. With deforestation, land leveling, and alterations of drainage patterns, deep silt or silt and sand deposits have obliterated the former environment of clean brooks that supported rich faunas including both stream-dwelling and burrowing crayfishes.

It is our belief that silt, in suspension, is not necessarily detrimental to crayfish populations. In fact, for some species [*Procambarus acutissimus* (Girard) and *P. a. acutus* (Girard), for example], tremendous populations occur in roadside ditches in which the water is so laden with finely divided particulate matter (principally clay particles) that a Secchi disk disappears at a depth of less than 5 cm below the surface. Furthermore, many streams in which the water is rendered opaque and reddish orange in color by the clay particles in suspension support large crayfish populations, but there the current is such that only in backwaters and eddies are there more than superficial deposits on the stream bed. Only in sluggish or eddy areas of streams or in lentic habitats receiving heavy silt loads are the crayfishes seriously affected. In such places, the effect is more probably due to a destruction of the habitat (obliteration of retreats under rocks and debris, and smothering of burrows) rather than to a direct serious adverse effect on the crayfish itself. Most, if not all, crayfishes are well equipped to strain, to a considerable extent, the water that passes into their gill chambers.

Unusual perhaps as a source of stream pollution, but nonetheless detrimental to benthic invertebrates, are ash wastes. Wastes from an incinerator were discharged into Indian Creek in DeKalb County, Georgia. Upstream from the incinerator, the stream was found to support 11 species of macroinvertebrates including the crayfishes, *Procambarus spiculifer* (LeConte) and *Cambarus latimanus* (LeConte). In the area immediately down stream from the incinerator, not only was the water blackish but also the rocks present on the stream bottom were covered by black ash deposits. At this locality, only five species of macroinvertebrates were represented, none of which were crayfish. In two man-hours spend in collecting at each of the two stations, 100 specimens were collected above the ash pollution source, whereas in the ash-littered locality only 30 specimens (3 caddis-fly larvae and 27 dipterans were obtained. The reduction in the fauna at the downstream locality was attributed to both organic and inorganic influences of the incinerator wastes (Anonymous, 1972a).

Among the most devastated areas of streams that we have visited have been those downstream from sawmill operations that utilize the streams for discarding sawdust. Not only does the sawdust have a smothering effect in the stream bed by covering the bottom and filling interstices between rocks, but it also deters the flow of water under and between the latter. The oxygen content of the water in such crevices is literally exhausted, and an anoxic environment is established—one in which few organisms characteristic of even a semihealthy stream can survive—and the crayfish fauna is nonexistent.

III. Temperature

While some interest has been demonstrated in the tolerance of crayfishes to warm temperatures, we are aware of no published data on the lower temperature range at which crayfishes can continue to live. Below 4°C, we have observed that crayfishes of a number of species which we have maintained in the laboratory enter a state of semitorpor, move sluggishly, and cease to feed. That some individuals can withstand 0°C for at least short periods of time became evident when a container in which a specimen of *Cambarus laevis* Faxon was being held overnight was placed too near the freezing unit in a refrigerator and became frozen. Much to our surprise, when the block of ice thawed the following day, the crayfish was alive.

The earliest experiments of which we are aware that were conducted to determine the tolerance of crayfishes to elevated temperatures were those of Park and his students (1940) who were concerned with adaptations of crayfishes living in lotic as opposed to lentic habitats. Data applicable to a single species are available only for the stream-dwelling *Orconectes propinquus* (Girard) which exhibited a mean time of death of 60.4 minutes at a mean temperature of 34.6°C, whereas the species collected from ponds, *Orconectes virilis* (Hagen) and *Cambarus d. diogenes* Girard, exhibited corresponding mean values (data combined for the two) of 80.4 minutes at 38.2°C. In summarizing findings from experiments conducted over six years for crayfishes collected in streams and ponds (pooled data irrespective of species occurring in each of the two habitats), Park (1945) indicated that the stream inhabitants showed a mean time of death of 65.6 minutes at a mean temperature of 37.15°C, those of the pond dwellers, 80.9 minutes at 39.88°C.

Bovbjerg (1952) studied the toleration of *O. propinquus* and *Fallicambarus fodiens (Cottle)* [=*Cambarus fodiens*] to increased temperatures and found that in animals that had not been acclimated to identical laboratory temperatures, the median survival time of *F. fodiens*, a pond dweller, is distinctly greater than that of *O. propinquus* which occurs in streams. After 30 individuals of each of the two species had been maintained in the laboratory at temperature of 18° to 28°C for five to six weeks and then subjected to temperatures of 34° to 35°, 16 individuals of each remained alive after eight days. As interpreted by Mobberly (1965), they exhibited a "12-hour median heat-tolerance of more than 35°C."

Mobberly (1965) was concerned with the median heat-tolerance exhibited by two populations of *Faxonella clypeata* (Hay) living in roadside ditches.

After acclimitization at 21 (±1.0)°C, representatives of both populations showed a 50% mortality at 40.30° and 41.00°C. "Individuals from both areas, when acclimated at 5°–6°C for seven days during the summer and winter, had a lower heat thermal threshold than control animals maintained at 21(±1.0)°C . . . When acclimated at 21(±1.0)°C for seven days *Faxonella clypeata* had a 12-hour median heat-tolerance at 36.5°C." During a 12-hour period, there was an 80% survival at 36°C, 94% at 35°C, and 100% at 34°C.

Spoor (1955) investigated the heat tolerance of the crayfish, *Orconectes rusticus* (Girard), a stream-dwelling species, and found that individuals taken from water in which the temperature was 22° to 26°C lived for no more than six hours at 37° but for at least 10 days at 35°C. The 12- and 24-hour median heat-tolerance limits proved to be 36.4° and 35.6°C, respectively. When acclimated to lower temperatures, seemingly the longer the period of acclimatization, the lower the tolerance limit became; tolerance could be regained (more quickly than lost) by a return to the original environmental temperature.

Bovbjerg (1952) exposed individuals of *O. propinquus* and *F. fodiens* to a constant temperature of 35°C and recorded the "median death" time in each series of experiments. If members of both species were utilized immediately after they were obtained from the field, both had a shorter median death time than those that had been maintained in the laboratory at temperatures ranging from 8° to 28°C for five to six weeks. (See Table I.)

While the above data relative to heat toleration have been presented in different forms and therefore cannot be concisely compared, it is obvious that the two stream-dwelling species (*O. propinquus* and *O. rusticus*) have a lower limit of toleration to elevated temperature than do *O. virilis*, *F. fodiens*, and *C. d. diogenes*. Adults of none of the five are able to tolerate, for extended periods, a temperature as high as 38°C, and the stream dwellers no more than 34° or 35°C. It is somewhat surprising that *F. clypeata*, a species typical of shallow, temporary pools in which the temperature must occasionally reach 38°C, demonstrated such a low tolerance limit. There are no

TABLE I

Range of Median Time (in Minutes) of Death in Several Experiments When Specimens Maintained at 35°C[a]

	O. propinquus	*F. fodiens*
Not acclimated	25–100	77–282
Acclimated	288–794	452–1032

[a] Data from Bovbjerg (1952).

comparable data for newly hatched young and small juvenile members of any of the species; consequently it is impossible to predict a safe thermal threshold for any species.

IV. Dissolved Oxygen

One of the earlier investigations of tolerations to lowered oxygen tensions by crayfishes was that of Park *et al.* (1940), and, in 1945, Park summarized the results of investigations of students in his advanced ecology classes. They found that stream-dwelling crayfishes in a closed container with an initial 3.1 mg/liter of oxygen had a mean time of death of 317.5 ± 23.1 minutes at an oxygen tension of approximately 1 mg/liter, whereas under the same conditions, pond-dwelling species had a mean time of death of 605.4 ± 50.4 minutes. The stream dwellers were *Orconectes propinquus* and *O. virilis*, and those from ponds included the latter, *O. immunis, Procambarus acutus* (= their *C. blandingii*) and *Cambarus d. diogenes*.

Burbanck *et al.* (1948) compared the toleration of the epigean *Orconectes n. neglectus* (Faxon) (= their *Cambarus rusticus*) and the troglobitic *Cambarus setosus* (Faxon) to lowered oxygen tensions. At the beginning of the experiments with the epigean form, the water contained a mean oxygen concentration of 4.07 cc/liter and, after a mean death time of 272.3 ± 21.5 minutes, the oxygen had been reduced to a mean figure of 0.2453 cc/liter. The respective values in experiments with the troglobite were a mean death time of 892.9 ± 35.0 minutes with a terminal oxygen concentration of 0.1978 cc/liter.

Hale (1969), in an investigation of the oxygen consumption in *C. latimanus* (= his *C. diogenes diogenes*) which is by no means restricted to streams, discovered that the mean minimal tolerance to dissolved oxygen is 0.26 ppm.

Boyce (1968) compared the oxygen requirements of *C. latimanus* and the stream-dwelling *P. spiculifer*. He found that the former died at concentrations ranging from 0.75 to 1.40 ppm, and the latter at 1.15 to 1.60 ppm.

In a few preliminary experiments conducted by one of us (Hobbs, unpublished), it was found that *P. spiculifer* dies at concentrations between 1.3 and 2.1 ppm and *Procambarus paeninsulanus* (Faxon) at 1.7 to 2.3 ppm. The latter species occurs in lotic and lentic habitats and in burrows. While data for the toleration of the burrowing *Procambarus advena* (LeConte) are no longer available, it was discovered that, although this crayfish dies at concentrations comparable to those just cited, it lives for an appreciably longer period of time.

Bovbjerg (1970) discovered that when *O. virilis* and *O. immunis* were subjected to water in which the oxygen concentration was less than 1 ppm, the respective mean survival times for the two were 4.2 ±0.40 and 1.6 ±0.83 hours; 75% of the *O. virilis* died within five hours as opposed to 20 hours for a comparable death rate in *O. immunis*. The significance of this latter finding is suggested in the fact that unless members of *O. virilis* can manage to expose their gills to the air–water interface or crawl out onto land, it is improbable that they can tolerate an overnight lowering of oxygen concentrations that frequently occurs in shallow lentic habitats with dense vegetation or rich organic substrates.

In 1952, Bovbjerg confined individuals of the stream-dwelling *O. propinquus* and the pond-dwelling *Fallicambarus fodiens* (= his *Cambarus fodiens*) that had been taken directly from the field to water having an initial oxygen concentration of "near 0.50 cc/l." In five experiments, the median time of death ranged from 110 to 312 minutes in the former and 365 to 1328 minutes in the latter. When the experiment was repeated utilizing animals that had been acclimatized to the same laboratory conditions, the median times of death ranged from 118 to 280 minutes for *O. propinquus*, and 284 to 1802 minutes for *F. fodiens*.

Wiens and Armitage (1961) were concerned with oxygen consumption of two crayfishes (*O. immunis*, occurring in temporary pools in roadside ditches, and *O. nais*, frequenting streams and permanent ponds) in response to variations in temperature and saturation. They concluded that, in addition to increased oxygen consumption with increasing temperature, there is a decrease in consumption with decreasing saturation as well as a decrease in metabolic rate with increase in body weight. They also stated that the two species "do not have a statistically different rate of oxygen consumption under experimental conditions of moderate temperatures and low oxygen saturations. The difference in rate appears to be due to failure of *O. nais* to regulate as well as does *O. immunis* under extreme conditions." (See also Armitage and Wiens, 1960.)

The Georgia Water Quality Control Board conducted an investigation of Lick Creek in Gordon County, Georgia, as well as in a tributary stream that received an effluent from a rug factory. The dissolved oxygen at a sampling station on the polluted tributary immediately above its confluence with Lick Creek ranged from 0.3 to 1.5 ppm, whereas the concentrations in Lick Creek immediately upstream from their confluence exhibited a concentration of 7.2 to 7.6 ppm. Twenty-four crayfish, *C. latimanus*, were collected at the latter locality, but not a single individual was found in that portion of the stream with the lowered oxygen content (Anonymous, 1971a).

Similar observations (Anonymous, 1971b) were made on Nance Spring Creek, a tributary to the Conasauga River in Whitfield County, Georgia. Upstream from the outfall of a sewer line, carrying the effluent from a rug factory, where the oxygen concentration ranged from 4.6 to 7.6 ppm,

Procambarus lophotus Hobbs and Walton and *Cambarus* species B were abundant. Downstream, where the oxygen concentration was only 0.4 to 0.7 ppm, no crayfishes were found.

A water quality survey of the Coosa Basin (Anonymous, 1970) revealed data on the distribution of certain crayfishes along much of the course of the Conasauga River. Upstream from the confluence of Coahulla Creek and the River, where the oxygen concentration is greater than 6 ppm, three crayfishes, *Orconectes spinosus* (Bundy), *Cambarus* species A and *Cambarus* species B occur in comparative abundance. Below the mouth of this Creek, where the oxygen concentration was measured at 3.6 ppm, *Cambarus* species A is absent, and downstream from the confluence of Drowning Bear Creek with the River, where the concentration diminishes to 1.6 ppm, only *Cambarus* species B remains. A short distance downstream, where the oxygen concentration was found to be only 0.7 ppm, there were no crayfishes present.

To be sure, one cannot be certain that the lowered oxygen level in any of the three streams just cited was alone responsible for the absence of one or more of the crayfishes, for the water also contained dyes and other wastes, some of which might have been toxic to these animals. Inasmuch as the concentrations of oxygen recorded in the streams approximate the lethal values cited in the above experiments, and, furthermore, in view of the fact that those wastes (also derived from rug factories) that were introduced into the Conasauga River through Coahulla Creek did nto exclude *O. spinosus* and *Cambarus* species B, it seems to us that the lowered oxygen levels in portions of the Conasauga Basin are probably primarily responsible for the absence of crayfishes in many, if not most, of the sectors in which they do not apparently exist.

In one locality on Okapilco Creek (a tributary to the Suwannee River in Brooks County, Georgia), a stream heavily polluted along its course by effluents from industries and municipal waste water treatment facilities, both *P. spiculifer* and *P. paeninsulanus* occurred abundantly in an area in which the oxygen concentration at the time of collecting was only 1.5 ppm. A few days earlier the concentration was 2.5 ppm (Anonymous, 1967), suggesting that the lower concentrations might have been of brief duration.

V. Hydrogen Ion

Few data are available that unequivocably can be attributed solely to the effects of hydrogen ion concentrations on crayfishes. Except for results obtained in experiments undertaken by Park *et al.* (1940), Park (1945), and Hobbs (unpublished), the data are based on effluents that contained other substances, some of which may have been toxic to the crayfishes.

Park *et al.* (1945) compared the toleration of the stream-dwelling *Orconectes propinquus* to acid and alkali with that of two species, *Cambarus d. diogenes* and *O. virilis*, that were collected from ponds. In subjecting the crayfishes to 0.2 *N* HCl, they found that the mean time of death of *O. propinquus* was 15.2 $\pm$ 0.77 minutes, whereas that of the pond-dwelling species was 26.2 $\pm$ 1.0. The same species were subjected to 0.2 *N* NaOH, and *O. propinquus* exhibited a mean time of death of 50.8 $\pm$ 2.12 minutes; that of the pond dwellers was 107.2 $\pm$ 5.65 minutes. Thus it was established that the pond dweller can withstand high acidities and high alkalinities for a longer period of time than can the stream-inhabiting *O. propinquus*.

Our observations on other crayfishes strongly suggest that those species that are restricted in their ecological distribution to streams do indeed have a more limited range of tolerance to water with low pH values than have those species that utilize shallow lentic habitats.

That concentrations of H_2SO_4 in water which lower the pH below 3.0 are highly toxic to the stream-dwelling *C. longulus* was demonstrated (Hobbs, unpublished data) by subjecting adult individuals to solutions maintained at pH 1.0, 2.0, 3.0, 4.0, and 5.0 for 10 days. None of the individuals survived at pH 1.0 or 2.0, but all survived at pH 3.0 and above.

Further evidence of the possible adverse effects of H_2SO_4 on crayfish populations are suggested by the following observations. The effluent (containing H_2SO_4) from a seed delinting plant, located on a tributary to the Oconee River in Barrow County, Georgia, passed through two holding ponds before being discharged into the adjacent stream. The pH of the stream approximately 300 feet below the second pond varied from 3.1 to 3.7. An unpolluted tributary, not unlike the stream carrying the effluent, joins the latter upstream from its confluence with Marbury Creek. The pH range observed in the unpolluted stream was 6.2 to 7.2, and that downstream from the confluence, 4.3 to 5.6. Collections were made in the three lotic habitats mentioned, and *C. latimanus* was found to occur abundantly in the stream in which the pH ranged from 6.2 to 7.2 but was absent in the other two localities. There is every reason to believe that in the absence of the contaminant(s), the same species would have been present in the other two localities (Anonymous, 1971c).

The study of Schwartz and Meredith (1962) on the Cheat River watershed in West Virginia includes a graphic account of the effect of acid mine wastes, and, to some extent, "lumber pollution," on the crayfish fauna. In these streams, "oxygen saturation was always 70% or more . . . [and] temperatures ranged from 0°C in winter to 20°C in summer." In waters having a pH range of 4.0 to 8.5 during the winter months, the crayfish population consisted of representatives of *Cambarus bartonii bartonii*, *C. b. carinirostris* Hay, and *Orconectes obscurus* (Hagen). [*Cambarus dubius* Faxon (= their *C. carolinus* (Erichson)) is a burrowing species and seldom invades open bodies of water;

it is therefore dismissed here.] In contrast, no crayfish was found in waters that had a pH of less than 4.0.

The effects of high caustic concentrations on crayfish populations are implied by observations on Swift Creek, a tributary to the Flint River in Upson County, Georgia. This stream was polluted by the effluent from a textile mill in Thomaston. The stream bed, volume of water, rate of flow, and shoreline cover were similar to other nearby tributaries in the basin. Conspicuously unlike them, in which the pH was only slightly higher than 7.0, and in which there exists a rich crayfish fauna consisting of populations of *C. latimanus* and *P. spiculifer*, Swift Creek has a pH of 11.4, and no crayfish has been found in it (Anonymous, 1972b). Elsewhere, the latter species has been found in waters in which the pH was 8.5 (unpublished).

VI. Toxic Chemicals

A. Metallic Ions

Helff (1931), in testing the antagonistic properties of four metallic ions (Na, Mg, K, and Ca) on *Procambarus clarkii* (Girard) (see Table II), discovered the sequence of toxic effects to be K, Ca, Mg, Na, but the toxicity of all four in general, with possible exception of the addition of potassium, was reduced when introduced to the crayfish in combination with one or more of the others. For details, Tables III and IV in Helff should be consulted.

TABLE II

Toxicity of Pure Salt Solutions[a]

Concentration of Solutions (*M*)	Duration of life (hours)			
	NaCl	$CaCl_2$	KCl	$MgCl_2$
0.5[b]	20.0	17.0	2.5	12.6
0.45	36.2	25.2	2.9	13.6
0.4	65.9	25.3	3.0	17.8
0.3	157.3(7[c])	32.4	3.9	24.0
0.2	160.2(7[c])	43.1	4.9	65.0
0.1	(10[c])	78.8(2[c])	6.5	99.5
0.05	(10[c])	95.2(1[c])	17.2	(10[c])
Dist. H_2O	(10[c])	(10[c])	(10[c])	(10[c])

[a]After Helff (1931).

[b]The $MgCl_2$ solutions used were twice the molecular strength of corresponding NaCl, $CaCl_2$, and KCl solutions indicated below.

[c]Denotes number living for 7 days when test was discontinued.

Penn (1956) reported that *P. clarkii* has been collected in littoral areas of Lake Pontchartrain where salinities were as high as 6‰.

Steeg (1942), in studying the resistance of young (16 to 25 mm carapace length) *P. clarkii* to sodium chloride solutions, discovered that 100% mortality results in 23 ppt, and, while able to tolerate concentrations of 15 ppt, as the salinity increases there is a corresponding increase in death rate. He suggested the possibility that individuals occurring naturally in low concentrations of salt might be able to withstand higher salinities than those individuals frequenting salt-free water.

Kendall and Schwartz (1954) found that some individuals of *O. virilis* (80 to 110 mm total length) were able to tolerate salinities of approximately 0, 6, 14, and 30 ppt for a maximum of 218, 216, 218, and 240 hours, respectively, in one test and 696, 696, 528, and 120 hours in a second one. In the second experiment one-half of the animals subjected to each of the four concentrations died in approximately 416, 96, 108, and 40 hours, respectively. Neglecting the results obtained in the 33 ppt salinity of the second test, they found that "no statistical differences existed in the rates of mortality ... The regressions of salinity and survival were best shown by the formulae $Y = 1.437 - 0.189\,X$ (Experiment 1) and $Y = 1.641 - 0.347\,X$ (Experiment 2)." In similar experiments with *Cambarus b. bartonii* (length not indicated), this crayfish was discovered to tolerate salinities of 0, 5.2, 13.6, and 27.5 ppt for 240, 252, 108, and 192 hours, respectively, but one-half of the test population died within 72 to 96 hours. A repetition of the experiment of 0, 5, 9, 15.4, and 33.6 ppt salinities resulted in survival periods of 624, 612, 228, and 120 hours during which one-half of the test animals died within 96 hours. "Salinities of 6 to 15 ppt were readily tolerated for a period of 27 days, while higher concentrations were detrimental. These salinity-survival relationships ... were best expressed by the regression formulas $Y = 1.314 - 0.073X$ for Experiment 1 and $Y = 1.473 - 0.191X$ for Experiment 2."

A comparative study of salinity tolerance of individuals of an inland population of *P. clarkii* with those of a coastal marsh population was conducted by Loyacano (1967). He found that newly hatched young are unable to withstand salinities of 15 to 30 ppt for as much as a week; those with a total length of 30 mm are able to tolerate concentrations up to 20 ppt but die within two or three days in 30 ppt. In contrast, individuals from 40 to 120 mm live at least a week in the latter concentration. It was discovered that growth varies inversely with salinity—individuals having a total length of 40 to 50 mm subjected to salinities of 0, 10, and 20 ppt for four weeks exhibited a weight increase of 4.4, 13.5, and 4.9%, respectively. Furthermore, growth in concentrations of "10 ppt was significantly greater ($< .05$) than that in 0 or 20 ppt." He concluded that adult *P. clarkii* are able to withstand salinities almost equivalent to that of seawater for a short time and that their tolerance to salinity is directly proportional to the size of the crayfish. He also indicated that representatives of the two populations "showed no significant difference ($<.05$) in tolerance to salinity, except that newly hatched young from coastal marsh were more resistant than those from the inland population."

LaCaze (1970) indicated that "experiments conducted in the marsh over a period of two years have established that crawfish [*Procambarus clarkii* (Girard)] will tolerate salt at the amounts given below:

Period	Salinity (ppt)
Egg laying and hatching September, October, November	4–8
Winter growth Dec. through March	4–7
Spring growth and breeding April through June	3–8

. . . . The changes in salinity were relatively gradual. Rapid changes in salinity, particularly during the egg laying and hatching period, would probably result in decreased crawfish production."

Miller (1965) summarized reports of the occurrence of crayfishes in brackish water and pointed out that *Pacifastacus leniusculus trowbridgii* (Stimpson) and *P. leniusculus leniusculus* (Dana) occur in, and the latter migrates annually to and from, brackish water, tolerating salinities as high as 13.7‰, and perhaps as high as 20‰. Many individuals encountered during his study were encrusted with barnacles. Both molting and egg laying occur in brackish water, and young, "approximately 11 millimeters long, were seen hiding beneath rocks high in the intertidal zone."

Hubschman (1967) studied the toxic effects of copper on the crayfish, *O. rusticus*, using solutions of $CuSO_4$. Following an exposure of 2.5 mg/liter for a period of 24 hours, adult crayfish died rather rapidly, and none were living after 15 days. He found that 3.0 mg/liter is sufficient to kill 50% of the adults exposed for 96 hours and concluded that "if this value were taken as a guide, severe damage could result even if the calculated amount was delivered over several weeks. Reduction by a factor of 10 may still result in destruction of the juvenile population and next year's brood stock." It was discovered that 50% mortality occurred among newly hatched young when they were exposed to a concentration of 1.0 mg/liter for only one-fiftieth the time required for the adult crayfish. "The recommended doses for algicidal work range up to 0.3 mg/liter in soft water and 1.0 mg/liter in hard water (Surber, 1961)." In view of Hubschman's findings, the potential harm that such concentrations might have on crayfish populations is obvious.

Coleman and Heath (1971) investigated the toxicity of $ZnSO_4$ to the crayfish, *C. b. bartonii*, and found that "96-hour tolerance limit [for 24 to 32 g crayfish] is 175 ppm. . . ."

B. Detergents

Surber and Thatcher (1963), in investigating the effects of alkyl benzene sulfonate (ABS) on aquatic invertebrates, found that when "young" *Orconectes rusticus*, having a total length of 26 to 32 mm, were subjected to an ABS concentration of 10 ppm for a 25-day period, 20% survived as compared with 48% of the controls.

C. Oils

The only account of the effects of petroleum products on crayfishes that has come to our attention is that of Bury (1972). He indicated that 1 to 4 days following a spillage of diesel fuel in Hayfork Creek, California, "many crayfish (*Astacus* sp.) [? = *Pacifastacus* sp.] were actively moving in the creek during daylight hours, a condition which had not been noticed in previous years," and 10 dead crayfish were found.

VII. Insecticides

The toxicity of ten insecticides on *Procambarus clarkii* was investigated by Muncy and Oliver (1963). Malathion, dimethoate, and mirex were not lethal in the concentrations investigated; however, seven additional compounds were. The median tolerance limits of the crayfish to them are presented in Table III.

TABLE III

TL_m (Median Tolerance Limit) Values for Seven Insecticides Tested on *Procambarus clarkii* (Girard)[a]

Insecticide	Number of crayfish tested	TL_m (ppm, technical grade) following exposures of		
		24-hour	48-hour	72-hour
Phosphamidon	120	20.0	6.0	5.5
Dibrom	64	6.0	4.0	4.0
Bidrin	128	5.5	4.0	3.0
Sevin	36	5.0	3.0	2.0
DDT	56	0.6	0.6	0.6
Endrin	56	0.4	0.3	0.3
Methyl parathion	80	0.05	0.04	0.04

[a]After Muncy and Oliver (1963).

Field experiments (primarily for tent caterpillars) with Bidrin and phosphamidon in which maximum concentrations in water in the test area did not exceed 0.08 and 0.03 ppm, respectively, produced no detectable mortality in *P. clarkii.*

In investigations utilizing *Procambarus hayi* (Faxon) and *Procambarus blandingii* (Harlan) (the latter probably either *P. acutus acutus* or *P. acutissimus*), Ludke *et al.* (1971) found that almost 100% mortality occurred in juvenile crayfishes exposed to 1 ppb of mirex for 144 hours although death was delayed five days. When these crayfishes were subjected to concentrations of 5 ppb for 6, 24, and 58 hours, mortality 10 days after the initial exposure was 26, 50, and 98%, respectively. Other crayfishes were exposed to 0.1 and 0.5 ppb mirex for 48 hours, and at the end of four days, 65 and 71% mortality occurred. Of the crayfishes exposed to a solution derived from placing 10 granules of bait (0.3% active ingredient) in two liters of tap water, 19 of the 20 were dead after seven days. Thirty-five additional crayfishes were exposed to solutions of 35 granules of mirex bait, and after 54 hours, 33 were dead. Duplicate samples of the solution indicated a concentration of 0.86 ppb mirex. Crayfishes that were fed one and two granules of mirex bait "suffered 55.5 and 100 percent mortality, respectively, in 4 days. . ." An additional eight crayfishes were placed separately in 250 ml of tap water with two granules of mirex bait. After six days, 50% of the animals so treated contained a 8.680 ppm residue. The hepatopancreas of four animals so treated contained a mirex residue of 62.795 ppm. Furthermore, it was reported that "when crayfish were placed into water in which Mirex bait was present, but inaccessible, the animals accumulated a residue (avg. = 1.45 ppm) 16,860-fold greater than that in the water (avg. = 0.86 ppb) into which the Mirex had leached." Additional data showed that "Mirex residues in crayfish bodies were from 940-fold to 27,210-fold greater than the concentration in the water in which the crayfish were held," and, after 8 days exposure, the hepatopancreas of four individuals that were still living revealed "a 126,603-fold greater quantity of Mirex than did the water." (See also Finley *et al.*, 1971).

Dowden and Bennett (1964) investigated the effect of chlordane on respiration in *P. clarkii* and found that "the mean oxygen consumption was very high at 1 ppm, very low at 2 ppm, increased at 4 and 8 ppm then levelled off thereafter. Crayfish in 8 and 16 ppm exhibited loss of nervous control of body and appendages."

Phillips *et al.* (1972) examined crayfish, "*Cambarus* sp.," occurring in tributaries of the Tennessee River in northern Alabama for the presence of DDT. Although quantitative data are not available, "DDT was found in all crayfish examined."

The persistence of DDT in the crayfish, *C. b. bartonii*, in northern Maine where aerial spray programs for controlling the spruce budworm were conducted, was studied by Diamond *et al.* (1968). Concentrations ranging from 2.5 to 0.5 ppm were found in specimens that had been collected the same year in which streams had been sprayed. Two years after treatment, residues in the crayfish "had declined to about 0.1 ppm but remained near this level though 9 yr after treatment." Contamination for such a long period of time was attributed to "highly persistent soil residues."

Freeman and Hall (1970), in studying the effects of DDT on the respiratory rate of a crayfish (species not cited) found "that a concentration as low as .5 ppm would rapidly affect the ability of a crayfish to carry on normal respiration."

Residues of DDT and its metabolites in pond-dwelling *Procambarus simulans* (Faxon) were investigated by Bridges *et al.* (1963). A pond in Colorado was treated with DDT to provide a concentration of approximately 0.22 ppm. One month later the crayfish exhibited a concentration of DDT and its metabolites of 1.82 ppm; within four months the concentration had dropped to 0.33 ppm, and during the succeeding 12 months vacillated between 1.04 and 0.10 ppm.

Certain physiological effects of DDT on the crayfish are discussed by Tobias *et al.* (1946).

VIII. Density of Polluted Waters

Among the greatest dangers to benthic animals, including crayfishes, are pollutants introduced in fluids that are denser than the water in which the animals are living. For example, acid mine wastes, with a pH below the tolerance limit of the crayfishes, have been discharged in shallow lakes in Illinois (Smith and Frey, 1971). Because of the density of the effluents, they plunge to the bottom and convert at least some of the otherwise holomictic lakes to meromictic ones. Such an alteration in the benthic region of a lake would prevent the seasonal migrations of segments of the crayfish population to deeper water (Momot and Gowing, 1972), and, in all probability, would seriously reduce or dissipate it.

IX. Enrichment

Although no quantitative data are available to support the belief that limited organic enrichment of a stream results in an increase in the size of a crayfish population, repeated observations in certain segments of creeks have provided us with such an impression.

Particularly in streams that are low in mineral content and that support an apparently impoverished fauna, the greatest concentration of crayfishes seems to occur under, and immediately downstream from, bridges. There, refuse of various sorts, ranging from entire, or parts of, domestic animals to vegetable wastes, continue to be disposed of, even in this age of environmental enlightenment.

While such concentrations of crayfishes seem to be more spectacular in these relatively unproductive streams, similar observations have been made in streams that support comparatively large populations. To some extent, the increase in numbers of individuals in the vicinity of bridges is undoubtedly due to the litter. Discarded automobile tires, mufflers, broken drain tiles, and cans offer retreats that are utilized by many species of stream-dwelling crayfishes. Nevertheless, the added nutrients included in the garbage surely must attract and aid in supporting larger numbers of individuals than similar stretches of the stream that are not so enriched.

In a small, swift, well-aerated stream in Augusta County, Virginia, not only was a surprisingly large population of *C. b. bartonii* discovered, but each seine haul netted, in addition to crayfish, an assortment of entrails, feathers, and other offal from a small-scale abattoir situated less than 100 yards upstream.

In sectors of streams in which organic enrichment results in obvious oxygen depletion, both the size and the composition of the crayfish population are strikingly altered; smaller populations of fewer species than occur in unpolluted sectors have frequently been found to exist in such polluted portions of streams (see the discussion of Dissolved Oxygen, Section IV).

References

Anonymous (1967). A Water Quality Study of Okapilco Creek. Georgia Water Quality Control Board, Unpublished rep.

Anonymous (1970). Coosa River Basin Study. Georgia Water Quality Control Board.

Anonymous (1971a). A Water Quality Investigation: Receiving Stream of Wastes from Ranger Rugs, Inc., Gordon County. Georgia Water Quality Control Board, Unpublished rep.

Anonymous (1971b). A Water Quality Investigation: Receiving Stream Wastes from Buck Creek Industries, Whitfield County. Georgia Water Quality Control Board, Unpublished rep.

Anonymous (1971c). A Water Quality Investigation: Receiving Stream of Wastes from Piedmont Acid Delinting Company, Barrow County. Georgia Water Quality Control Board.

Anonymous (1972a). A Biological Investigation of Indian and Snapfinger Creeks, DeKalb County. Georgia Water Quality Control Board.

Anonymous (1972b). Flint River Basin Study. Georgia Water Quality Control Board.

Armitage, K. B., and Wiens, A. W. (1960). Role of Oxygen Consumption in Ecological Distribution of Two Species of Crayfish. *Bull. Ecol. Soc. Amer.* **41**(3), 73 (Abstr.).

Bovbjerg, R. V. (1952). Comparative Ecology and Physiology of the Crayfish *Orconectes propinquus* and *Cambarus fodiens*. *Physiol. Zool.* **25**(1), 34–56.

Bovbjerg, R. V. (1970). Ecological Isolation and Competitive Exclusion in two crayfish (*Orconectes virilis* and *Orconectes immunis*). *Ecology* **51**(2), 225–236.

Boyce, J. L. (1969). An Ecological Study of *Cambarus latimanus* (LeConte) and *Procambarus spiculifer* (LeConte) in the Yellow River of the Altamaha Basin of Georgia with Particular Emphasis on Respiration and Tolerance to Low Oxygen. Unpublished Thesis, Emory Univ.

Branson, B. A., and Batch, D. L. (1972). Effects of Strip Mining on Small-Stream Fishes in East-Central Kentucky. *Proc. Biol. Soc. Washington* **84**(59), 507–518.

Bridges, W. R., Kallman, B. J., and Andrews, A. K. (1963). Persistence of DDT and its Metabolites in a Farm Pond. *Trans. Amer. Fish. Soc.* **92**(4), 421–427.

Burbanck, W. D., Edwards, J. P., and Burbanck, M. P. (1948). Toleration of Lowered Oxygen Tension by Cave and Stream Crayfish. *Ecology* **29**(3), 360–367.

Bury, R. B. (1972). The Effects of Diesel Fuel on a Stream Fauna. *Calif. Fish and Game* **58**(4), 291–295.

Coleman, M. A., and Heath, A. G. (1971). Toxicity of $ZnSO_4$ to the Crayfish *Cambarus bartonii*. *ASB Bull.* **18**(2), 29–30.

Diamond, J. B., Kadunce, R. E., Getchell, A. S., and Blease, J. A. (1968). Persistence of DDT in Crayfish in a Natural Environment. *Ecology* **49**(4), 759–762.

Dowden, B. F., and Bennett, H. J. (1964). Some Effects of a Chlorinated Hydrocarbon Insecticide on the Respiration of *Procambarus clarki*. Mimeographed.

Finley, M. T., Ludke, J. L., and Lusk, C. (1971). Toxicity of Mirex to Several Crustaceans. *ASB Bull.* **18**(2), 34 (Abstr.).

Freeman, G., and Hall, W. (1970). The Effects of Small Concentrations of DDT on the Respiratory Rate of Crayfish. *J. Tenn. Acad. Sci.* **46**(4), 117 (Abstr.).

Hale, H. E. (1969). Oxygen Consumption in *Cambarus diogenes diogenes* (Girard, 1852) as Determined by an Improved Technique. Georgia State College, Mimeographed.

Helff, O. M. (1931). Toxic and Antagonistic Properties of Na, Mg, K, and Ca Ions on Duration of Life of *Cambarus clarkii*. *Physiol. Zool.* **4**(3), 380–393.

Hobbs, H. H., Jr. (1972). Biota of Freshwater Ecosystems. Identification Manual No. 9, Crayfishes (Astacidae) of North and Middle America. Environmental Protection Agency.

Hubschman, J. H; (1967). Effects of Copper on the Crayfish *Orconectes rusticus* (Girard). I. Acute Toxicity. *Crustaceana* **12** (Part 1), 33–41.

Kendall, A., and Schwartz, F. J. (1964). Salinity Tolerances of Two Maryland Crayfishes. *Ohio J. Sci.* **64**(6), 403–409.

LaCase, C. (1970). Crawfish Farming. Louisiana Wild Life and Fish. Comm.

Loyacano, H. (1967). Some Effects of Salinity on Two Populations of Red Swamp Crawfish, *Procambarus clarki*. *Proc. 21st Annu. Conf. Southeast. Ass. Game and Fish Comm.*, **21**, 423–435.

Ludke, J. L., Finley, M. T., and Lusk, C. (1971). Toxicity of Mirex to Crayfish, *Procambarus blandingi*. *Bull. Environ. Contam. and Toxicol.* **6**(1), 89–96.

Miller, G. C. (1965). Western North American Crawfishes (*Pacifastacus*) in Brackish Water Environments. *Res. Briefs, Fish Comm. Oregon* **11**(1), 42–50.

Mobberly, W. C., Jr. (1965). Lethal Effect of Temperature on the Crawfish Faxonella clypeata *Proc. Louisiana Acad. Sci.* **28**, 45–51.

Momot, W. T., and Gowing, H. (1972). Differential Seasonal Migration of the Crayfish *Orconectes virilis* (Hagen) in Marl Lakes. *Ecology* **53**(3), 479–483.

Muncy, R. J., and Oliver, A. D., Jr. (1963). Toxicity of Ten Insecticides to the Red Crawfish, *Procambarus clarki* (Girard). *Trans. Amer. Fish. Soc.* **92**(4), 428–431.

Park, T. (1945). A Further Report on Toleration Experiments by Ecology Classes. *Ecology* **26**(3), 305–308.

Park, T., Gregg, R. E., and Lutherman, C. Z. (1940). Toleration Experiments by Ecology Classes. *Ecology* **21**(1), 109–111.

Penn, G. H. (1956). The Genus *Procambarus* in Louisiana (Decapoda, Astacidae). *Amer. Midl. Natur.* **6**(2), 406–422.

Phillips, J. F., Bourne, J. R., Lindsey, C., and Dickson, W. J. (1972). An Investigation of Crayfish from Selected Streams of the Tennessee River Valley for the Presence of Chlorinated Hydrocarbon Residues. *J. Alabama Acad. Sci.* **41**(3), 131 (Abstr.).

Schwartz, F. J., and Meredith, W. G. (1962). Crayfishes of the Cheat River Watershed in West Virginia and Pennsylvania. Part II. Observations Upon Ecological Factors Relating to Distribution *Ohio J. Sci.* **62**(5), 260–273.

Smith, R. W., and Frey, D. G. (1971). Acid Mine Pollution Effects on Lake Biology. U.S. Environ. Protection Agency, Water Pollut. Contr. Res. Ser.

Spoor, W. A. (1955). Loss and Gain of Heat-Tolerance by the Crayfish. *Biol. Bull.* **108**(1), 77–87.

Steeg, W. A. (1942). Tolerance of Immature *Cambarus clarkii* in Sodium Chloride Solutions. *Tulane Biol.* **6**, 1–4.

Surber, E. W. (1961). Improving Sport Fishing by Control of Aquatic Weeds. *Bur. Sport Fishing. Wildl. Circ.* **128**, 4–13.

Surber, E. W., and Thatcher, T. O. (1963). Laboratory Studies of the Effects of Alkyl Benzine Sulfonate (ABS) on Aquatic Invertebrates. *Trans. Amer. Fish. Soc.* **92**(2), 152–160.

Tobias, J. M., Kollros, J. J., and Savit, J. (1946). Acetyl-choline and Related Substances in the Cockroach, Fly and Crayfish and the Effect of DDT. *J. Cell. Comp. Physiol.* **28**(2), 159–182.

Wiens, A. W., and Armitage, K. B. (1961). The Oxygen Consumption of the Crayfish *Orconectes immunis* and *Orconectes nais* in Response to Temperature and to Oxygen Saturation. *Physiol. Zool.* **34**(1), 39–54.

CHAPTER 8

Clams and Mussels (Mollusca: Bivalvia)

SAMUEL L. H. FULLER

I. Introduction

The North American freshwater bivalved molluscan fauna consists of three elements. The vast majority of native species belongs to the Unionacea (the freshwater mussels or naiades) and Sphaeriacea (the pill and finger nail clams), two superfamilies of worldwide distribution. Also representing the Sphaeriacea, *Corbicula manilensis* (Philippi) is an introduced Asiatic species which poses economic and biotic threats. Third, there is minor representation of several groups of essentially marine bivalves. These animals reflect phylogenetically diverse, more or less unsuccessful attempts to colonize continental waters. None of them is a legitimate freshwater species, although several can tolerate extremely low salinities. For instance, I have taken *Polymesoda caroliniana* (Bosc) in the company of freshwater mussels in tidewaters of the St. Johns river system of northeastern Florida. Nevertheless, for the purposes of this volume, these faunal elements are considered extralimital. For the student interested in pursuing the natural histories of these species, some references are included in the bibliography: Hopkins (1970), Hopkins and Andrews (1970), Morrison (1955), Wolfe (1971), and van der Schalie (1933).

Like the essentially marine elements, in contrast to the Unionacea and the other Sphaeriacea, and as an alien, *Corbicula* will receive rather short shrift here. *Corbicula* is itself a pollutant, and its own pollution biology has not been of great interest to students who would prefer to see it eliminated! It is the threat which *Corbicula* poses to the native fauna as a competitor for space that has attracted attention. The threat depends upon this species' free-swimming larva, unknown among native bivalves, and its ability to exploit virtually any substrate. Thus the range and population sizes of *Corbicula* have exploded since the first wild populations were discovered on the West Coast about 35 years ago.

Matters of geographic and ecologic distribution, normal ecology and life cycle, and economic and biotic problems may be surveyed through Sinclair's (1971) bibliography, which, though not exhaustive, provides an excellent overview of the *Corbicula* story. Little has been added to it since, but this pest has recently made its way into the Atlantic drainage (Sickel, 1973; Fuller and Powell, 1973), and W. M. Richardson *et al.* (1970) have discovered a natural enemy.

My treatments of the Unionacea and Sphaeriidae, also, display certain limitations in scope. I am especially mindful of the general lack of chemical data. Unfortunately, we do not know the limits of the tolerance by a single species for a single chemical parameter. On the other hand, the two groups share an indifference to natural variations in water chemistry over a range much wider than might aprioristically be supposed; for example, unionid mussels are not "supposed" to live in soft water or at low pH, but they can and do (Harman, 1969). Under the circumstances, I see no advantage in tabulating chemical data, and not many are discussed. Interested students can find copious information of this sort in papers by Cvancara and Harrison (1965), Dawley (1947), Imlay and Paige (1972), Morrison (1932), Patrick *et al.* (1967), Shoup *et al.* (1941), Wurtz and Roback (1955), and others.

II. Unionacea

A. Taxonomy and Systematics

The vast majority of the North American freshwater bivalve fauna consists of the mussels, Margaritiferidae and Unionidae. The somber beauty of their shells has occasioned enormous interest among naturalists since the first quarter of the last century. The rather large size of many of these animals might lead one to believe that distinguishing among our many score species would be an easy matter. Unfortunately, their variable morphologies and the uncritical gaze of most observers have combined to produce many times more names than are warranted. The resulting taxonomic confusion remains incompletely resolved to this day, and there is no one work which may be used to identify all species reliably. As Ortmann (1924) would have it, "we are able to control the identification" of perhaps a majority, but certainly not all, of our fauna. Correlating modern taxonomy with older nomenclature is an additional difficulty, particularly when early usage was not shored up by reference to published concepts or figures which can be reevaluated today.

In sharp contrast to the difficulties inherent in species determinations, most of the nearctic naiad genera recognized by Heard and Guckert (1971) and Valentine and Stansbery (1971) are readily identified with one or more of a number of works—Clench (1959), Haas *et al.* (1969), Heard (1968), Ortmann (1912), Pennak (1953), Valentine and Stansbery (1971), or Walker (1918).

The van der Schalies (1950) recognized six major zoogeographic subdivisions of the North American naiades: the Pacific, Ozarkian, Mississippian, Cumberlandian, Apalachicolan, and Atlantic faunas. Somewhere between the rather antipodal viewpoints of Ingram (1948) and Hannibal (1912) lies a true taxonomic appreciation of the Pacific drainage naiades. The Ozarkian fauna has never been monographed; Call's (1895) is a relevant paper, but it is out of date and was published before certain characteristically Ozarkian elements were distinguished (see, for example, Ortmann and Walker, 1912). Much of this fauna is common to the Mississippian and Cumberlandian regions, as well. The papers by Ortmann (1918) and Neel and Allen (1964) form a good introduction to the latter fauna, while a number of papers are very helpful for the Mississippian, which has received most attention of the six: Baker (1928), Call (1900), Heard and Burch (1966), Goodrich and van der Schalie (1944), Murray and Leonard (1962), Parmalee (1967), Simpson (*in* Baker, 1898), Starrett (1971), and Valentine and Stansbery (1971). The works by Starrett and by Murray and Leonard are particularly useful in coordinating early and recent nomenclature.

The fauna of the Canadian interior basin is an outgrowth of the Mississippian and has been wonderfully monographed by Clarke (1973). The taxonomy of Atlantic drainage naiades of Canada and the United States, including peninsular Florida, is in very good condition in view of papers by Athearn and Clarke (1962), Clarke and Berg (1959), Clarke and Rick (1963), Fuller (1971, 1972, 1973), Harman (1970b), Johnson (1970, 1972), and Shelley (1972). The Appalachicolan fauna, also, is quite well understood taxonomically (Athearn, 1964; Clench and Turner, 1956; Fuller and Bereza, 1973; Johnson, 1967, 1968). Westward from the Appalachicolan region, however, knowledge of the mussel fauna is very imperfect. Two papers by van der Schalie (1938c, 1939b) amount to the only introduction of any scope to the rich fauna of the Alabama river system. Grantham's (1969) treatment of the naiades of the adjoining state of Mississippi is helpful. A gaggle of papers (Frierson, 1897, 1902, 1911; Vaughan, 1892; Vanatta, 1910; Shira, 1913; Coker, 1915) may aid in understanding the fauna of Louisiana, which has never been considered as a unit. Strecker (1931) provided an excellent overview of Texan mussels. There are few species known from Arizona (Stearns, 1883; Taylor, 1966; Bequaert and Miller, 1973) or New Mexico (Cockerell, 1902; Henderson, 1933). Certain essentially Mexican elements—notably *Popenaias* (also in peninsular Florida) and *Cyrtonaias* (Heard and Guckert, 1971)—occur in the extreme western Gulf drainage of the United States. On the excellent chance that other exotics will be discovered there, I mention the great works on Mexican and Central American naiades by Fischer and Crosse (1894) and von Martens (1900), as well as an informative paper by Pilsbry (1910).

Until the reevaluation by Heard and Guckert (1971), there had been little improvement on the system of naiad classification developed by Ortmann, the main features of which are embodied particularly well in his 1912 and 1919 papers. At this time I prefer a variation on Heard and Guckert which recognizes two families, Margaritiferidae and Unionidae, with four subfamilies in the latter: Ambleminae, Unioninae, Lampsilinae, and Anodontinae.

B. Life Cycle

Many of the ecologically significant aspects of mussel biology are revealed by a skeletal account of their normal life cycle. Its great weaknesses lie in the prerequisites and consequences of the period of larval parasitism on a vertebrate host. Stein (1971) has made the point emphatically, and further details may be gained from the better naiad natural histories, including Lefevre and Curtis (1912), Coker *et al.* (1921), Baker (1928), and Pennak (1953).

Sperms are shed into the water from the male gonad. So far as is known, sperm are aggregated in volvocoid bodies (Edgar, 1965), which are drawn into the female on her inhalant current. Utterback (1931) observed a female *Lampsilis ovata* (Say 1817) fanning sperm bodies into her mantle cavity with her mantle flaps. Wherever water currents are reduced (see Section II, F,2), especially in ponds and lakes, fertilization of ova is less likely, mussel populations are often small, and individuals may be scattered too far apart to breed effectively (Wilson and Clark, 1912b). The level of glochidial infection on fishes may be very low in lakes (Evermann and Clark, 1918).

The fertilized ova are incubated for varying periods of time in varying portions (the marsupium) of the gills which are permanently modified to that end. During this time they are liable to attack by bacteria and protozoans (see Sections II,D,2 and 4). Mature larvae are minute bivalved creatures (glochidia), microspined and/or hooked (Arey, 1924). In spite of great mortality in the marsupium (Ellis, 1929), glochidia are shed in enormous numbers, sometimes reckoned in the millions (Merrick, 1930). In the Margaritiferidae and in all unionid subfamilies but (at least most of) the Lampsilinae, the glochidia are discharged through the excurrent aperture, individually, in small aggregates, or in large masses. Glochidia are often suspended in the water on mucous lacework (Matteson, 1955; Yokley, 1972), the more readily to come into contact with suitable host fishes (see Section II,D,16), and fish may actually devour masses of glochidia (e.g., Chamberlain, 1934). Certain fish feed upon mussels and are thus infected. Howard (1914c) believed that potential hosts are attracted to mussel beds to feed on them and associated foodstuffs, including fortuitous egg masses.

Among the Lampsilinae the glochidia are discharged through minute pores in the marsupium directly outside into the water near the incurrent aperture (Ortmann, 1910). In advanced genera, such as *Lampsilis* itself, the marsupium occupies the postbasal portion of the outer female demibranchs and may be protruded through the valves, where it is apt to be touched, even attacked, by curious or predatory fish which have been attracted by the colors, piscine shapes, and undulatory movements of the mantle flaps. Aspects of this sophisticated device for increasing the chances of glochidial infection have been observed and/or discussed by numerous authors, including Kirtland (1851), Wilson and Clark (1912b), Utterback (1915–1916), Coker *et al.* (1921), Baker (1922), and Howard and Anson (1922). Flaps have been figured by Welsh (1961), Harman (1970), Kraemer (1970), and Fuller (1971).

Since mussels do not have free-swimming larvae, the glochidial hosts serve to disperse them. Glochidia have been detected alive in plankton (Kofoid, 1910; Clark and Stein, 1921), and those of *Margaritifera falcata* (Gould 1850), for example, can remain alive as plankters for as many as

11 days (Murphy, 1942), but it is highly unlikely that planktonic glochidia contribute significantly to mussel dispersal or even to successful host infection. The vast majority of glochidia fail to infect at all, and the vast majority of these fall to the bottom, where they are subject to predation (see Sections II,D,5 and 10) and smothering.

The glochidia of Anodontinae are unusually large, hooked, and equipped with a "thread gland," which is thought to aid in host infection; significantly, these glochidia are usually found in successful infection only on tougher tissues, such as fins, and even under scales. *Megalonaias gigantea* (Barnes 1823), the Washboard (Ambleminae), and *Elliptio dilatata* (Rafinesque 1820), the Lady Finger or Spike (Unioninae), possess thread glands, and each of them can infect fins almost as readily as the more delicate tissue of gills (Howard, 1914c).

Tactile stimuli are sufficient to induce glochidial attachment on host tissue (Arey, 1921), but chemical stimuli are required to prolong attachment until host tissue can encyst the glochidia by epithelial proliferation (Howard, 1914c; Arey, 1921; Heard and Hendrix, 1964; Lukacsovics and Labos, 1965). Profound qualitative changes take place during encystment, and some glochidia increase greatly in size (Coker and Surber, 1911; Murphy, 1942). Having reared glochidia through successful metamorphosis in a solution of salts, sugars, and amino acids, Ellis and Ellis (1926) concluded that the parasitic life is not absolutely necessary, but does provide nourishment (see, also, Arey, 1932b; Blystad, 1923; Ellis, 1929; Jones, 1950; and Yokley, 1972), protection against bacterial and protozoan attack, and the opportunity of dispersal. That the Ellises managed this without glochidial mortality reminds one of the enormous loss of larvae in nature, else our waterways would be choked with mussels.

There are a few mussels—*Anodonta imbecilis* Say 1829, a Floater or Paper Shell (Howard, 1914d; Clark and Stein, 1921; E. Allen, 1924); *Strophitus undulatus* (Say 1817), the Squaw Foot (Lefevre and Curtis, 1911, 1912); and *Obliquaria reflexa* Rafinesque 1820, the Three-horned Warty Back (Lefevre and Curtis, 1912; Howard, 1914c; Utterback, 1915–1916)—whose parasitism, allegedly or definitely, is only facultative. Perhaps this harkens back to early times before larval incubation and parasitic dispersal (see Howard, 1953, and Sellmer, 1967). In those trying times the widespread hermaphroditism known among mussels (van der Schalie, 1970; Heard, 1970a) must have been most advantageous (see Tomlinson, 1966).

After remaining encysted for varying lengths of time, glochidia help initiate rupture of their cysts (Arey, 1932a) and drop from the host. If they land on suitable substrate, development ensues. A "suitable substrate" is firm (but yielding) and stable (Negus, 1966); shifting sands and fine muds are, as a rule, inimical to mussels, young and old alike. Some juvenile mussels have a byssus, which provides purchase on solid objects (Kirtland, 1840; Sterki, 1891a,b; Frierson, 1903; Isely, 1911; Coker, 1912; Howard, 1914c). Read and Oliver (1953) observed juveniles only on gravel shoals, clamped to algal filaments or attached to rocks by their byssi. As requirements for successful mussel reproduction, Isely (1911) included sand- and silt-free riffles, abundant food and dissolved oxygen, and holdfasts for the byssi. He added that mussels radiate to other habitats as they grow larger. Possibly the difference between adult and juvenile habitats is a device meant to reduce competition for space (d'Eliscu, 1972). "Competition" among mussels—even for such essentials as food and space—seems a rather passive affair, depending primarily on degrees of success in reproduction. Perhaps it is really superior reproductive ability that accounted for the "competitive" succession of mussel families during geological time that was cited by Bănărescu (1971).

C. Food and Feeding

W. R. Allen (1914) believed that a suspension of any finely divided, decaying tissue will serve as food for mussels. Churchill (1916) showed that mussels can utilize nutrient in solution, including fat (Churchill, 1915; Coker *et al.*, 1921). There seem to be no interspecific differences in feeding among naiades; identifiable stomach contents are almost invariably mud, desmids, diatoms, and other unicellular algae (see Section II,D,3) (Evermann and Clark, 1918; Coker *et al.*, 1921). W. R. Allen (1921) found nannoplankton very important in the mussel diet, and Read and Oliver (1953) felt that mussels prefer zooplankton to phytoplankton. Churchill and Lewis (1924) concluded that microorganisms, especially Protozoa (see Section II,D,4), and detritus form the sole food of mussels and noted that larger populations develop below areas where disintegration of rich vegetation is occurring (see Section II,F,6). Mussel food may include rotifers and flagellates, also (Wilson and Clark, 1912a). Living and active diatoms have been observed at the posterior ends of mussels' alimentary canals (Wilson and Clark, 1912a; Coker *et al.*, 1921). Finally, several unnatural substances have sustained mussel life: strained beef and beef heart (Salbenblatt and Edgar, 1964), "baby food beef" (Edgar, 1965), and trout fry food (Imlay and Paige, 1972).

Thus a heterogeneous collection of observations on mussel food has accumulated over the years, and the mussel diet is reasonably well understood: it consists primarily of detritus and animal plankters. (It might usefully be emphasized that the prevalent notion that mussels feed mainly on diatoms is a myth.)

There are very few observations on mussel feeding activity. W. R. Allen (1921) wrote that sand and mud are usually sorted from foodstuffs, but

Churchill and Lewis (1924) and Coker *et al.* (1921) felt that rejection of a potential food is accomplished, not by ciliary sorting, but by refusal to feed. There is unresolved difference of opinion here.

D. Biotic Associations

Reproduction, habitat, and food are commonly discussed aspects of a natural history of any group of organisms, but their relationships with other plants and animals are often treated rather superficially. It happens, however, that mussel symbioses and other biotic relationships are critically important ingredients of their normal and pollution ecologies—and of any attempt to understand the value of mussels as indicator organisms.

1. Viruses (Virulenta)

Pauley (1968a,b) described the "spongy" disease of the foot that has been detected in *Margaritifera falcata*. Watery lesions develop, and the affliction can involve the reduction of epithelium to necrotic, squamous tissue, which may disappear altogether. These remarks are offered here on the chance that they represent a viral infection.

2. Bacteria (Schizomycetes)

Particularly under conditions of siltation (see Section II,F,3), (unspecified) bacteria may attack glochidia while still in the marsupium (Ellis, 1929). Imlay and Paige (1972) found that growth of *Amblema plicata* (Say 1817), the Three Ridge or Blue Point, can be retarded by sufficient quantities of (unspecified) bacteria. Apparently, bacteria pose a threat to mussels only when their populations expand upon suitable environmental disturbance, doubtless particularly with organic enrichment (see Section II,F,6).

3. Algae (particularly Chlorophyta)

Very little specific information is available on the important relationship between mussels and algae. Vinyard (1955) felt that there are possibly undescribed genera and species of algae growing on "clams and snails," suggested that some of these associations may prove highly specific, and pointed out the previous lack of literature on the subject. Wilson and Clark (1912a) noted abundant multicellular algae, especially the green alga *Cladophora*, on the shells of living mussels, and W. R. Allen (1914), alleging that it may have a nutritional role, remarked the "private garden" of diatoms and other algae which often adorn the posterior ends of shells. Wilson and Clark (1912b) suggested that such algae may help aerate stagnant water near the mussel and provide it with added buoyancy. All these ideas remain untested, and the one indisputably important role that algae play in mussels' lives, though commonly overestimated, is as food (see Section II,C).

4. Protozoans (Protozoa)

Other than as an extremely important food (see Section II,C), protozoans have rather little to do with mussels. Coker *et al.* (1921) observed the familiar genera *Paramecium* and *Vorticella* in the mantle cavity and on the mantle, respectively, and members of a small group of thigmotrichous ciliates are regularly associated with mussels (Kelly, 1902). These include *Heterocinetopsis uniodarum* Jarocki and Raabe, which Antipa and Small (1971) tentatively considered parasitic, and several species of *Conchophthirius*. Combining their data with Kelly's (1902), Antipa and Small found *C. curtus* Engelmann present, as a commensal, in 81.5% of the mussel species examined for it at that time. They felt that mussels whose shells gape in life are especially susceptible to protozoan infestation—one would thus expect less infection among Ambleminae and Unioninae than among Anodontinae and Lampsilinae—and Wilson and Clark (1912a) thought *Conchophthirius* universally distributed among species of their acquaintance. Kidder (1934) considered *C. curtus* and *C. anodontae* (Ehrenberg) both commensals. J. H. Penn (1958) noted heavy infections with *C. curtus* in all individual mussels examined and interpreted the species as a world-wide "parasite." The more recent of these papers include lengthy bibliographies, and information about specific associations between these thigmotrichs and their mussel hosts may be gained from any of the papers cited above. Environmental disturbances could scarcely increase the size, but perhaps the variety, of protozoan infections of mussels.

5. Flatworms (Platyhelminthes: Turbellaria)

Howard and Anson (1922) reported a member of the genus *Stenostomum* (Rhabdocoela) which was preying upon glochidia that had fallen to the stream bed. The occasional occurrence of "planarians" (probably Tricladida) within mussels was thought unimportant by Kelly (1902).

6. Flukes (Platyhelminthes: Trematoda)

Freshwater mussels are afflicted with members of several trematode families, most of which are classified among the digenetic flukes, that is, in the order Digenea in Hyman's (1951) arrangement, which first subdivides the Trematoda at ordinal level. Perhaps the most familiar of these is *Allocreadium ictaluri* Pearse (Allocreadiidae), which, argued Hopkins (1934), is the organism whose irritating presence leads to pearl formation in mussels (see Section II,D,15); Seitner (1951) wrote a life history of *A. ictaluri*. This is one of the "distomids," references to which are so common in the naiad literature. The old family Distomidae has been dismembered and its elements, scattered among modern digenetic families (Hyman, 1951). Gasterostomes of the genus *Bucephalus* (Bucephalidae) affect mussels quite differently, often damaging or destroying gonad tissue (Kelly, 1902; Wilson and Clark, 1912a,b; Lefevre and Curtis, 1912). It is the rare paper (e.g., Gentner and Hopkins, 1966) which correlates individual mussel species with any variety of "distomids" identified and classified according to recent concepts and nomenclature. Other trematode host-parasite relationships of varying taxonomic reliability can be found in Leidy (1858), Fischthal (1954), and Coil (1954). Allocreadiids and bucephalids can interfere with mussel lives so profoundly under the best of circumstances that it is hard to imagine that they could do a great deal more damage in a disturbed environment.

More common is information about members of the other group of flukes which infect mussels, the family Aspidogastridae in the order Aspidobothrea. In North America these parasites fall into the genera *Cotylogaster*, *Cotylaspis*, and *Aspidogaster*. Hendrix and Short (1965) showed that all *Cotylaspis* infecting Nearctic naiades are *C. insignis* Leidy. Each of the other two genera includes only one species parasitic upon North American mussels (Stromberg, 1970): *A. conchicola* von Baer and *Cotylogaster occidentalis* Nickerson. The latter is rarely recorded from Nearctic mussels (Kelly, 1926), but the former is common and may cause disease involving severe modification or destruction of affected host tissue (Pauley and Becker, 1968). The following works list most of the published records associating specific mussels with *C. insignis* and *A. conchicola*: Gentner and Hopkins (1966), Hendrix (1968), Hendrix and Short (1965), Kelly (1902), Kofoid (1899), Leidy (1858, 1859), Monticelli (1892), Najarian (1955, 1961), Pauley and Becker (1968), Stromberg (1970), Stunkard (1917), Osborn (1898a,b), Utterback (1916), van Cleave and Williams (1943), Vidrine (1973), and Wilson and Clark (1912b).

7. *Roundworms* (Nematoda)

Wilson and Clark (1912a) detected roundworms in the alimentary canals of mussels. Coker *et al.* (1921) confirmed the phenomenon and pronounced the worms parasites. Nothing is known about nematodes' potential for damage to mussels.

8. *Rotifers* (Rotifera)

Rotifers can play an apparently insignificant role in the diet of mussels (Wilson and Clark, 1912a).

9. *Bryozoans* (Entoprocta and Ectoprocta)

Wilson and Clark (1912a) noted a member of the ectoproct genus *Plumatella*—probably *P. repens* (Linnaeus) or a close relative—growing on various *Anodonta* and *Amblema plicata*. I have seen the entoproct *Urnatella gracilis* Leidy on mussels, mostly *Elliptio complanata* (Lightfoot), in the Potomac

River, Maryland and Virginia, and the ectoproct *Pottsiella erecta* (Potts) on *Uniomerus tetralasmus* (Say 1831) in numerous waterways of the southeastern states. Are these consistently specific associations? An indeterminate bryozoan was reported by Williams (1969) on "baldies" (i.e., shells which are nearly or quite denuded of periostracum) in Kentucky Lake, an impoundment on the Tennessee River. In view of the common association of divers bryozoans with healthy mussels, no given significance can reasonably be attached to Williams' observations.

10. Aquatic Earthworms (Annelida: Oligochaeta)

Chaetogaster limnaei von Baer, a naid worm, was reported within mussels by Kelly (1902), who thought it might be a predator. *Chaetogaster diaphanus* (Gruithuisen) definitely preys upon glochidia which have fallen to the bottom (Howard and Anson, 1922). The latter species, at least, plays a probably minor role in regulation of naiad population sizes. My taxonomic usage here has followed Brinkhurst (*in* Brinkhurst and Jamieson, 1971).

11. Leeches (Annelida: Hirudinea)

Leeches are often found attached to dead or living mussel shells, but perhaps it is nothing more than the solid substrate provided by the shell which attracts the leech. Leeches are commonly "parasitic" in mussels and have been thought to eat their mucous (Wilson and Clark, 1912a; Coker *et al.*, 1921). *Placobdella montifera* Moore has been observed in mussels, but was not noticed to feed upon them (Moore, 1912). Kelly (1902) thought unimportant the occasional occurrence of leeches within mussels. The only leeches which I have found actually inside (i.e., within the mantle cavity) living mussels are *P. parasitica* (Say) and *P. montifera*. The former association is rare; the latter is regularly encountered, but there seems to be no pattern of species which serve as "hosts." The advantages secured by *P. montifera*—other than clandestine shelter—are unclear. In any event, although leech populations can reach epidemic proportions, these animals seem to pose no threat to mussels, even though the relationship is more common in lentic situations. Finally, glochidia have been observed attached to leeches (Seshaiya, 1941), but successful metamorphosis is doubtful. My taxonomic usage here has followed Sawyer (1972).

12. Copepods (Arthropoda: Crustacea: Copepoda)

Wilson (1916) concluded that piscine immunity to glochidia can be induced by copepod infection, but that the opposite is true, as well. However, Cope's (1959) data indicated that this is true neither everywhere nor with all species—which is hardly surprising in view of the fact that glochidial infection can induce immunity to itself (Arey, 1932c).

13. Crayfish (Crustacea: Decapoda)

Wilson and Clark (1912a) observed crayfish feeding upon dead mussels and wondered if the crustaceans might not have killed them. I am not aware that the possibility has been verified.

14. Insects (Arthropoda: Insecta)

Wilson and Clark (1912a) reported midge larvae (Diptera: Chironomidae) in the mantle cavity of an unspecified mussel. Doubtless this was an accidental intrusion; the record, to my knowledge, has not been duplicated.

15. Water Mites (Arthropoda: Acari)

The nonmarine aquatic mites include a family, the Unionicolidae, many of whose members are symbiotic with freshwater mussels. These mites are *Najadicola ingens* (Koenike) and numerous species of *Unionicola*. The symbiosis may be parasitic or commensalistic, depending upon the species of mite and/or the stage in its life cycle. The given mite may exploit numerous mussel species, and the given mussel may harbor more than one species of mite. Some instances of apparently well developed specificity were revealed by Mitchell and Wilson (1965) and Davids (1973), whereas *U. formosa* Dana and Whelpley, for example, parasitizes several different *Anodonta* (Mitchell, 1957). The genus *Anodonta* seems to support a characteristic assemblage of *Unionicola* species (*ibid.*), at least some of which may be present in great numbers, particularly in naturally lentic or even in impounded habitats, where *Anodonta* are peculiarly well suited to survival.

The proliferation of water mites and the damage they can do can be indirect, adverse effects of man's impact upon formerly lotic waterways. A heavy mite infestation may lead to shredding of portions of the gills, a favorite locus of attack, or even to death (Davids, 1973). Mite eggs can form the nuclei of pearls (Wilson and Clark, 1912a). A sense of aspects of the biology and host relationships of *Najadicola ingens* is sought in papers by Humes and Jamnback (1950), Humes and Russell (1951), and Humes and Harris (1952). Information on host relationships and natural histories of several species of *Unionicola* is given by Dana and Whelpley (1836), Evermann and Clark (1918), Kelly (1902), Leidy (1884), Marshall (1926), Mitchell (1955, 1957, 1965), Mitchell and Wilson (1965), Murray and Leonard (1962), Utterback (1916), Wilson and Clark (1912b), and Wolcott (1898, 1899). Sadly, in view of superior species concepts and nomenclature developed in the later papers many records in the earlier ones must be considered suspect.

16. Fishes (Chordata: Vertebrata: Pisces)

A few kinds of fishes prey heavily upon mussels (Baker, 1916), but the practice commonly results in parasitism by glochidia. The relationship between predator and prey is mutualistic at worst. Catfish often eat benthic mollusks, including mussels (Kendall, 1910), and *Aplodinotus grunniens* Rafinesque, the Freshwater Drum, feeds on little else (Forbes and Richardson, 1908). Drum host glochidia of at least 11 species of mussels (Table I). They are most readily infected when crushing the shells of larvigerous females of fragile species, such as members of the lampsiline genera *Leptodea*, *Proptera*, *Ellipsaria*, and *Truncilla.*

Other than their difficulties at the hands of man, the most important biotic relationship involving mussels is the almost universal parasitism of their glochidia upon fishes. Especially under the crowded conditions of fish hatcheries (Murphy, 1942) glochidial infestation may produce host mortality (Ellis, 1929), but in nature the glochidium is normally a "good" parasite. Glochidial infection induces in the fish an immunity which strengthens with repeated infections. In certain cases, immunity can provide protection against attack by copepods, as well (see Section II, D, 12).

Disruption of the relationship between mussel and fish usually depends on destruction of the mussel habitat (see, in particular, Sections II, F, 1 and 2) and/or on elimination of the host. The latter difficulty concerns us here. This is the most subtle and poorly known aspect of naiad pollution biology. We are largely or completely ignorant of the identities of the potential fish hosts for most mussels. The information in Table I provides (generally inadequate) knowledge of these relationships for only perhaps one fifth or so of the Nearctic naiad fauna. Here is an area where the most informative and practical kind of observation and research can be accomplished in field and laboratory. Accurate knowledge would allow prediction that a given mussel will be threatened when its glochidial hosts are disturbed by waterway alteration.

Conner (1905) was first to identify partners in a Nearctic mussel-host relationship. At about this time, the United States Bureau of Fisheries Laboratory at Fairport, Iowa, became the center of research on the artificial propagation of freshwater mussels, whose seemingly inexhaustible supplies were dwindling (Kunz, 1898; Simpson, 1899; Smith, 1899; Coker, 1916, 1919). Commencing with the pioneering studies by Lefevre and Curtis (1910a,b, 1911, 1912), the staff and associates of the Laboratory produced a long series of papers on naiad natural history, which included much information on glochidial hosts of commercial species (Table I, in part), as well as some of lesser value. Many unpublished data on noncommercial species were lost when the Laboratory was destroyed by fire in 1917 (Coker *et al.*, 1921). From the mussels' point of view, so to speak, this was a tragedy on the order of the TVA damming of the Tennessee River, the current excesses of our Army Corps of Engineers, and three centuries' agrarian malpractice in the eastern United States. The Fairport Laboratory was rebuilt, but, with depletion of the mussel and stocks the advent of plastic buttons shortly there-

TABLE I

CERTAIN NEARCTIC FRESHWATER MUSSELS (UNIONACEA) AND THEIR KNOWN AND/OR IMPLICATED GLOCHIDIAL HOST FISHES[a]

Mussel	Host fish	References
Margaritiferidae	Salmonidae	
Margaritifera falcata (Gould)	*Oncorhynchus tschawytscha* (Walbaum), Chinook Salmon	Davis (1946)
	Salmo gairdneri Richardson, Rainbow Trout	Davis (1934), Murphy (1942), K. A. Wilson and Ronald (1967)
	S. trutta Linnaeus, Brown Trout	Murphy (1942)
	Salvelinus fontinalis (Mitchill), Brook Trout	Murphy (1942)
	Cyprinidae	
	Rhinicthys osculus (Girard), Speckled Dace	Murphy (1942)
	Richardsonius egregius (Girard), Lahontan Redside	Murphy (1942)
	Catostomidae	
	Catostomus tahoensis Gill and Jordan, Tahoe Sucker	Murphy (1942)
M. margaritifera (Linnaeus)	Salmonidae	
	Salmo trutta (Linnaeus)	Clarke and Berg (1959)
	Salvelinus fontinalis (Mitchill)	Clarke and Berg (1959)
Unionidae:		
Ambleminae	Lepisosteidae	
Amblema plicata (Say)	*Lepisosteus platostomus* Rafinesque, Shortnose Gar	Coker *et al.* (1921), Howard and Anson (1922)
	Esocidae	
	Esox lucius Linnaeus, Northern Pike	Coker *et al.* (1921), C. B. Wilson (1916)
	Catostomidae	
	Carpiodes velifer (Rafinesque), Highfin Carpsucker	Howard (1914c)
	Ictaluridae	
	Ictalurus punctatus (Rafinesque), Channel Catfish	Howard (1914c)
	Pylodictis olivaris (Rafinesque), Flathead Catfish	Howard (1914c)
	Percicthyidae	
	Morone chrysops (Rafinesque), White Bass	Coker *et al.* (1921), C. B. Wilson (1916)
	Centrarchidae	
	Ambloplites rupestris (Rafinesque), Rock Bass	Stein (1968)
	Lepomis cyanellus Rafinesque, Green Sunfish	Stein (1968)
	L. gibbosus (Linnaeus), Pumpkinseed	Coker *et al.* (1921), Stein (1968)

[a]See pp. 238–239 and 273.

TABLE I (*continued*)

Mussel	Host fish	References
	L. gulosus (Cuvier), Warmouth	Coker *et al.* (1921), Howard (1914c), Pearse (1924), Stein (1968)
	L. macrochirus Rafinesque, Bluegill	Howard (1914c), Stein (1968)
	Micropterus salmoides (Lacépède), Largemouth Bass	Coker *et al.* (1921), Howard (1914c), Lefevre and Curtis (1912), Reuling (1919)
	Pomoxis annularis Rafinesque, White Crappie	Coker *et al.* (1921), Howard (1914c), Surber (1913), C. B. Wilson (1916)
	P. nigromaculatus (Lesueur), Black Crappie	Coker *et al.* (1921), Howard (1914c)
	Percidae	
	Stizostedion canadense (Smith), Sauger	Coker *et al.* (1921), Howard (1914c), Surber (1913), C. B. Wilson (1916)
Fusconaia ebena (Lea)	Clupeidae	
	Alosa chrysochloris (Rafinesque), Skipjack Herring	Coker (1919), Coker *et al.* (1921), Howard (1914c, 1917), Surber (1913), C. B. Wilson (1916)
	Centrarchidae	
	Lepomis cyanellus Rafinesque	Coker *et al.* (1921)
	Micropterus salmoides (Lacépède)	Howard (1914c)
	Pomoxis annularis Rafinesque	Howard (1914c)
	P. nigromaculatus (Lesueur)	Howard (1914c)
F. flava (Rafinesque)	Centrarchidae	
	Lepomis macrochirus Rafinesque	Howard (1914c)
	Pomoxis annularis Rafinesque	Coker *et al.* (1921), Howard (1914c), C. B. Wilson (1916)
	P. nigromaculatus (Lesueur)	Surber (1913), C. B. Wilson (1916)
Megalonaias gigantea (Barnes)	Amiidae	
	Amia calva Linnaeus, Bowfin	Howard (1914c)
	Anguillidae	
	Anguilla rostrata (Lesueur), American Eel	Coker *et al.* (1921), Surber (1915) C. B. Wilson (1916)
	Clupeidae	
	Alosa chrysochloris (Rafinesque)	Coker *et al.* (1921), C. B. Wilson (1916)
	Dorosoma cepedianum (Lesueur), Gizard Shad	Coker *et al.* (1921), Howard (1914c)
	Catostomidae	
	Carpiodes velifer (Rafinesque)	Howard (1914c)
	Ictaluridae	
	Ictalurus melas (Rafinesque), Black Bullhead	Coker *et al.* (1921), Howard (1914c)

TABLE I (*continued*)

Mussel	Host fish	References
	I. nebulosus (Lesueur), Brown Bullhead	Coker *et al.* (1921)
	I. punctatus (Rafinesque)	Coker *et al.* (1921), Howard (1914c)
	Pylodictis olivaris (Rafinesque)	Coker *et al.* (1921), Howard (1914c)
	Percicthyidae	
	Morone chrysops (Rafinesque)	Coker *et al.* (1921), Howard (1914c), C. B. Wilson (1916)
	Centrarchidae	
	Lepomis macrochirus Rafinesque	Coker *et al.* (1921), Howard (1914c)
	Micropterus salmoides (Lacépède)	Howard (1914c)
	Pomoxis annularis Rafinesque	Coker *et al.* (1921)
	P. nigromaculatus (Lesueur)	Coker *et al.* (1921), Howard (1914c)
	Percidae	
	Stizostedion canadense (Smith)	Howard (1914c)
	Sciaenidae	
	Aplodinotus grunniens Rafinesque, Freshwater Drum	Coker *et al.* (1921), Howard (1914c), Surber (1913, 1915), C. B. Wilson (1916)
Quadrula metanevra (Rafinesque)	Centrarchidae	
	Lepomis cyanellus Rafinesque	Surber (1913), C. B. Wilson (1916)
	L. macrochirus Rafinesque	Coker *et al.* (1921), Howard (1914c), Surber (1913)
	Percidae	
	Stizostedion canadense (Smith)	Coker *et al.* (1921), Howard (1914c)
Q. nodulata (Rafinesque)	Ictaluridae	
	Ictalurus punctatus (Rafinesque)	Coker *et al.* (1921), C. B. Wilson (1916)
	Pylodictis olivaris (Rafinesque)	Coker *et al.* (1921)
	Centrarchidae	
	Lepomis macrochirus Rafinesque	Howard (1914c)
	Micropterus salmoides (Lacépède)	Howard (1914c)
	Pomoxis annularis Rafinesque	Coker *et al.* (1921), Surber (1913), C. B. Wilson (1916)
	P. nigromaculatus (Lesueur)	Howard (1914c)
Q. pustulosa (Lea)	Acipenseridae	
	Scaphirhynchus platorhynchus (Rafinesque), Shovelnose Sturgeon	Coker *et al.* (1921)
	Ictaluridae	
	Ictalurus melas (Rafinesque)	Coker *et al.* (1921), Howard (1913, 1914c)
	I. nebulosus (Lesueur)	Coker *et al.* (1921), Howard (1914c)
	I. punctatus (Rafinesque)	Coker *et al.* (1921), Howard (1913, 1914c)

TABLE I (*continued*)

Mussel	Host fish	References
	Pylodictis olivaris (Rafinesque)	Coker *et al.* (1921), Howard (1913, 1914c), C. B. Wilson (1916)
	Centrarchidae	
	Pomoxis annularis Rafinesque	Coker *et al.* (1921), Surber (1913), C. B. Wilson (1916)
Q. quadrula (Rafinesque)	Ictaluridae	
	Pylodictis olivaris (Rafinesque)	Howard and Anson (1922)
Unionidae: Unioninae		
Elliptio complanata (Lightfoot)	Percidae	
	Perca flavescens (Mitchill), Yellow Perch	Lefevre and Curtis (1912), Matteson (1948)
E. crassidens (Lamarck)	Clupeidae	
	Alosa chrysochloris (Rafinesque)	Howard (1914c, 1917)
E. dilatata (Rafinesque)	Clupeidae	
	Dorosoma cepedianum (Lesueur)	C. B. Wilson (1916)
	Ictaluridae	
	Pylodictis olivaris (Rafinesque)	Howard (1914c)
	Centrarchidae	
	Pomoxis annularis Rafinesque	Howard (1914c), C. B. Wilson (1916)
	P. nigromaculatus (Lesueur)	Howard (1914c)
	Percidae	
	Perca flavescens (Mitchill)	Howard (1914c)
Plethobasus cyphyus (Rafinesque)	Percidae	
	Stizostedion canadense (Smith)	Surber (1913), C. B. Wilson (1916)
Pleurobema cordatum (Rafinesque)	Cyprinidae	
	Notropis ardens (Cope), Rosefin Shiner	Yokley (1972)
	Centrarchidae	
	Lepomis macrochirus Rafinesque	Coker *et al.* (1921), Surber (1913)
Unionidae: Anodontinae		
Alasmidonta calceola (Lea)	Percidae	
	Etheostoma nigrum Rafinesque, Johnny Darter	Morrison (*in* Clarke and Berg, 1959)
	Cottidae	
	Cottus bairdi Girard, Mottled Sculpin	Morrison (*in* Clarke and Berg, 1959)
A. marginata (Say)	Catostomidae	
	Catostomus commersoni (Lacépède), White Sucker	Howard and Anson (1922)
	Hypentelium nigricans (Lesueur), Northern Hog Sucker	Howard and Anson (1922)
	Moxostoma macrolepidotum (Lesueur), Shorthead Redhorse	Howard and Anson (1922)

TABLE I (*continued*)

Mussel	Host fish	References
	Centrarchidae	
	Ambloplites rupestris (Rafinesque)	Howard and Anson (1922)
	Lepomis gulosus (Cuvier)	Howard and Anson (1922)
Anodonta beringiana Middendorf	Salmonidae	
	Oncorhynchus nerka (Walbaum), Sockeye Salmon	Cope (1959)
	O. tschawytscha (Walbaum)	Cope (1959)
	Gasterosteidae	
	Gasterosteus aculeatus Linnaeus, Threespine Stickleback	Cope (1959)
A. californiensis Lea	Poeciliidae	
	Gambusia affinis (Baird and Girard), Mosquitofish	d'Eliscu (1972)
A. cataracta Say	Cyprinidae	
	Cyprinus carpio Linnaeus, Carp	Lefevre and Curtis (1910b)
A. grandis Say	Lepisosteidae	
	Lepisosteus spatula Lacépède, Alligator Gar	Coker *et al.* (1921), C. B. Wilson (1916)
	Clupeidae	
	Alosa chrysochloris (Rafinesque)	Surber (1913), C. B. Wilson (1916)
	Dorosoma cepedianum (Lesueur)	C. B. Wilson (1916)
	Cyprinidae	
	Cyprinus carpio Linnaeus	Lefevre and Curtis (1910b), Morrison (*in* Clarke and Berg, 1959)
	Notemigonus chrysoleucas (Mitchill), Golden Shiner	Lefevre and Curtis (1910b), Read and Oliver (1953)
	Ictaluridae	
	Ictalurus natalis (Lesueur), Yellow Bullhead	C. B. Wilson (1916)
	Gasterosteidae	
	Culaea inconstans (Kirtland), Brook Stickleback	Morrison (*in* Clarke and Berg, 1959)
	Percicthydae	
	Morone chrysops (Rafinesque)	C. B. Wilson (1916)
	Centrarchidae	
	Ambloplites rupestris (Rafinesque)	Lefevre and Curtis (1910b), Tucker (1928), C. B. Wilson (1916)
	Lepomis cyanellus Rafinesque	Tucker (1928)
	L. macrochirus Rafinesque	Lefevre and Curtis (1910b), Morrison (*in* Clarke and Berg, 1959), Penn (1939), C. B. Wilson (1916)
	L. megalotis (Rafinesque), Longear Sunfish	Penn (1939)
	Micropterus salmoides (Lacépède)	Morrison (*in* Clarke and Berg, 1959), Penn (1939), C. B. Wilson (1916)

TABLE I (*continued*)

Mussel	Host fish	References
	Pomoxis annularis Rafinesque	Lefevre and Curtis (1910b), Morrison (*in* Clarke and Berg, 1959), C. B. Wilson (1916)
	P. nigromaculatus (Lesueur)	C. B. Wilson (1916)
	Percidae	
	Etheostoma exile (Girard), Iowa Darter	Morrison (*in* Clarke and Berg, 1959)
	E. nigrum Rafinesque	Hankinson (1908), Morrison (*in* Clarke and Berg, 1959)
	Perca flavescens (Mitchill)	Lefevre and Curtis (1910b)
	Sciaenidae	
	Aplodinotus grunniens Rafinesque	C. B. Wilson (1916)
A. imbecilis Say	Cyprinidae	
	Semotilus atromaculatus (Mitchill), Creek Chub	Clarke and Berg (1959)
	Centrarchidae	
	Lepomis cyanellus Rafinesque	Tucker (1927)
A. implicata Say	Clupeidae	
	Alosa pseudoharengus (Wilson), Alewife	Davenport and Warmuth (1965), Johnson (1946)
	Catostomidae	
	Catostomus commersoni (Lacépède)	Davenport and Warmuth (1965)
	Percicthyidae	
	Morone americana (Gmelin), White Perch	Davenport and Warmuth (1965)
	Centrarchidae	
	Lepomis gibbosus (Linnaeus)	Davenport and Warmuth (1965)
Anodontoides ferussacianus (Lea)	Petromyzontidae	
	Petromyzon marinus Linnaeus, Sea Lamprey	K. A. Wilson and Ronald (1967)
	Cottidae	
	Cottus bairdi Girard	Morrison (*in* Clarke and Berg, 1959)
Arcidens confragosa (Say)	Anguillidae	
	Anguilla rostrata (Lesueur)	C. B. Wilson (1916)
	Clupeidae	
	Dorosoma cepedianum (Lesueur)	Surber (1913), C. B. Wilson (1916)
	Centrarchidae	
	Ambloplites rupestris (Rafinesque)	Surber (1913)
	Pomoxis annularis Rafinesque	Surber (1913), C. B. Wilson (1916)
	Sciaenidae	
	Aplodinotus grunniens Rafinesque	C. B. Wilson (1916)

TABLE I (*continued*)

Mussel	Host fish	References
Lasmigona complanata (Barnes)	Cyprinidae	
	Cyprinus carpio Linnaeus	Lefevre and Curtis (1910b)
	Centrarchidae	
	Lepomis cyanellus Rafinesque	Lefevre and Curtis (1912)
	Micropterus salmoides (Lacépède)	Lefevre and Curtis (1910b)
	Pomoxis annularis Rafinesque	Lefevre and Curtis (1912)
L. costata (Rafinesque)	Cyprinidae	
	Cyprinus carpio Linnaeus	Lefevre and Curtis (1910b)
Strophitus undulatus (Say)	Cyprinidae	
	Fundulus zebrinus Jordan and Gilbert, Rio Grande Killifish	Ellis and Keim (1918)
	Semotilus atromaculatus (Mitchill)	Howard (R. L. Barney *in* Baker, 1928)
	Centrarchidae	
	Lepomis cyanellus Rafinesque	Ellis and Keim (1918)
	Micropterus salmoides (Lacépède)	Howard (R. L. Barney *in* Baker, 1928)
Unionidae: Lampsilinae		
Actinonaias carinata (Barnes)	Anguillidae	
	Anguilla rostrata (Lesueur)	Coker *et al.* (1921)
	Ictaluridae	
	Noturus gyrinus (Mitchill), Tadpole Madtom	Coker *et al.* (1921)
	Percicthyidae	
	Morone chrysops (Rafinesque)	Coker *et al.* (1921), Surber (1913), C. B. Wilson (1916)
	Centrarchidae	
	Ambloplites rupestris (Rafinesque)	Lefevre and Curtis (1910b)
	Lepomis cyanellus Rafinesque	Coker *et al.* (1921), Lefevre and Curtis (1912), C. B. Wilson (1916)
	L. macrochirus Rafinesque	Coker *et al.* (1921), C. B. Wilson (1916)
	Micropterus dolomieui Lacépède, Smallmouth Bass	Coker *et al.* (1921), Howard and Anson (1922)
	M. salmoides (Lacépède)	Coker *et al.* (1921), Lefevre and Curtis (1910b, 1912), Reuling (1919), C. B. Wilson (1916)
	Pomoxis annularis Rafinesque	Coker *et al.* (1921), Lefevre and Curtis (1912), C. B. Wilson (1916)
	P. nigromaculatus (Lesueur)	Coker *et al.* (1921)
	Percidae	
	Perca flavescens (Mitchill)	Coker *et al.* (1921), Lefevre and Curtis (1910b)

TABLE I (*continued*)

Mussel	Host fish	References
	Stizostedion canadense (Smith)	Coker *et al.* (1921), Pearse (1924)
Carunculina parva (Barnes)	Centrarchidae	
	Lepomis cyanellus Rafinesque	Mermilliod (1973)
	L. gulosus (Cuvier)	C. B. Wilson (1916)
	L. humilis (Girard), Orangespotted Sunfish	Mermilliod (1973)
	L. macrochirus Rafinesque	Mermilliod (1973)
	Pomoxis annularis Rafinesque	Mermilliod (1973)
Cyprogenia irrorata (Lea)	Cyprinidae	
	Carassius auratus (Linnaeus)	Chamberlain (1934)
Lampsilis orbiculata (Hildreth)	Percidae	
	Stizostedion canadense (Smith)	Coker *et al.* (1921), Surber (1913), C. B. Wilson (1916)
	Sciaenidae	
	Aplodinotus grunniens Rafinesque	Coker *et al.* (1921), C. B. Wilson (1916)
L. ovata (Say) (including *L. ventricosa*) (Barnes)	Centrarchidae	
	Lepomis macrochirus Rafinesque	Coker *et al.* (1921)
	Micropterus dolomieui Lacépède	Coker *et al.* (1921)
	M. salmoides (Lacépède)	Coker *et al.* (1921), Lefevre and Curtis (1912), Reuling (1919)
	Pomoxis annularis Rafinesque	Coker *et al.* (1921), C. B. Wilson (1916)
	Percidae	
	Perca flavescens (Mitchill)	Coker *et al.* (1921)
	Stizostedion canadense (Smith)	Coker *et al.* (1921), C. B. Wilson (1916)
L. radiata luteola (Lamarck)	Ictaluridae	
	Noturus gyrinus (Mitchill)	Coker *et al.* (1921)
	Percicthyidae	
	Morone chrysops (Rafinesque)	Coker *et al.* (1921), Corwin (1920)
	Centrarchidae	
	Ambloplites rupestris (Rafinesque)	Evermann and Clark (1918, 1920)
	Lepomis macrochirus Rofinesque	Coker *et al.* (1921), Everman and Clark (1918, 1920), Howard (1922)
	Micropterus dolomieui Lacépède)	Coker *et al.* (1921), Corwin (1920)
	M. salmoides (Lacépède)	Coker *et al.* (1921), Arey (1923), Howard (1914b, 1922), Reuling (1919)
	Pomoxis annularis Rafinesque	Coker *et al.* (1921), Howard (1922)
	P. nigromaculatus (Lesueur)	Coker *et al.* (1921), Howard (1922)
	Percidae	
	Perca flavescens (Mitchill)	Coker *et al.* (1921), Corwin (1920), Pearse (1924)
	Stizostedion canadense (Smith)	Coker *et al.* (1921), Corwin (1920)
	S. v. vitreum (Mitchill), Walleye	Coker *et al.* (1921), Corwin (1920, 1921).

TABLE I (*continued*)

Mussel	Host fish	References
L. teres (Rafinesque)	Acipenseridae	
	Scaphirhynchus platorhynchus (Rafinesque)	Coker *et al.* (1921), Surber (1913), C. B. Wilson (1916)
	Lepisosteidae	
	Lepisosteus osseus (Linnaeus), Longnose Gar	Coker *et al.* (1921), Jones (1950), Reuling (1919), C. B. Wilson (1916)
	L. platostomus Rafinesque	Coker *et al.* (1921), Howard (1914a), Howard and Anson (1922, Jones (1950), Reuling (1919), C. B. Wilson (1916)
	Centrarchidae	
	Lepomis cyanellus Rafinesque	Coker *et al.* (1921), Surber (1913)
	L. gulosus (Cuvier)	C. B. Wilson (1916)
	L. humilis (Girard)	Coker *et al.* (1921), Surber (1913)
	Micropterus salmoides (Lacépède)	Coker (1919), Coker *et al.* (1921), C. B. Wilson (1916)
	Pomoxis annularis Rafinesque	Coker *et al.* (1921), Surber (1913), C. B. Wilson (1916)
	P. nigromaculatus (Lesueur)	Coker *et al.* (1921), Surber (1913)
Leptodea fragilis (Rafinesque)	Sciaenidae	
	Aplodinotus grunniens Rafinesque	Howard (1913), C. B. Wilson (1916)
Ligumia recta (Lamarck)	Anguillidae	
	Anguilla rostrata (Lesueur)	Coker *et al.* (1921)
	Centrarchidae	
	Lepomis macrochirus Rafinesque	Clarke and Berg (1959), Coker *et al.* (1921), Lefevre and Curtis (1912), C. B. Wilson (1916)
	Micropterus salmoides (Lacépède)	Lefevre and Curtis (1912)
	Pomoxis annularis Rafinesque	Clarke and Berg (1959), Coker *et al.* (1921), Lefevre and Curtis (1912), C. B. Wilson (1916)
	Percidae	
	Stizostedion canadense (Smith)	Pearse (1924)
L. subrostrata (Say)	Centrarchidae	
	Lepomis cyanellus Rafinesque	Lefevre and Curtis (1912)
	L. macrochirus Rafinesque	Lefevre and Curtis (1912)
	Micropterus salmoides (Lacépède)	Lefevre and Curtis (1912)
Obovaria olivaria (Rafinesque)	Acipenseridae	
	Scaphirhynchus platorhynchus (Rafinesque)	Coker *et al.* (1921), Howard (1914a)
Ellipsaria lineolata (Rafinesque)	Centrarchidae	
	Lepomis cyanellus Rafinesque	Surber (1913), C. B. Wilson (1916)

TABLE I (*continued*)

Mussel	Host fish	References
	Percidae	
	Stizostedion canadense (Smith)	Surber (1913)
	Sciaenidae	
	Aplodinotus grunniens Rafinesque	Coker (1919), Coker *et al.* (1921), Howard (1914a), Howard and Anson (1922), C. B. Wilson (1916)
Proptera alata (Say)	Sciaenidae	
	Aplodinotus grunniens Rafinesque	Howard (1913), C. B. Wilson (1916)
Proptera laevissima (Lea)	Centrarchidae	
	Pomoxis annularis Rafinesque	Surber (1913), C. B. Wilson (1916)
	Sciaenidae	
	Aplodinotus grunniens Rafinesque	Coker and Surber (1911), Howard and Anson (1922), Surber (1912, 1913), C. B. Wilson (1916)
P. purpurata (Lamarck)	Sciaenidae	
	Aplodinotus grunniens Rafinesque	Surber (1913, 1915), C. B. Wilson (1916)
Truncilla donaciformis (Lea)	Percidae	
	Stizostedion canadense (Smith)	Surber (1913), C. B. Wilson (1916)
	Sciaenidae	
	Aplodinotus grunniens Rafinesque	Howard (1913, 1914a), Howard and Anson (1922), Surber (1912, 1913), C. B. Wilson (1916)
T. truncata Rafinesque	Percidae	
	Stizostedion canadense (Smith)	C. B. Wilson (1916)
	Sciaenidae	
	Aplodinotus grunniens Rafinesque	C. B. Wilson (1916)

after, the mussel fishery declined, the Bureau lost interest, and most additional information about mussel hosts has trickled down the decades in scattered, shorter papers.

As an aid to the interested researcher and fisheries biologist alike, and because of my belief in the importance to mussels of such information in the days to come, I have assembled in Table I all information on glochidial hosts that is known to me. Insofar as possible, the tabulated data have been brought taxonomically up to date. For the fishes, Trautman (1957) and Jordan *et al.* (1930) have been invaluable, and the vernacular and scientific nomenclature in Bailey *et al.* (1970) has been followed precisely. Simpson (1900), Frierson (1927), Murray and Leonard (1962), Johnson (1970), and Starrett (1971) have been used in the updating of naiad nomenclature.

The information in Table I reveals some points of interest and significance. The Centrarchidae (sunfishes, basses) are identified or implicated as hosts for more than half the mussel species listed. The 11 centrarchids involved come to over one third of the species given by Bailey *et al.* (1970) for the United States and Canada, and they are a higher representation than exists for any other family. About one-fifth of all known mussel hosts are centrarchids, and about two-fifths of known records involve this family. The White Crappie, *Pomoxis annularis* Rafinesque, a centrarchid, hosts more kinds of glochidia (16 species) than any other fish will tolerate. Clearly, the Centrarchidae are the most important host family. Fortunately, the family contains many durable and widespread species and theoretically provides a host fish reservoir of some permanence. Of course, we cannot assume that the mussel fauna will survive simply because the centrarchids do, and, of course, only a limited variety of mussel glochidia is recorded from centrarchids. Incidentally, the importance of the Drum (Sciaenidae) and of *Stizostedion canadense* (Smith), the Sauger (Percidae), is scarcely less.

Not mentioned in Table I are several less specific records. Morrison (*in* Clarke and Berg, 1959) and Read and Oliver (1953) listed indeterminate species of *Notropis* (Cyprinidae) as hosts for a Floater, *Anodonta grandis* Say 1829. Young (1911) gave the Banded Killifish, *Fundulus diaphanus* (Lesueur) (Cyprinidontidae), as host for at least one unspecified mussel. Williams (1969) did likewise for the Spotted Sucker, *Minytrema melanops* (Rafinesque) (Catostomidae), in Kentucky Lake. According to C. L. Hubbs (*in* Jordan *et al.*, 1930), *Lepomis euryorus* McKay is a hybrid between the centrarchids *L. cyanellus* Rafinesque, the Green Sunfish, and *L. gibbosus* (Linnaeus), the Pumpkinseed; Coker *et al.* (1921) gave *euryorus* as a glochidial host for *Amblema plicata*. On the Blackstripe Topminnow, *Fundulus notatus* (Rafinesque) (Cyprinidontidae), Shira (1913) found a glochidium smaller than, but very similar to, that of a Pocketbook, *Proptera capax* (Green 1832).

Fishes have enormous influence on mussel distribution. For example, Wilson and Danglade (1914) noted that above the Falls of St. Anthony at Minneapolis, Minnesota, on the Mississippi River there were Muckets (representing the lampsiline genera *Lampsilis* and *Actinonaias*), but none of the "*Quadrula*-group" (primarily the amblemine genera *Amblema, Quadrula*, and *Megalonaias*). They felt that no host fishes for the latter group had been able to surmount the Falls. Similarly, Wilson and Danglade believed that the St. Louis River gorge at Carlton, Minnesota, had blocked all glochidial hosts because no mussels whatsoever were found in the river above it. A third example of this sort concerns Cumberland Falls on the Cumberland River in Kentucky; only a few members of the rich naiad fauna in this river occur above the Falls (Wilson and Clark, 1914; Neel and Allen, 1964).

Anthropogenic changes in the fish fauna are a real and present danger to mussels. Athearn (1967) emphasized the adverse effects of wholesale destruction of "trash" fish, many of which are mussel hosts (Table I); Heard (1970) told how *Anodonta imbecilis* and another Floater, *A. peggyae* Johnson 1965, are being destroyed in Lake Talquin, an impoundment on the Ochlockonee River, Florida, by rotenone treatments intended to eliminate a "pest" fish, *Dorosoma cepedianum* (Lesueur), the Gizzard Shad. After each intensive application, the shore would be littered with dead and dying bivalves of these and other species. Ironically, it is *A. imbecilis*—otherwise an enormously tolerant species—that is being eliminated. It is an additional irony that this Shad is a glochidial host for at least four species of mussels (Table I), including representatives of *Anodonta* and of *Megalonaias*, another genus known from the Ochlockonee system. Although this is the least important point, it must finally be added that Lake Talquin happens to be the type locality of *A. peggyae*. An "improvement" in the fish fauna from the angler's point of view is not necessarily in the best interests of all organisms concerned.

One last point about mussels and fishes: the really successful mussels—notably *Amblema plicata*, *Anodonta grandis*, and the Washboard, *Megalonaias gigantea* (Barnes 1823)—parasitize large number of host fishes (15, 19, and 16 species, respectively).

17. *Amphibians* (Vertebrata: Amphibia)

There is a record (van der Schalie, 1937) of a small mussel in the stomach of a frog, and Howard (1951) suggested that the mudpuppy, *Necturus maculosus* Rafinesque, may devour *Simpsoniconcha ambigua* (Say 1825). Perhaps it is in this way that the latter's glochidia get on *Necturus*, its only recorded host (Howard, 1914c, 1915, 1951). In addition, Seshaiya (1941) described successful glochidial metamorphoses on tadpoles. The limited information indicates that mussels derive little harm and much good from their associations with Amphibia.

18. *Reptiles* (Vertebrata: Reptilia)

Turtles will occasionally (and insignificantly) feed on mussels (Coker *et al.*, 1921).

19. *Birds* (Vertebrata: Aves)

As predators on young fishes, waterfowl interfere with naiad reproduction (Baker, 1922), and some birds prey directly upon adult mussels. Simpson (1899) recounted how crows would drop mussels through distances sufficient to break their shells. Coker *et al.* (1921) confirmed Simpson's observation and added that shorebirds will feed upon mussels stranded in shallow waters. Snyder and Snyder (1969) gave detailed accounts of the ways Limpkins, Everglade Kites, and Boat-tailed Grackles employ to open mussel shells. These records are few and bird predation upon mussels, light.

20. Mammals (Vertebrata: Mammalia)

A few aquatic mammals feed more or less heavily on mussels: otter, mink, and muskrat (Coker, 1912; Evermann and Clark, 1918; Simpson, 1899; Parmalee, 1967; van der Schalie, 1938b; Williams, 1969; Wilson and Clark, 1912a,b). There is a small number of more detailed observations on the habits of muskrat. Wilson and Clark (1914) noted their preference for Pigtoes (*Fusconaia* or *Pleurobema*) in the Cumberland River. Muskrats have severely restricted the ranges of several species in shallow waters: *Anodonta* spp. (Headlee, 1906), *Lasmigona costata* (Rafinesque 1820) (Zetek, 1918), and *L. costata* and *Strophitus undulatus* (van Cleave, 1940). Raccoon are occasional mussel predators (Simpson, 1899), and hogs can root mussel beds to pieces (Kirtland, 1851; Meek and Clark, 1912).

Aboriginal man consumed mussels in great numbers (Matteson, 1953; J. L. Murphy, 1971; Ortmann, 1909a; Roscoe, 1967; Stansbery, 1966b; van der Schalie and Parmalee, 1960; Wilson and Clark, 1912a). Documentations are few, but modern man will occasionally feed on mussels (e.g., Wilson and Clark, 1912b).

Sadly, modern man's associations with freshwater mussels have not all been of the culinary variety. The ways in which his activities and his wastes have affected mussels are, against a backdrop of "normal" chemical and physical needs, the main theme of the next pages of this chapter.

E. Chemical and Physical Parameters

1. Temperature

A modest fund of not usually very specific information is available on the effects of temperature upon freshwater mussels. The blood of the Eurasian *Anodonta cygnea* (Linnaeus 1758) freezes at -0.078°C (Potts, 1954); the slight difference between that figure and the freezing point of water cannot be of great comfort to the mussel. Salbenblatt and Edgar (1964) contributed the useful information that temperature tolerances do indeed vary among species; specifically, 29°C was lethal for most *Anodontoides ferussacianus* (Lea 1834) tested, whereas most *Anodonta grandis* and Muckets, *Lampsilis radiata luteola* (Lamarck 1819), survived.

Edgar (1965) observed sperm release by *Anodontoides ferussacianus* stimulated by a drop from 27° to 22°C. Having disproved the significance in this matter of dissolved oxygen, pH, and carbon dioxide, Chamberlain (1934) confirmed only temperature as the determining factor in the discharge of egg masses by the Arkansas Fan Shell, *Cyprogenia aberti* (Conrad 1850). Sudden drops in temperature promote egg mass abortion (Matteson, 1948). Low temperature dulls glochidial response (see Section II, B) to opportunities for infection (Arey, 1921). Warmer water often shortens the period of glochidial encystment (Young, 1911). Whereas the glochidia of northern populations of *Anodonta grandis* would ordinarily overwinter in the marsupia before host infection in the spring, G. H. Penn (1939) described autumnal release of *grandis* glochidia in New Orleans.

Low temperature represses development of the mussel's digestive style (W. R. Allen, 1921). Oxygen consumption rises with temperature (Lukacsovics and Salánki, 1964). Hobden (1970b) showed that semipurified digestive gland catalase exhibits optimal activity at 10°C in *Elliptio complanata*. Mantle flap movement in *Lampsilis radiata luteola* varies directly with change in temperature (Grier, 1926).

There is disagreement about what inspires mussels, ordinarily very sedentary, to move about. Movement was ascribed to changes in water depth by Wilson and Clark (1912a), van der Schalie (1938b), and Grantham (1969). Evermann and Clark (1918) stated that in winter mussels *sometimes* move farther offshore and burrow more deeply into the substrate, and van Cleave (1940) noted some White Heel Splitters, *Lasmigona complanata* (Barnes 1823), which moved to deeper waters with a 10°F drop in temperature. Bovjerg (1957), however, was convinced that only starvation regularly stimulates mussel activity.

2. Light

W. R. Allen (1923) offered the following observations and speculations on the relationship between mussels and light. Sudden reductions in light, as a shadow passing over a mussel, cause it to contract its mantle "siphons," possibly because shadows are so often cast by predators. Perhaps the dappling among weeds induces the same response, which would explain why only a few species of mussels are commonly found in weed beds. On the other hand, bright light stimulates movement, probably to start the mussel on his quest for deeper water (i.e., less light) away from the shallows with their hazards, dessication, predation, and the like. Although this pattern of motion would be appropriate to most mussels, the Lilliput Shell, *Carunculina parva* (Barnes 1823), for example, may be extremely abundant in a few inches' depth, following the water's rise and fall, seeking its margin (Clench and Turner, 1956; Grantham, 1969; Isely, 1925; Murray and Leonard, 1962; Utterback, 1915–1916). One wonders in what way turbidity (see Section II, F, 3) and great depth (see Section II, F, 2) disturb naiad phototaxes.

3. Hardness

In central New York state, Clarke and Berg (1959) found no unionid mussels in waters of less than 47 ppm hardness (as $CaCO_3$), but Harman (1969) found several species at levels as low as 21 ppm. Although hardness (especially the availability of calcium) is essential to the welfare of mussels, there seems to be no published information about exact levels at which hardness determines the presence or absence of a given mussel species. Soft, poorly buffered waters may experience rapid changes in pH, which can, in turn, do harm to mussels (*ibid.*). At least some American species can combat this by drawing upon shell materials for buffers (Ellis *et al.*, 1931), and the Eurasian *Anodonta cygnea* (Linnaeus 1758) can take calcium from its valves when the diet is poor in this element (de Waele, 1930).

Calcium is antagonistic to metals, and alkaline waters (Section II, E,4) precipitate them as insoluble, harmless hydrates (Wurtz, 1962).

4. Alkalinity

New York state's Tioga River was "practically devoid of molluscan life" wherever Harman (1970) found about 15 ppm or less of total alkalinity. Similarly, Pennak (1953) stated that Anodontinae rarely live at levels of "bound CO_2" less than 15 mg/liter, yet Morrison (1932) found an anodontine species, *Anodonta c. cataracta* (Say 1817), a Floater or Paper Shell, in water with only 2.6 ppm of "fixed CO_2."

5. pH

In a study of the Big Muddy River, a naturally acidic stream in Illinois, Jewell (1922) reported nine mussel species and concluded that their distributions had to do with substrate type, not water quality. Harman (1969) implied that pH has less to do with mussels than is commonly supposed. Morrison reported mussels living throughout a broad range (5.6 to 8.3) of pH. Ellis *et al.* (1931) showed that mussel blood pH varies widely in the subneutral range. Nevertheless, the blood is readily modified by the surrounding waters (*ibid.*). Thus, when Matteson (1955) introduced mussels into lake waters of pH 4.4 to 6.1, the intolerable acidity induced a response akin to aestivation: the valves clamped shut, and the mussels gradually lost weight. Matteson suggested that declining pH of water in the mantle cavity can have a lethal effect. In fact, low pH reflects a chemical regime featuring so much dissolved carbon dioxide that mussels' gas exchange is probably impeded. Despite the prevailing subneutral pH of mussel blood, some physiological functions proceed best in the basic pH range; for example, Hobden (1970b) showed that semipurified digestive gland catalase of *Elliptio complanata* has optimal activity at pH 7.8.

Ellis' (1936) data suggest that low pH allows suspended materials to remain in suspension, thereby doubtless interfering with some aspects of mussels' biological activities (see, for example, Section II, F, 3). Wurtz (1962) pointed out that acidic waters bring metals (including toxins) into solution.

6. Arsenic

Ellis (1937) reported that 16 ppm of sodium arsenite in "hard water" are fatal to *Amblema plicata* in 3–16 days.

7. Cadmium

Solutions of $CdCl_2$ above 0.001*M* significantly inhibit respiration of *Anodonta cygnea* (Lukacsovics and Salánki, 1964).

8. Chlorine

Reporting on the mussels of Turtle River, North Dakota, Cvancara and Harrison (1965) recorded none where the river flows through a belt of naturally saline soils and chloride concentration rises to 87 ppm and above. Ironically, the heightened chloride level was accompanied by unusually high dissolved oxygen and low turbidity (as ppm of SiO_2).

Destruction of mussels by chloride-laden oil field brine was first recorded by Shira (1913). Williams (1969) reviewed the tale of oil brine damage in the Green River of Kentucky. In 1958 the chloride concentration shot from fewer than 10 to more than 1000 ppm. Mussel bed recruitment became possible in the afflicted area only where high flow and dissolved oxygen levels obtained just below dams. *Amblema plicata* and *Megalonaias gigantea*, two very valuable commercial species, withstood brines best among mussels. Much of the magnificent mussel fauna (Ortmann, 1926) was badly damaged or destroyed, but remnants remain, as at Munfordville, where Stansbery (1965b) secured over 60 species some years ago.

9. Copper

Imlay (1971) reported that several months' exposure to 25 ppb of copper was lethal to certain unspecified mussels. Wurtz (1962) considered copper second only to zinc (Section II, E, 16) in toxicity among metals.

10. Iron

Hobden (1970a) presented a detailed account of iron concentration by *Elliptio complanata*. There was little metabolic turnover or excretion of the iron. Hobden wondered if excretion was naturally inadequate or uptake, hyperactive. He thought that the high concentrations might be an emergency store, but noted that no known enzyme system needs so much of this trace element.

11. Mercury

Yokley (1973) noted that certain mussels—*Elliptio crassidens* (Lamarck 1819), the Elephant Ear, and *Cyclonaias tuberculata* (Rafinesque 1820), the Purple Warty Back—can accumulate in the soft tissues not quite 3 μg of mercury, apparently without ill effect. Evidently there is no published information on lethal and sublethal effects upon mussels by this much publicized element. Wurtz (1962) considered mercury's (and silver's) toxicity to be exceeded only by zinc's (Section II, E, 16) and copper's (Section II,E,9).

12. Nitrogen

Starrett (1971) furnished an account of the possible effects of nitrogen on mussels in the Illinois River. Nitrogen (as ammonia) exceeded 6.0 ppm throughout the upper river, where (in 1966) no mussels had occurred. Mussels reappeared where ammonia nitrogen was at or below 6.0 ppm. Noting the deleterious effect of this substance upon fish, Starrett assumed that the adverse effect on mussels had been nothing more than interference with their glochidial hosts. Emphasizing that they are not hosts, he listed these four as the only common fishes in the upper Illinois: *Notropis atherinoides* Rafinesque, the Emerald Shiner; *Ictalurus melas* (Rafinesque), the Black Bullhead; *Carassius auratus* (Linnaeus), the Goldfish; and *Cyprinus carpio* Linnaeus, the Carp. Starrett was unaware that at least the latter three *are* glochidial hosts (Table I). Moreover, the glochidia in question represent several species designated by Starrett as still alive in the lower river. Perhaps ammonia nitrogen has a direct effect upon mussels, after all.

Of course, nitrogen (chiefly as nitrate) plays an important role in the eutrophication of waterways (see Section II,F,6).

13. Oxygen

In Imlay's (1971) experiments, adults and juveniles of several unspecified "riffle species" of mussels required 2.5 ppm of dissolved oxygen for survival at laboratory temperatures corresponding to those of the summer, when oxygen levels are usually least. *Amblema plicata* survived for 10 weeks at 0 ppm of oxygen; this is a "pool" species which is tolerant of other adversities, as well, including muddy bottoms and other conditions associated with impoundment (see Section II,F,2) (Baker, 1922; Imlay, 1972a; Isely, 1925; Howard, 1914c; Murray and Leonard, 1962; Parmalee, 1967; Starrett, 1971; Williams, 1969; Wilson and Clark, 1914; Wilson and Danglade, 1914). However, all species examined by Imlay required 6 ppm of dissolved oxygen for normal growth. Ellis (1931b) found that mussels become inactive when oxygen tension is no greater than one-fifth of saturation. It is a curious irony that gravidity hinders respiration, by rendering portions of the gills unsuitable for gas exchange (Howard, 1914c; W. R. Allen, 1921; Matteson, 1955).

In Imlay's (1971) experiment, *Amblema plicata* doubtless survived total lack of oxygen because, as a powerful species whose valves fit closely, it is able to clamp shut very tightly. Shells of members of the genus *Anodonta* usually gape even when closed, so the animals cannot rely upon a supply of water hermetically sealed within the mantle cavity, plus the lowered metabolic rate of dormancy, to see them safely through periods of marginal oxygen. Anodontae must survive these conditions on physiological aptitude alone. This is one of the reasons, of course, why they can so successfully exploit impoundments (see Section II, F, 2).

Lukacsovics (1966) found that *Anodonta cygnea* experienced 95% mortality within three days under hypoxial conditions, but American species—with the possible exception of West Coast forms (see Hannibal, 1912)—appear more durable. *Anodonta implicata* Say 1829, for example, evolves enough metabolic oxygen to ensure survival when the dissolved oxygen of the surrounding water is exhausted (Eddy and Cunningham, 1934). Similarly, Hiestand (1938) demonstrated that *A. imbecilis* can respire normally at down to about 0.73 ml/liter of oxygen and concluded that this species either uses very little or can very delicately adjust metabolic rate. Larger mollusks have the lower metabolic rates (*ibid.*); perhaps this is one reason—in addition to increased buoyancy—why many anodontae are so corpulent. These remarkable faculties are not limited to *Anodonta* among anodontine genera: Cole (1926) found *Anodontoides ferussacianus* active as far as one foot below the surface of organic silt where the dissolved oxygen was only 6% of that in the stream flowing above.

W. R. Allen (1923) made the disconcerting point that at lower levels of oxygen, before the mussel is stimulated to close its valves tightly, it tends to gape in an effort to maximize the passage of water to the gills. At this time, of course, the animal is extremely vulnerable.

Grantham (1969) found no mussels alive where dissolved oxygen occasionally dropped as low as 3 mg/liter. Ellis (1931a) felt that mussels would not survive at oxygen levels below 5 mg/liter.

Badman and Chin (1973) studied the metabolic responses of a Pigtoe, *Pleurobema*, to anaerobic conditions.

14. Phosphorus

Starrett (1971) felt that in the Illinois River there was in 1966 no apparent correlation between mussel abundance and the total phosphate load, including industrial wastes and agricultural run-off. Thus it seems that the phosphorus threat to mussels lies in the contribution which phosphates make to organic enrichment (Section II,F,6).

15. Potassium

Lukacsovics and Salánki (1968) found *Anodonta cygnea* more sensitive to K^+ than to any other cation, and Imlay (1971) added several observations on the relation of potassium to mussels. The natural absence of mussels from certain areas (for example, in the western Mississippi basin) is probably due to the presence of potassium in toxic quantities (see Section II,E,8). The lowest lethal level is between 4 and 7 ppm. Potassium is a common pollutant from industry, particularly paper mills (see Section II,F,5); irrigation return water; and petroleum brine. Further details concerning the important role of potassium in the life of mussels were summarized by Imlay (1973).

16. Zinc

Wurtz (1962) considered zinc the most toxic of heavy metals, followed by copper (Section II,E,9), mercury (Section II,E,11), and silver. Mullican *et al.* (1960) implied that levels of zinc averaging 65 ppm had at least aided in the extermination of mussels in afflicted portions of the Nolichucky River in Tennessee.

F. Adverse Anthropogenic Effects on Mussels

There is no question that the mussel fauna has suffered severely from anthropogenic substances and activities (see Section II,D,20). In a comparison of the evidence from Indian shell middens and from modern collections, Stansbery (1965a) concluded that seven Unioninae had disappeared from the Scioto River of Ohio during historic times. Stansbery (1970b) listed 41 of 103 Ohio River drainage naiades as rare and endangered, including eight species—all in the riffle-adapted lampsiline genus *Dysnomia* (see Stansbery, 1971)—which he thought probably extinct. Dineen (1971) found in the St. Joseph River, Indiana, only about one half of the species recorded by Wenninger (1921) and van der Schalie (1936). Parmalee (1967) felt that at least 12 species of the original mussel fauna were missing in Illinois, but, after a more detailed study of the Illinois River in 1966, Starrett (1971) concluded that more than 25 species had disappeared since 1900, with five more excessively rare and none living in the upper river (see Section II,E,12). Other equally distressing illustrations of degradation appear in the pages to come. The story is an old one (see Ortmann, 1909b, and Kirtland, 1851), recently reviewed by Stansbery (1970b).

1. Channelization

Wilson and Clark (1912a) pointed out that canals need not be inferior mussel habitats if their walls and floors are stable. Kirtland (1851) thought them superior places, in fact, but deplored how they tended to accumulate "filth." Channelization, of course, straightens and deepens winding waterways until they are virtual canals. Wilson and Clark (1912b) remarked the total loss of mussel fauna in those portions of the Yellow and Kankakee Rivers of Indiana which had been straightened and dredged. In perhaps the earliest indictment of channelization, they told how these practices changed the rivers into motionless pools (see "Dams," below) alternating with unbroken stretches where silt and sand constantly scud along the bottom. No mussels live in rolling sand and on shifting bars (Williams, 1969). Williams noted, also, the utter destruction of mussels and their habitats by gravel dredging in the Ohio River, and Baker (1922) made the same point about the fauna in the Big Vermilion River of Illinois.

2. Dams

Riffle conditions are usually best for juvenile mussels, but adults, especially of the heavy-shelled commercial species, often do better in quieter waters (Wilson and Clark, 1912c; Danglade, 1914). In theory, then, one might expect that impounded waters behind dams would provide an especially good mussel habitat. For a variety of reasons, this is not the case, even though certain mussels (especially *Anodonta*, other advanced Anodontinae, and lightly built lampsiline species) are peculiarly suited to impoundment conditions.

Probably the most critical adverse effect of impoundment upon mussels is disruption of their reproductive processes. The deeper, cooler waters of western Lake Erie delay maturation of certain Lampsilinae (Stansbery, 1967). Discussing the commercial species of Kentucky Lake, Williams (1969) noted that many individual fishes bore glochidial infections, but the level of infection on each was very low (see Scruggs, 1960). Williams found no mussels less than four years old, few less than 13, and no recruitment in the preimpoundment beds, by then under 55 feet of water. On the other hand, he discovered that many females were gravid. Therefore, the minimal reproduction was not due to problems with adult fertility or glochidial host availability, but, perhaps, to host sickness, loss of glochidia in the substrate (see Section II,F,3), or attacks upon larvae by microorganisms (see Ellis, 1929; "Bacteria," Section II,D,2; and "Protozoans," Section II,D,4).

Impoundment can profoundly disturb the fish fauna, driving out glochidial hosts in the process. Ellis (1931a) reported breeding grounds for fish and plankton were disappearing around Lake Keokuk, a Mississippi River impoundment, due to weed clearing for pleasure boating, as well as construction of roads and riprap along the margins. Isom and Yokley (1968a) noted that the mussel fauna of Bear Creek, a Tennessee River tributary, now differs from Ortmann's (1925) records; since the Creek was then still pristine, they attributed the faunal changes to failure of glochidial hosts to traverse the lower portion of the Creek, impounded by back-up of the Tennessee. These authors (1968b) interpreted a decline in the mussel fauna of the Duck River—again judged against Ortmann's (1924) records—in the same way. A specific example is Yokley's (1972) forecasting the doom of *Fusconaia ebena* (Lea 1831), the Ebony Shell, in Kentucky Lake because its host, *Notropis ardens* (Cope), the Rosefin Shiner, no longer tolerates this impoundment. Similarly, Coker (1929) reported that *Lampsilis teres* (Rafinesque 1820), aptly y-clept the Yellow Sand Shell, had been found in 1926 abundant on sand bars thrown up between wing dams in Lake Keokuk. Five years later Ellis (1931b) found that these bars had muddied and become largely devoid of *teres*, in spite of an increase in gars (*Lepisosteus*), this species' best hosts (Table I). *Lampsilis teres* was the species most readily killed by silt during Ellis' (1936) experiments.

Thus adult and juvenile mussels, also, can be hurt by impoundment. Baker (1922) noted that parasitism upon *Lampsilis radiata luteola* was heavier above than below the dam across the Big Vermilion River at Homer Park, Illinois, even though this was then a very low dam. Again in contrast to Ortmann's (1918) previous records, Isom (1971) determined a net loss of 60 species from the Tennessee River in the Fort Loudon Reservoir area; he attributed this disaster to periodic oxygen sag (Section II,E,13). Perhaps because of low oxygen and temperatures, but also because mussel species seek their individual optimal water depths (W. R. Allen, 1923), mussels are rarely found at great depths—notwithstanding Reigle's (1967) record of *Anodonta grandis* alive, but stunted, at 102 feet in Lake Michigan. Parasitism, low temperature, oxygen sag, and the water pressure at great depths are undoubtedly inimical to mussels and their glochidial hosts, but siltation (Section II,F,3) is the most important of the many adverse effects associated with damming. Saving this one rather graphic illustration, siltation problems are discussed below: Scruggs (1960) found that *Pleurobema cordatum* (Rafinesque 1820), a Pigtoe, was able to reproduce in Wheeler Reservoir on the Tennessee River and that adults were surviving impoundment. Populations were in decline, however, not because of commercial fishery pressure, but because most juvenile mussels did not survive the silt accumulating on the bed of the old river channel. Wilson and Danglade (1914) recounted allegations that mussels were buried one and three feet deep in the floors of the Shell River and Lake Bemidji, Minnesota, respectively. These reports have been neither confirmed nor complemented since.

Spillways below dams holding back shallow impoundments often support rich naïad populations. Baker (1922) reported 28 species below Homer Park dam, but 10 fewer just above. Doubtless the difference has much to do with the adverse impoundment effects behind the dam, but Baker emphasized the salubrious features below: the Big Vermilion's original stable bottom, with the advantages of highly oxygenated water laden with nutrients from the impoundment. However, the promise of the spillway is usually ultimately unfulfilled. The one below Magnolia Springs in Jenkins County, Georgia, supports large populations of numerous naiad species, including the rare Atlantic Pigtoe, *Fusconaia masoni* (Conrad 1834). The Springs are used as a fish hatchery, and, judged by the quantity of vegetation in the spillway, the water falling over its dam is nutrient-rich. Here is induction of impoundment conditions in one part of a waterway by the nice union of eutrophication and actual ponding in another. The spillway has changed little enough since W. J. Clench, K. J. Boss, and I visited in 1961, but the day will come when it is choked with weeds, and only *Anodonta* and the abundant *Uniomerus tetralasmus* will survive. Another instance of snatching a little bit of victory from the jaws of overwhelming defeat is Williams' (1969) ironic account of how, in the wake of petroleum brine pollution (see Section II,E,8), mussel beds recruited successfully only in the fast waters below dams in the Green River.

Dams oppress mussels in additional ways; some are rather subtle, but as effectively lethal as the foregoing examples. Wing dams, for instance, create novel water currents, which move sands over mussel beds and sweep juveniles onto these shifting bars (see Williams, 1969), where they, too, are smothered (Grier, 1922).

Clench, Boss, Fuller, and H. D. Athearn once searched for mussels on a shoal below a rather high dam spanning the Ogeechee River of Georgia. Only the painfully cold water released at the foot of the dam was flowing through the shallows below. Some effects of low temperatures upon mussels have already been discussed (Section II,E,1). The incomparable Athearn found several living mussels; the rest of us found none.

Long ago Simpson (1899) warned that waterways will occasionally dry out below dams. Grier (1922) noted that the high sand bars thrown up by wing dams can become an excellent habitat for many valuable species, but these will be left to dessicate when the water falls. Most mussels experience difficulty in escaping the consequences of falling waters (Grier, 1922; Parmalee, 1955). Coker (1915) made the point that flooding—which, of course, is *not* always checked by dams—and other fluctuations in water level can interrupt naiad growth; this is particularly disadvantageous to the mussel industry (Section II,F,4). Dessication is the most important consequence of reduced water level; most mussels weather it poorly. For all their resistance to environmental turmoil, the advanced Anodontinae, with their thin and poorly fitted valves, are among these. In general, the Lampsilinae are sensitive and mobile; they move to deeper waters rather readily. *Carunculina parva* (see Section II,E,2) and *Proptera laevissima* (Lea 1830) (Isely, 1925; Riggs and Webb, 1956) are able examples. Like most Ambleminae, *Amblema plicata* is resistant to dessication (Wilson and Clark, 1914; Howard, 1914c), probably because of its thick and closely fitting valves. This is the more remarkable because Prosser and Weinstein (1950) demonstrated that blood amounts to less than 10% of the volume of *A. plicata*. A unionine species, *Uniomerus tetralasmus*, is that mussel most notoriously resistant to dessication, as reported by many authors, signally Simpson (1892), Strecker (1908), and Isely (1914). This species, incidentally, manages to survive in probably more types of demanding habitat than any other Nearctic mussel. Perhaps this has to do with the fact that it possesses the highest blood cell count, including phagocytes, among species studied by Dundee (1953).

Mussels' true mettle has been severely tested by impoundment. For example, Bates (1962) noticed that only eight species had invaded the post-impoundment Kentucky Lake shallows, and only two of them had been listed by Ortmann (1925) or van der Schalie (1939a) as members of the pre-impoundment mussel fauna of the lower Tennessee River: *Quadrula quadrula* (Rafinesque 1820), the Maple Leaf, and *Truncilla donaciformis* (Lea 1928), the Fawn's Foot. The six pioneers were three species of *Anodonta*, *Proptera laevissima*, *Leptodea fragilis* (Rafinesque 1820), and *Carunculina parva*.

3. Silt

Anything which obstructs current—dams, weed beds, and so forth—slows them and creates eddies, which allows settling of suspended silt (Ellis, 1931a). Wing dams (Section II,F,2) are particularly reprehensible because they set up powerful currents which shift bars, erode stream banks, and increase silt (Ellis, 1931b). These phenomena disturb the existing stream bed and render it unstable. Silt limits light penetration (Ellis, 1936), dulling the sensitivity of mussels' phototactic responses and reducing the production of mussel foodstuffs. Suspended silt causes mussels to remain closed almost half again as much as otherwise (*ibid.*); Stansbery (1970a) showed that siltation can retard growth of *Amblema plicata*. On the other hand, mussels can feed in water so choked by silt that they are invisible in this suspension (Churchill and Lewis, 1924). Williams (1969) observed coal dust in mussels' digestive tracts, but observed no ill effects. Nevertheless, Headlee (1906) showed that choking the gills with sediment has a terminal effect, and Ellis (1936) saw dying mussels with quantities of silt in their mantle cavities and gills. In spite of their ability to secrete copious mucous in order to remove silt, most of the mussels in Ellis' (1936) experiments died in one quarter to one inch of silt, and all were doomed when fully covered. Silt carries organic materials from suspension into the stream bed, causing them to remain longer and in higher concentrations (*ibid.*). Thus the adverse effects of enrichment (Section II,F,6) are increased and localized. Lefevre and Curtis

(1912) reviewed divers aspects of the negative effects of silt upon mussels. As a classic example, we have Stansbery's (1964) report that since the impoundment of much of the Mussel Shoals in the Tennessee River there has been a net loss of 30 species from the fauna outlined by Ortmann (1925). Much recent work concerning the effect of silt upon aquatic invertebrate organisms is reviewed by Chutter (1969).

4. The Mussel Industry

Those readers who never saw Burt Lancaster brail for mussels in "The Kentuckian" or never read Mannix' (1965) article in *True* ("The Man's Magazine"), those unfortunates, I say, are probably unaware of the commercial interest taken in mussels during the last century. Contemporary accounts of the early pearl and mussel shell fisheries are available (Kunz, 1894, 1898a,b; Carlander, 1954; Temté, 1968; Coker, 1919; Smith, 1899; Vertrees, 1913), as well as papers which consider ways to conserve the mussel resource for, and in spite of, the industry (Simpson, 1899; Coker, 1914; Smith, 1919; van der Schalie, 1938a; Krumholz *et al.*, 1970; Jorgensen and Sharp, 1971). Aside from the outright destruction of animals killed for their shells and pearls, harvesting has these deleterious effects: reduction in breeding stock to the point where reproduction does not offset mortality; destruction and disruption of the stream bed; abortion of gravid females when disturbed (J. F. Boepple *in* Simpson, 1899); waste deaths of juveniles below useful and legal limits; and deaths of adults which cannot rebury themselves after being needlessly uprooted (see Imlay, 1972b). Fishing pressure and the use of plastic in button manufacture nearly terminated the shell industry after World War I, but the use of spheres of American mussel nacre as cultured pearl nuclei by the Japanese revitalized the harvest during the 1960's. The use of scuba and other diving gear enables shellers to collect entire beds. However localized and imperfectly lethal, this may be the greatest threat to commercially valuable mussels today.

5. Wood Products Wastes

Simpson (1899) mentioned the adverse effect upon mussels of sawdust as a false streambed. Similarly, Wilson and Danglade (1914) noted bark dislodged from logs driven downstream coating the bottom of the Prairie River of Minnesota. Ortmann (1918) cited the damage done mussels by wastes from wood pulp and paper mills (see Section II,E,15) in the upper Tennessee River drainage. Mackie and Qadri (1973) found mollusks, including unionid mussels, limited by wastes from the same sorts of mills in Ottawa River, Canada. Finally, all flora and fauna have been completely destroyed for 15 miles downstream to the Gulf from a paper mill on the Fenholloway River of Florida (Heard, 1970b).

6. *Organic Enrichment*

A certain amount of eutrophication may be of service to mussels (van der Schalie, 1938b), presumably by increasing the amount of available food (Section II,C). Churchill and Lewis (1924) recorded an instance of unusual concentration of mussels below an area where rich vegetation disintegrated, and Coker *et al.* (1921) found healthy, even exaggerated, growth of mussels in water that was influenced by sewage. Nevertheless, too great organic enrichment, from whatever source(s), induces conditions akin to those of ponding and impoundment (see Section II,F,2). There is proliferation of submerged vegetation, which contributes to a softer bottom with the products of its own decay mixed with silt which it strains from the water; the current slackens; and (Grantham, 1969) there develop increased bacterial oxygen demand, increased carbon dioxide, and lowered pH. These conditions are inimical to the great majority of mussels. In addition to phosphates, nitrates, and other ordinary ingredients of eutrophication, many sources of organic pollution include toxic substances, as well. Thus, for example, Mackie and Qadri (1973) found that mussels (and other mollusks) were severely limited by slaughter house wastes in the Ottawa River, Canada. Similarly, sewage has been responsible for total destruction of mussels in portions of the Kankakee River of Indiana (Wilson and Clark, 1912b) and the Salt Fork of the Vermilion River of Illinois (Baker, 1922).

7. *Acid Mine Waste*

Primarily in the form of mine drainage, acidic waters can have a direct, extraordinarily adverse effect upon mussels. A 1% acetate solution causes the mantle margin to contract, and the effect may be visible after several hours (W. R. Allen, 1923). Only the *intensity* of the response declines with lower concentrations (*ibid.*). Williams (1969) found only "baldies" (mussels with little or no periostracum) in Kentucky Lake, and Simpson (1899) stated that acid can eat holes through the shell to the animal. Yokley (1973) noted widespread mussel mortality due to acid in Kentucky Lake. I have seen streams where practically all macroinvertebrate life had been exterminated by acid mine wastes in the Slippery Rock Creek drainage of Western Pennsylvania, and as many as 60 years ago such effects were being recorded for portions of the Cumberland River in Kentucky (Wilson and Clark, 1914; Stansbery, 1969). Some mussels remain comparatively free of damage by acids at low concentrations; for example, Wilson and Clark (1914) noticed that *Amblema plicata* was free of corrosion in the Cumberland.

8. *Pesticides*

I am unaware of studies on mortality caused by pesticides in mussels. Bedford *et al.* (1971) found that *Lampsilis radiata luteola* and *Anodonta grandis* are excellent monitors of chlorinated hydrocarbons, including DDT, aldrin, and methoxychlor; they implied that at least one of these substances had been responsible for observed deaths. Miller *et al.* (1966) showed that *Elliptio complanata* concentrated diazinon and parathion at levels greatly in excess of their concentrations in the surrounding water and metabolized them very slowly. Fikes and Tubb (1971) found that *Amblema plicata* concentrated 20 ppt ("t" for "trillion"!) dieldrin at measurable levels about 2500 times background levels. Starrett (1971) discovered unexpectedly small concentrations of organochlorine pesticides (DDT, DDE, heptachlor epoxide, and dieldrin), probably because of their adsorption in stream-bed mud (see Ellis, 1936) and/or because of a shift in agricultural usage toward organophosphate varieties. Uptake of dieldrin and DDT by *E. dilatata* and *Anodonta grandis* under experimental conditions was studied by Zabik and Bedford (1972).

9. Radionuclides

As in the case of pesticides (Section II,F,8), I am unaware of studies about the mortality and sublethal effects of radioactive materials upon mussels, but their ability to indicate the presence of substances occurring in extremely low concentrations in nature by accumulating them in readily detectable quantities has received a great deal of inquiry. Lee and Wilson (1969) found that the contrast between Sr/Ca ratios in ancient and modern mussel shells is indicative of different paleohydrologic conditions. On the other hand, Garder and Skulberg (1965) found that uptake in the Eurasian *Anodonta piscinalis* Nilsson 1822 varies with physiological and seasonal conditions. These authors and Nelson (1962, 1964) showed that isotopes of phosphorus, strontium, and cesium have especially great affinity for mussel bodies and/or shells. Nelson (*ibid.*) found that Tennessee River drainage mussels can concentrate Sr^{90} at levels from 2500 to 9000 times normal. Brungs (1967) and Harvey (1969) found that mussels concentrate Zn^{65} (Section II,E,16) more heavily than any other isotopes studied by them. Finally, Nelson (1967) studied mussel concentration of *non*radioactive isotopes of several elements. These are only a few examples of the rich and useful literature that is abuilding in this area, including studies by Harrison (1969) and Short *et al.* (1969).

10. Miscellaneous Pollutants

Gas works wastes, especially tar and oily scum, were reported by Wilson and Clark (1912b) and by Baker (1928) as very damaging to mussels. Wilson and Clark (1912a) cited dye from knitting mills as toxic to mussels. Williams (1969) could find no living mussels for six miles below an industrial area bordering the lower Tennessee River in Kentucky, but he did not discriminate among the pollutants. Coker *et al.* (1921) reviewed the adverse effects of deforestation, including irregular stream flow, high water temperatures, and lowered dissolved oxygen; Grier (1922) recommended reforestation to strengthen stream banks against sand flow and silt runoff. The activities of private conchologists, research malacologists, commercial collectors for biological supply houses, and ecologists conducting biotic surveys all pose threats similar to those inherent in the pearl and shell industry (Section II, F,4). Stansbery (1970b) and van der Schalie (1938a) have reviewed numerous polluting agents, including some rather minor ones.

G. Mussels as Indicator Organisms

The presence of unusually large populations of mussels can be indicative of pollution only in the case of minor degrees of organic enrichment (Section, II,F,6). The absence of mussels can logically be an indication of environmental disruption only when and where their former presence can be demonstrated. It is very rare that we can quantitatively and/or qualitatively correlate the composition and size of the mussel fauna with a specific disruption, be it chemical or physical (see Ingram, 1956). Indubitably, we have much to learn. On the other hand, there can be no doubt that mussels have extraordinary value as qualitative indicators of pesticides, radionuclides, and, presumably, other trace substances in nature. Finally, organisms with ecologically narrow habitats are particularly sensitive to change. For example, several Nearctic Margaritiferidae prefer softer, cleaner waters and peculiar substrates (Bjork, 1962; Hendelberg, 1960; McMillan, 1966; Roscoe and Redelings, 1964; Stansbery, 1966a; Stober, 1972). These characteristics may be correlated with glochidial host predilections (Table I). As indicators, the Margaritiferidae must be thought both less useful and more sensitive than the Unionidae.

III. Sphaeriidae

Until about two decades ago the great majority of literature on Nearctic Sphaeriidae was taxonomic, and most biological data involved such mainstays of "good, old-fashioned natural history" as notes on ranges and habitats. While these observations were often extremely informative, it was only recently that the reasons behind them were sought with any frequency. There is still an insufficiency of this more sophisticated information (but see Ingram, 1956), which accounts for the brevity and other shortcomings of this section almost as much as do the predilections and other weaknesses of the writer!

Thanks to the monographic treatments by Herrington (1962) and Burch (1972), sphaeriid taxonomy is not perfect, perhaps, but quite stable, particularly when compared to that of most Unionacea. Older nomenclature has been brought up to the present in the light of these works.

Most of the biological literature prior to Herrington's revision is found in his bibliography and is not repeated in mine unless cited in the paragraphs below. Also, some relevant papers (Crowther, 1894; Monk, 1928; Ingram, 1941; Ingram *et al.*, 1953) were not mentioned by Herrington. Some recent European literature (Dance, 1970; Meier-Brook, 1970; Wolff, 1970; Kuiper and Wolff, 1970) contains information on sphaeriids which are Nearctic, as well as Eurasian.

Heard (1962, 1964, 1965b) has provided a great deal of information on aspects of "normal" reproduction and life cycles of many Nearctic sphaeriids, and Herrington (1962), as already noted, has assembled a host of data on "normal" habitats. By and large, the Sphaeriidae are a eurytopic lot, but some have exploited unusual habitats and may be more or less restricted to them. For example, Thomas (1959, 1963, 1965) has found that *Sphaerium partumeium* (Say 1822) exhibits some unusual adaptations (e.g., precocious production of young) to the evaporation of ephemeral pools, which are its usual habitat. Neither the habitat nor, to the best of my knowledge, the adaptation is encountered widely in this genus.

The case of *Sphaerium partumeium* is illustrative of the great success achieved by the family. Like many Unionacea (Section II,B), sphaeriids are hermaphroditic, even self-fertile (Thomas, 1959), and only a single individual may be required to extend a species' ecologic and geographic ranges. Differing from the Unionacea (see van der Schalie, 1945), Sphaeriidae are subject to mechanical distribution as a key element in their vagility. Resistance to dessication (see Ingram, 1941) and hermaphroditism are thus especially advantageous.

There are some other odd sphaeriid life styles. *Pisidium conventus* Clessin 1877 is most abundant in deep, cool, highly oxygenated lake waters (Rawson, 1953; Heard, 1963; Hamilton, 1971). This habitat is so narrow that disappearance of *P. conventus*—as of certain Margaritiferidae (Section II,G)—would surely indicate an important change in water quality. *P. idahoense* Roper 1890, on the other hand, shows the same predilection for great depths, but it is able to survive lengthy periods with little or no oxygen (Juday, 1908; Cole, 1921). Certainly this species has the lesser value as an indicator organism. *Pisidium ultramontanum* Prime 1865, also, is an indicator, but of another sort: the pattern of its geographical distribution casts light on ancient drainage patterns (Taylor, 1960).

The Sphaeriidae include at least one clear example of the sort of indicator that responds positively to a measure of pollution. *Sphaerium transversum* (Say 1829) was believed by earlier authorities (e.g., R. E. Richardson, 1928) to increase in numbers under the influence of sewage and not to react adversely until such time as the macroinvertebrate fauna had been greatly simplified. My experience of this creature in numerous eutrophic waterways corroborates this point of view precisely. *Sphaerium transversum* belongs to the group *Musculium*, variously regarded as congeneric or subgeneric to *Sphaerium. Musculium* has had a reputation as a tolerator of impoundment (Ellis, 1931a). Zetek (1918) reported an oxbow "alive" with *transversum*, and I have found them by the tens of thousands in luxuriant weed beds. In the only paper on the reaction of a single Nearctic sphaeriid to so much as a single form of pollution, Ingram *et al.* (1953) noted a similar, though not so well developed, tolerance of enrichment conditions by *S. striatinum* (Lamarck 1818). Silt was the substrate of the only *striatinum* I have ever collected. Certainly such sphaeria as these are the sphaeriid analog of the unionid genus *Anodonta*.

Grantham (1969) listed a number of authorities who have thought of the Sphaeriidae as tolerant of polluted, nearly septic conditions. There is no doubt that certain Sphaeriidae have real value as indicators of environmental disturbance, especially conditions of impoundment and eutrophication. Our knowledge is insufficiently detailed to define sphaeriid "pollution indication" in chemical terms, so, as with the Unionacea, I have refrained from tabulating probably meaningless and possibly misleading water chemistry data. Some can be found in papers on American sphaeriids by Filice (1958, 1959), Gillespie (1969), Thut (1969), Tuthill and Johnson (1969), Tuthill and Laird (1963–1964), and Zumoff (1973). In evaluating the relative worth of sphaeriid groups as indicators in terms of their responses to ecological pressure, it may be instructive to recall that Morrison (1932) stated that the lowest pH levels tolerated by *Sphaerium*, "*Musculium*," and *Pisidium* in northeastern Wisconsin lakes were 6.8, 5.9, and 5.1, respectively. As illustrated above, the group *Pisidium* shows perhaps the greatest tolerance among sphaeriids in other ways, as well. I would add that in my experience *P. compressum* Prime 1851 and *P. casertanum* (Poli 1791) respond almost as favorably to organic enrichment as does *Sphaerium transversum*. Significantly, Herrington (1962) regarded these two as the most common pisidia in North America.

As with unionaceans, once again, the parasites of sphaeriids have been generally ignored by malacologists. Once again, it is tempting to suppose that a correlation can eventually be demonstrated between degree of parasitism and amount of environmental stress. Wolcott (1899) recorded *Unionicola crassipes* (Müller), perhaps a questionable determination (see Section II,D,15), as a symbiont of *Sphaerium simile* (Say 1816). Several authors have recorded parasitism of sphaeriids by "distomids" (see Section II,D,6) and other digenetic trematodes (Goodchild, 1939a,b; Gentner and Hopkins, 1966; Groves, 1945; H. E. Henderson, 1938; Parker, 1932; Steelman, 1939; Vickers, 1940).

Eupera cubensis (Prime 1865) is my final sphaeriid topic. Heard (1965a) has an elegant taxonomy and partial natural history of this species, which occurs primarily in lower portions of river systems of the Atlantic drainage of the Carolinas to the Caribbean drainage of South America. It is especially likely to be encountered in submerged rootstocks and other tangles, which offer refuge and an avenue to the surface: *Eupera* is often found out of water crawling on roots (Walker, 1915; Athearn *in* Clench and Turner, 1956; Hubricht, 1966). Heard records it attached, often with a byssus, to the "rims of rusty (but not fresh) beer cans." This species is regularly associated with *Sphaerium transversum*, *Pisidium casertanum*, and *P. compressum* in the Savannah River of Georgia and South Carolina. In enriched areas *E. cubensis* is the one of the four which is lacking if any are lacking at all.

Acknowledgments

One does not put together an eclectic piece like this without a little help! My thanks go to Ruth Brown, Librarian at the Academy of Natural Sciences of Philadelphia; to two of her associates, Martha Pilling and Tamsen, my sister; to Arlene Mogilefsky, Librarian in the Academy's Department of Limnology; and to my colleague Daniel Bereza. Nor has the forbearance of Micki, my wife, gone unnoticed of late months.

References

Allen, E. (1924). The existence of a short reproductive cycle in *Anodonta imbecillis*. *Biol. Bull.* **46**, 88–94.

Allen, W. R. (1914). The food and feeding habits of freshwater mussels. *Biol. Bull.* **27**, 127–147.

Allen, W. R. (1921). Studies of the biology of freshwater mussels. Experimental studies of the food relations of certain Unionidae. *Biol. Bull.* **40**, 210—241.

Allen, W. R. (1923). Studies of the biology of freshwater mussels. II. The nature and degree of response to certain physical and chemical stimuli. *Ohio J. Sci.* **23**, 57–82.

Antipa, G. A., and Small, E. B. (1971). The occurrence of thigmotrichous ciliated Protozoa inhabiting the mantle cavity of unionid molluscs of Illinois. *Trans. Amer. Micros. Soc.* **90**, 463–472.

Arey, L. B. (1921). An experimental study on glochidia and the factors underlying encystment. *J. Exp. Zool.* **33**, 463–499.

Arey, L. B. (1923). Observations on an acquired immunity to a metazoan parasite. *J. Exp. Zool.* **38**, 377–381.

Arey, L. B. (1924). Glochidial cuticulae, teeth, and the mechanics of attachment. *J. Morphol. Physiol.* **39**, 323–335.

Arey, L. B. (1932a). The formation and structure of the glochidial cyst. *Biol. Bull.* **62**, 212–221.

Arey, L. B. (1932b). The nutrition of glochidia during metamorphosis. *J. Morphol.* **53**, 201–221.

Arey, L. B. (1932c). A microscopical study of glochidial immunity. *J. Morphol.* **53**, 367–379.

Athearn, H. D. (1964). Three new unionids from Alabama and Florida and a note on *Lampsilis jonesi*. *Nautilus* **77**, 134–139.

Athearn, H. D. (1967). Changes and reductions in our freshwater molluscan populations. *Annu. Rep. 1967 Amer. Malacolog. Un.* 44–45.

Athearn, H. D., and Clarke, A. H., Jr. (1962). The freshwater mussels of Nova Scotia. *Nat. Mus. Can. Bull. No. 183, Contrib. Zool., 1960–1961*, 11–41.

Badman, D. G., and Chin, S. L. (1973). Metabolic responses of the fresh-water bivalve, *Pleurobema coccineum* (Conrad), to anaerobic conditions. *Comp. Biochem. Physiol.* **44B**, 27–32.

Bailey, R. M., Fitch, J. E., Herald, E. S., Lachner, E. A., Lindsey, C. C., Robins, C. R., and Scott, W. B. (1970). A list of common and scientific names of fishes from the United States and Canada. *Spec. Publ. Amer. Fish. Soc.* No. 6, 1–150.

Baker, F. C. (1898). The Mollusca of the Chicago area. The Pelecypoda. *Bull. Natur. Hist. Surv. Chicago Acad. Sci.* No. 3, 1–130.

Baker, F. C. (1916). The relation of mollusks to fish in Oneida Lake. *Tech. Publ. N. Y. State College Forestry* **4**, 1–366.

Baker, F. C. (1922). The molluscan fauna of the Big Vermilion River, Illinois. *Ill. Biol. Monogr.* **7**, 1–126.

Baker, F. C. (1928). The fresh water Mollusca of Wisconsin. Part II. Pelecypoda. *Bull. Wis. Geol. Natur. Hist. Surv.* No. 70, 1–495.

Bănărescu, P. (1971). Competition and its bearing on the fresh-water faunas. *Rev. Roum. Biol. Ser. Zool.* **16**, 153–164.

Bates, J. M. (1962). The impact of impoundment on the mussel fauna of Kentucky Lake Reservoir, Tennessee River. *Amer. Midl. Natur.* **68**, 232–236.

Bedford, J. W., Roelefs, E. W., and Zabik, M. J. (1968). The freshwater mussel as a biological monitor of pesticide concentrations in a lotic environment. *Limnol. Oceanogr.* **13**, 118–126.

Bequaert, J. C., and Miller, W. B. (1973). "The Mollusks of the Arid Southwest with an Arizona Check List," pp. 1–271. Univ. of Arizona Press, Tucson.

Bjork, S. (1962). Investigations on *Margaritifera margaritifera* and *Unio crassus*. *Acta Limnol.* **4**, 1–109.

Blystad, C. N. (1923). Significance of larval mantle of fresh-water mussels during parasitism, with notes on a new mantle condition exhibited by *Lampsilis luteola*. *Bull. U. S. Bur. Fish.* **39**, 203–219. Separately issued as Bur. Fish. Document No. 950.

Bovjerg, R. V. (1957). Feeding related to mussel activity. *Proc. Iowa Acad. Sci.* **64**, 650–653.

Brinkhurst, R. O., and Jamieson, B. G. M. (1971). "Aquatic Oligochaeta of the World," pp. 1–860. Univ. of Toronto Press, Toronto.

Brungs, W. A., Jr. (1967). Distribution of cobalt 60, zinc 65, strontium 85, and cesium 137 in a freshwater pond. U. S. Publ. Health Serv. Publ. No. 999-RH-24, pp. 1–52.

Burch, J. B. (1972). Freshwater sphaeriacean clams (Mollusca: Pelecypoda) of North America. U. S. Environmental Protection Agency, Biota of Freshwater Ecosystems, Identification Manual No. 3, pp. 1–31.

Burch, J. B. (1973). Freshwater unionacean clams (Mollusca: Pelecypoda) of North America. U. S. Environmental Protection Agency, Biota of Freshwater Ecosystems, Identification Manual No. 11, pp. 1–176.

Call, R. E. (1895). A study of the Unionidae of Arkansas, with incidental reference to their distribution in the Mississippi valley. *Trans. Acad. Sci. St. Louis* **7**, 1–65.

Call, R. E. (1900). A descriptive illustrated catalogue of the Mollusca of Indiana, *Annu. Rep. Indiana Dep. Geol. Natur. Resources, 24th* 335–535.

Carlander, H. B. (1954). "History of fish and fishing in the upper Mississippi River," pp. 1–96. Upper Mississippi River Conservation Committee, Davenport, Iowa.

Chamberlain, T. K. (1934). The glochidial conglutinates of the Arkansas Fanshell, *Cyprogenia aberti* (Conrad). *Biol. Bull.* **66**, 55–61.

Churchill, E. P., Jr. (1915). The absorption of fat by freshwater mussels. *Biol. Bull.* **29**, 68–87.

Churchill, E. P., Jr. (1916). The absorption of nutriment from solution by freshwater mussels. *J. Exp. Zool.* **21**, 403–430.

Churchill, E. P., Jr., and Lewis, S. I. (1924). Food and feeding in fresh-water mussels. *Bull. U.S. Bur. Fish.* **39**, 439–471. Separately issued as Bur. Fish. Document No. 963.

Chutter, F. M. (1969). The effects of silt and sand on the invertebrate fauna of streams and rivers. *Hydrobiologia* **34**, 57–76.

Clark, H. W., and Stein, S. (1921). Glochidia in surface towings. *Nautilus* **35**, 16–20.

Clarke, A. H., Jr. (Ed.) (1970). Rare and endangered mollusks of North America. *Malacologia* **10**, 1–56.

Clarke, A. H., Jr. (1973). The freshwater molluscs of the Canadian interior basin. *Malacologia* **13**, 1–509.

Clarke, A. H., Jr., and Berg, C. O. (1959). The freshwater mussels of central New York with an illustrated key to the species of northeastern North America. *Cornell Univer. Agr. Exp. Sta. Mem.* **367**, 1–79.

Clarke, A. H., Jr., and Rick, A. M. (1963). Supplementary records of Unionacea from Nova Scotia with a discussion of *Anodonta fragilis* Lamarck. *Nat. Mus. Can. Bull. No. 199, Contributions Zool., 1963*, 15–27.

Clench, W. J. (1959). Mollusca. *In* "Fresh-water Biology" (W. T. Edmondson, Ed.), pp. 1117–1160. Wiley, New York.

Clench, W. J., and Turner, R. D. (1956). Freshwater mollusks of Alabama, Georgia, and Florida from the Escambia to the Suwanee River. *Bull. Florida State Mus., Biol. Sci.* **1**, 97–239.

Coçkerell, T. D. A. (1902). *Unio popeii*, Lea, in New Mexico. *Nautilus* **16**, 69–70.

Coil, W. H. (1954). Two new rhopalocercariae (Gorgoderinae) parasitic in Lake Erie mussels. *Proc. Helminthol. Soc. Washington* **21**, 17–29.

Coker, R. E. (1912). Mussel resources of the Holston and Clinch Rivers of eastern Tennessee. *Rep. U. S. Comm. Fish. for 1911 and Spec. Papers*, pp. 1–13. Separately issued as Bur. Fish. Document No. 765.

Coker, R. E. (1914). The protection of fresh-water mussels. *Rep. U. S. Comm. Fish. for 1912 and Spec. Papers*, pp. 1–23. Separately issued as Bur. Fish. Document No. 793.

Coker, R. E. (1915). Mussel resources of the Tensas River of Louisiana. *U. S. Bur. Fish. Econ. Circ.* No. 14, 1–7.

Coker, R. E. (1916). The Fairport fisheries biological station: its equipment, organization, and functions. *Bull. U. S. Bur. Fish.* **34**, 383–405. Separately issued as Bur. Fish. Document No. 829.

Coker, R. E. (1919). Fresh-water mussels and mussel industries of the United States. *Bull. U. S. Bur. Fish.* **36**, 13–89. Separately issued as Bur. Fish. Document No. 865.

Coker, R. E. (1929). Keokuk Dam and the fisheries of the upper Mississippi River. *Bull. U. S. Bur. Fish.* **45**, 87–139. Separately issued as Bur. of Fish Document No. 1063.

Coker, R. E., Shira, A. F., Clark, H. W., and Howard, A. D. (1921). Natural history and propagation of fresh-water mussels. *Bull. U. S. Bur. Fish.* **37**, 77–181. Separately issued as Bur. Fish. Document No. 893.

Coker, R. E., and Surber, T. (1911). A note on the metamorphosis of the mussel *Lampsilis laevissimus*. *Biol. Bull.* **20**, 179–182.

Cole, A. E. (1921). Oxygen supply of certain animals living in water containing no dissolved oxygen. *J. Exp. Zool.* **33**, 293–320.

Cole, A. E. (1926). Physiological studies on fresh-water clams. Carbon-dioxide production in low oxygen tensions. *J. Exp. Zool.* **45**, 349–359.

Conner, C. H. (1905). Glochidia of *Unio* on fishes. *Nautilus* **18**, 142–143.

Cope, O. B. (1959). New parasite records from stickleback and salmon in an Alaska stream. *Trans. Amer. Microsc. Soc.* **78**, 157–162.

Corwin, R. S. (1920). Raising freshwater mussels in enclosures. *Trans. Amer. Fish. Soc.* **49**, 81–84.

Corwin, R. S. (1921). Further notes on raising freshwater mussels in enclosures. *Trans. Amer. Fish. Soc.* **50**, 307–311.

Crowther, H. (1894). Biology of *Sphaerium corneum*. *J. Conchol.* **7**, 417–421.

Cvancara, A. M., and Harrison, S. S. (1965). Distribution and ecology of mussels in the Turtle River, North Dakota, *Proc. N. D. Acad. Sci.* **19**, 128–146.

Dana, J. D., and Whelpley, J. (1836). On two American species of the genus *Hydrachna*. *Amer. J. Sci. Arts* **30**, 354–359.

Dance, S. P. (1970). *Pisidium lilljeborgii* Clessin, in the River Teifi, West Wales. *J. Conchol.* **27**, 177–181.

Danglade, E. (1914). The mussel resources of the Illinois River. *Rep. U. S. Comm. Fish for 1913*, Appendix VI, pp. 1–48. Separately issued in Bur. Fish. Document No. 804.

Davenport, D., and Warmuth, M. (1965). Notes on the relationship between the freshwater mussel *Anodonta implicata* Say and the alewife *Pomolobus pseudoharengus* (Wilson). *Limnol. Oceanogr.* **10**, R74–R78.

Davids, C. (1973). The relations between mites of the genus *Unionicola* and the mussels *Anodonta* and *Unio*. *Hydrobiologia* **41**, 37–44.

Davis, H. S. (1934). Care and diseases of trout. *U. S. Bur. Fish. Investigational Rep.* **1**, No. 22, 1–69.

Davis, H. S. (1946). Care and diseases of trout. *U. S. Fish Wildl. Serv. Res. Rep.* No. 12, 1–98.

Dawley, C. (1947). Distribution of aquatic mollusks in Minnesota. *Amer. Midl. Natur.* **38**, 671–697.

d'Eliscu, P. N. (1972). Observation of the glochidium, metamorphosis, and juvenile of *Anodonta californiensis* Lea, 1857. *Veliger* **15**, 57–58.

de Waele, A. (1930). Le sang d'*Anodonta cygnea* et la formation de la coquille. *Acad. Roy. Belg. Classe Sci. Mem. Coll. in-4°, Deuxieme Ser.* **10**, 1–52.

Dineen, C. F. (1971). Changes in the molluscan fauna of the Saint Joseph River, Indiana, between 1959 and 1970. *Proc. Indiana Acad. Sci.* **80**, 189–195.

Dundee, D. S. (1953). Formed elements of the blood of certain fresh-water mussels. *Trans. Amer. Microsc. Soc.* **72**, 254–270.

Eddy, N. W., and Cunningham, R. B. (1934). The oxygen consumption of the fresh-water mussel, *Anodonta implicata*. *Proc. Penn. Acad. Sci.* **8**, 140–143.

Edgar, A. L. (1965). Observations on the sperm of the pelecypod *Anodontoides ferussacianus* (Lea). *Trans. Amer. Microsc. Soc.* **84**, 228–230.

Ellis, M. M. (1929). The artificial propagation of freshwater mussels. *Trans. Amer. Fish. Soc.* **59**, 217–223.

Ellis, M. M. (1931a). A survey of conditions affecting fisheries in the upper Missouri River. *U. S. Bur. Fish. Circ.* No. 5, 1–18.

Ellis, M. M. (1931b). Some factors affecting the replacement of the commercial fresh-water mussels. *U. S. Bur. Fish. Fish. Circ.* No. 7, 1–10.

Ellis, M. M. (1936). Erosion silt as a factor in aquatic environments. *Ecology* **17**, 29–42.

Ellis, M. M. (1937). Detection and measurement of stream pollution. *Bull. U. S. Bur. Fish.* **48**, 365–437. Separately issued as Bur. Fish. Bulletin No. 22.

Ellis, M. M., and Ellis, M. D. (1926). Growth and transformation of parasitic glochidia in physiological nutrient solutions. *Science* **64**, 579–580.

Ellis, M. M., and Keim, M. (1918). Notes on the glochidia of *Strophitus edentulus pavonius* (Lea) from Colorado. *Nautilus* **32**, 17–18.

Ellis, M. M., Merrick, A. D., and Ellis, M. D. (1931). The blood of North American fresh-water mussels under normal and adverse conditions. *Bull. U. S. Bur. Fish.* **46**, 509–542. Separately issued as Bur. Fish. Document No. 1097.

Evermann, B. W., and Clark, H. W. (1918). The Unionidae of Lake Maxinkuckee. *Proc. Indiana Acad. Sci. 1917* 251–285.

Evermann, B. W., and Clark, H. W. (1920). "Lake Maxinkuckee. A physical and biological survey, Volume II. Biology," pp. 1–512. Indiana Department of Conservation, Indianapolis.

Fikes, M. H., and Tubb, R. A. (1971). *Amblema plicata* as a pesticide monitor. *In* Jorgensen and Sharp (1971, pp. 34–37).

Filice, F. P. (1958). Invertebrates from the estuarine portion of San Francisco Bay and some factors influencing their distributions. *Wasmann J. Biol.* **16**, 159–211.

Filice, F. P. (1959). The effect of wastes on the distribution of bottom invertebrates in the San Francisco Bay estuary. *Wasmann J. Biol.* **17**, 1–17.

Fischer, P., and Crosse, H. (1880–1902). Études sur les mollusques terrestres et fluviatiles du Mexique et du Guatemala. *In* "Mission scientifique au Mexique et dans l'Amerique Centrale, recherches zoologiques pour servir à l'histoire de la faune de l'Amerique Centrale et du Mexique," (M. Milne-Edwards, Ed.), Part 7 (Mollusques), Volume 2, pp. 1–731. The section on Unionidae (pp. 505–622) was published in 1894.

Fischthal, J. H. (1954). *Cercaria tiogae* Fischthal, 1953, a rhopalocercous form from the clam, *Alasmidonta varicosa* (Lamarck). *Trans. Amer. Microsc. Soc.* **73**, 210–215.

Forbes, S. A., and Richardson, R. E. (1908). "The fishes of Illinois," pp. 1–357. Natural History Survey of Illinois, State Laboratory of Natural History, Urbana.

Frierson, L. S. (1897). Conchological notes from Louisiana. *Nautilus* **11**, 3–4.

Frierson, L. S. (1902). Collecting Unionidae in Texas and Louisiana. *Nautilus* **16**, 37–40.

Frierson, L. S. (1903). Observations on the byssus of Unionidae. *Nautilus* **17**, 76–77.

Frierson, L. S. (1911). A comparison of the Unionidae of the Pearl and Sabine Rivers. *Nautilus* **24**, 134–136.

Frierson, L. S. (1927). "A Classified and Annotated Check List of the North American Naiades," pp. 1–111. Baylor Univ. Press, Waco, Texas.

Fuller, S. L. H. (1971). A brief field guide to the fresh-water mussels (Mollusca: Bivalvia: Unionacea) of the Savannah river system. *Ass. Southeast. Biol. Bull.* **18**, 137–146.

Fuller, S. L. H. (1972). *Elliptio marsupiobesa*, a new fresh-water mussel (Mollusca: Bivalvia: Unionidae) from the Cape Fear River, North Carolina. *Proc. Acad. Natur. Sci. Philadelphia* **124**, 1–10.

Fuller, S. L. H. (1974). *Fusconaia masoni* (Conrad 1834) (Bivalvia: Unionacea) in the Atlantic drainage of the southeastern United States. *Malacol. Rev.* **6**, 105–117.

Fuller, S. L. H., and Bereza, D. J. (1973). Recent additions to the naiad fauna of the eastern Gulf drainage (Bivalvia: Unionoida: Unionidae). *Ass. Southeast. Biol. Bull.* **20**, 53–54.

Fuller, S. L. H., and Powell, C. E., Jr. (1973). Range extensions of *Corbicula manilensis* (Philippi) in the Atlantic drainage of the United States. *Nautilus* **87**, 59.

Garder, K., and Skulberg, O. (1965). Radionuclide accumulation by *Anodonta piscinalis* Nilsson (Lamellibranchiata) in a continuous flow system. *Hydrobiologica* **26**, 151–169.

Gentner, H. W., and Hopkins, S. H. (1966). Changes in the trematode fauna of clams in the Little Brazos River, Texas. *J. Parasitol.* **52**, 458–461.

Gillespie, D. M. (1969). Population studies of four species of molluscs in the Madison River, Yellowstone National Park. *Limnol. Oceanogr.* **14**, 101–114.

Goodchild, C. G. (1939a). *Cercaria donecerca* n. sp. (gorgoderid cercaria) from *Musculium partumeium* (Say), 1822. *J. Parasitol.* **25**, 133–135.

Goodchild, C. G. (1939b). *Cercaria conica* n. sp. from the clam *Pisidium abditum* Haldeman. *Trans. Amer. Microsc. Soc.* **58**, 179–184.

Goodrich, C., and van der Schalie, H. (1944). A revision of the Mollusca of Indiana. *Amer. Midl. Natur.* **32**, 257–326.

Grantham, B. J. (1969). The fresh-water pelecypod fauna of Mississippi. Doctoral dissertation, Univ. of Southern Mississippi, Hattiesburg, pp. 1–243.

Grier, N. M. (1922). Final report on the study and appraisal of mussel resources in selected areas of the upper Mississippi River. *Amer. Midl. Natur.* **8**, 1–33.

Grier, N. M. (1926). Notes on the naiades of the upper Mississippi drainage: III. On the relation of temperature to the rhythmical contractions of the "mantle flaps" in *Lampsilis ventricosa* (Barnes). *Nautilus* **39**, 111–114.

Groves, R. E. (1945). An ecological study of *Phyllodistomium solidum* Rankin, 1937 (Trematoda: Gorgoderidae). *Trans. Amer. Microsc. Soc.* **64**, 112–132.

Haas, F., and contributors (1969). Superfamily Unionacea. *Treatise Invertebrate Paleontol. Part N*, Vol. 1 (of 3), Mollusca 6, Bivalvia: *N*411–*N*470.

Hamilton, A. L. (1971). Zoobenthos of fifteen lakes in the Experimental Lakes Area, northwestern Ontario. *J. Fish. Res. B. Can.* **28**, 257–263.

Hannibal, H. (1912). A synopsis of the Recent and Tertiary land and freshwater Mollusca of the Californian Province. *Proc. Malacolog. Soc. London* **10**, 112–165.

Harman, W. N. (1969). The effect of changing pH on the Unionidae. *Nautilus* **83**, 69–70.

Harman, W. N. (1970a). New distribution records and ecological notes on central New York Unionacea. *Amer. Midl. Natur.* **84**, 46–58.

Harman, W. N. (1970b). *Anodontoides ferussacianus* (Lea) in the Susquehanna River watershed in New York state. *Nautilus* **83**, 114–115.

Harrison, F. L. (1969). Accumulation and distribution of ^{54}Mn and ^{65}Zn in freshwater clams. *In*: *Symp. Radioecol.* (D. J. Nelson and F. C. Evans *et al.*, Eds.), pp. 198–220. U. S. At. Energy Comm. Div. of Tech. Inform. Extension, Oak Ridge.

Harvey, R. S. (1969). Uptake and loss of radionuclides by the freshwater clam *Lampsilis radiatata* (Gmel.). *Health Phys.* **17**, 149–154.

Headlee, T. J. (1906). Ecological notes on the mussels of Winona, Pike, and Center Lakes of Kosciusko County, Indiana. *Biol. Bull.* **11**, 305–318.

Heard, W. H. (1962). Distribution of Sphaeridae (Pelecypoda) in Michigan, U.S.A. *Malacologia* **1**, 139–160.

Heard, W. H. (1963). The biology of *Pisidium* (*Neopisidium*) *conventus* Clessin (Pelecypoda; Sphaeriidae). *Papers Mich. Acad. Sci. Arts Lett.* **48**, 77–86.

Heard, W. H. (1964). Litter size in the Sphaeriidae. *Nautilus* **78**, 47–49.

Heard, W. H. (1965a). Recent *Eupera* (Pelecypoda: Sphaeriidae) in the United States. *Amer. Midl. Natur.* **74**, 309–317.

Heard, W. H. (1965b). Comparative life histories of the North American pill clams (Sphaeriidae: *Pisidium*). *Malacologia* **2**, 381–411.

Heard, W. H. (1968). Mollusca. *In* "Keys to Water Quality Indicative Organisms (Southeastern United States)" (F. K. Parrish Ed.), pp. G1–G26. *Fed. Water Pollut. Contr. Administration, Washington, D.C.*

Heard, W. H. (1970a). Hermaphroditism in *Margaritifera falcata* (Gould). *Nautilus* **83**, 113–114.

Heard, W. H. (1970b). Eastern freshwater mollusks. (II) The south Atlantic and Gulf drainages. *In* Clarke (1970, pp. 23–27).

Heard, W. H., and Burch, J. B. (1966). Key to the genera of freshwater pelecypods (mussels and clams) of Michigan. *Circ. Mus. Zool., Univ. Michigan* No. 4, 1–14.

Heard, W. H., and Guckert, R. H. (1971). A re-evaluation of the Recent Unionacea (Pelecypoda) of North America. *Malacologia* **10**, 333–355.

Heard, W. H., and Hendrix, S. S. (1964). Behavior of unionid glochidia. *Annu. Rep. 1964 Amer. Malacolog. Un.* 2–4.

Hendelberg, J. (1960). The fresh-water pearl mussel, *Margaritifera margaritifera* (L.). *Inst. Freshwater Res. Drottningholm, Sweden*, Rep. No. 41, 149–171.

Henderson, H. E. (1938). The cercaria of *Crepidostomum cornutum* (Osborn). *Trans. Amer. Microsc. Soc.* **57**, 165–172.

Henderson, J. (1933). *Lampsilis* at old New Mexican camp sites. *Nautilus* **46**, 107.

Hendrix, S. S. (1968). New host and locality records for two aspidogastrid trematodes, *Aspidogaster conchicola* and *Cotylaspis insignis*. *J. Parasitol.* **54**, 179–180.

Hendrix, S. S., and Short, R. B. (1965). Aspidogastrids from northeastern Gulf of Mexico river drainages. *J. Parasitol.* **51**, 561–569.

Herrington, H. B. (1962). A revision of the Sphaeriidae of North America (Mollusca: Pelecypoda). *Misc. Publ. Mus. Zool. Univ. Michigan* No. 118, 1–74.

Hiestand, W. A. (1938). Respiration studies with fresh-water molluscs: I. Oxygen consumption in relation to oxygen tension. *Proc. Indiana Acad. Sci.* **47**, 287–292.

Hobden, D. J. (1970a). Aspects of iron metabolism in a freshwater mussel. *Can. J. Zool.* **48**, 83–86.

Hobden, D. J. (1970b). The catalase of a freshwater bivalve. *Can. J. Zool.* **48**, 201–203.

Hopkins, S. H. (1934). The parasite inducing pearl formation in American fresh-water Unionidae. *Science* **79**, 385–386.

Hopkins, S. H. (1970). Studies on brackish water clams of the genus *Rangia* in Texas. *Proc. Nat. Shellfish. Ass.* **60**, 5–6.

Hopkins, S. H., and Andrews, J. D. (1970). *Rangia cuneata* on the East Coast: thousand mile range extension, or resurgence? *Science* **167**, 868.

Howard, A. D. (1913). The catfish as a host for fresh-water mussels. *Trans. Amer. Fish. Soc.* **42**, 65–70.

Howard, A. D. (1914a). Some cases of narrowly restricted parasitism among commercial species of fresh water mussels. *Trans. Amer. Fish. Soc.* **44**, 41–44.

Howard, A. D. (1914b). A new record in rearing fresh-water pearl mussels. *Trans. Amer. Fish. Soc.* **44**, 43–47.

Howard, A. D. (1914c). Experiments in propagation of fresh-water mussels of the *Quadrula* group. *Rep. U. S. Comm. Fish. for 1913*, Appendix IV, pp. 1–52. Separately issued as Bur. Fish. Document No. 801.

Howard, A. D. (1914d). A second case of metamorphosis without parasitism in the Unionidae. *Science* **40**, 353–355.

Howard, A. D. (1915). Some exceptional cases of breeding among the Unionidae. *Nautilus* **29**, 4–11.

Howard, A. D. (1917). An artificial infection with glochidia on the river herring. *Trans. Amer. Fish. Soc.* **46**, 93–100.

Howard, A. D. (1922). Experiments in the culture of fresh-water mussels. *Bull. U. S. Bur. Fish.* **38**, 63–89. Separately issued as Bur. Fish. Document No. 916.

Howard, A. D. (1951). A river mussel parasitic on a salamander. *Natur. Hist. Misc.* No. 77, 1–6.

Howard, A. D. (1953). Some viviparous pelecypod mollusks. *Wasmann J. Biol.* **11**, 233–240.

Howard, A. D., and Anson, B. J. (1922). Phases in the parasitism of the Unionidae. *J. Parasitol.* **9**, 68–82.

Hubricht, L. (1966). Habitat of *Eupera singleyi* Pilsbry. *Nautilus* **80**, 33.

Humes, A. G., and Harris, S. K. (1952). The clam hosts of *Najadicola ingens* (K.) (Acarina) in a Quebec lake. *Can. Field-Natur.* **66**, 83–84.

Humes, A. G., and Jamnback, H. A. (1950). *Najadicola ingens* (Koenike), a water-mite parasitic in fresh-water clams. *Psyche* **57**, 77–87.

Humes, A. G., and Russell, H. D. (1951). Seasonal distribution of *Najadicola ingens* (K.) (Acarina) in a New Hampshire pond. *Psyche* **58**, 111–119.

Hyman, L. H. (1951). "The Invertebrates: Platyhelminthes and Rhynchocoela, Vol. II, The Acoelomate Bilateria", pp. 1–550. McGraw-Hill, New York.

Imlay, M. J. (1971). Bioassay tests with naiads. *In* Jorgensen and Sharp (1971, pp. 38–41).

Imlay, M. J. (1972a). Reproduction of *Amblema costata* (Rafinesque) in Moose River, Minnesota. *Nautilus* **85**, 146.

Imlay, M. J. (1972b). Greater adaptability of freshwater mussels to natural rather than to artificial displacement. *Nautilus* **86**, 76–79.

Imlay, M. J., and Paige, M. L. (1972). Laboratory growth of freshwater sponges, unionid mussels, and sphaeriid clams. *Progr. Fish-Cultur.* **34**, 210–216.

Ingram, W. M. (1941). Survival of fresh-water mollusks during periods of dryness. *Nautilus* **54**, 84–87.

Ingram, W. M. (1948). The larger freshwater clams of California, Oregon, and Washington. *J. Entomol. Zool.* **40**, 72–92.

Ingram, W. M. (1956). The use and value of biological indicators of pollution: fresh water clams and snails. *In* "Biological Problems in Water Pollution" (C. M. Tarzwell, Ed.), pp. 94–135. Robert A. Taft Sanitary Eng. Center, Cincinatti, Ohio.

Ingram, W. M., Ballinger, D. G., and Gaufin, A. R. (1953). Relationship of *Sphaerium solidulum* Prime to organic pollution. *Ohio J. Sci.* **53**, 230–235.

Isely, F. B. (1911). Preliminary note on the ecology of the early juvenile life of the Unionidae. *Biol. Bull.* **20**, 77–80.

Isely, F. B. (1914). Experimental study of the growth and migration of fresh-water mussels. *Rep. U. S. Comm. Fish. for 1913*, Appendix III, pp. 1–24. Separately issued as Bur. Fish. Document No. 792.

Isely, F. B. (1925). The fresh-water mussel fauna of eastern Oklahoma. *Proc. Okla. Acad. Sci.* **4**, 43–118.

Isom, B. G. (1971). Mussel fauna found in Fort Loudon Reservoir, Tennessee River, Knox County, Tenessee. *Malacolog. Rev.* **4**, 127–130.

Isom, B. G., and Yorkley, P., Jr. (1968a). Mussels of Bear Creek watershed, Alabama and Mississippi, with a discussion of area geology. *Amer. Midl. Natur.* **79**, 189–196.

Isom, B. G., and Yokley, P., Jr. (1968b). The mussel fauna of Duck River in Tennessee, 1965. *Amer. Midl. Natur.* **80**, 34–42.

Jewell, M. E. (1922). The fauna of an acid stream. *Ecology* **3**, 22–28.

Johnson, R. I. (1946). *Anodonta implicata* Say. *Occas. Pap. Mollusks* **1**, 109–116.

Johnson, R. I. (1967). Additions to the unionid fauna of the Gulf drainage of Alabama, Georgia and Florida (Mollusca: Bivalvia). *Breviora* No. 270, 1–21.

Johnson, R. I. (1968). *Elliptio nigella*, overlooked unionid from Apalachicola river system. *Nautilus* **82**, 20, 22–24.

Johnson, R. I. (1970). The systematics and zoogeography of the Unionidae (Mollusca: Bivalvia) of the southern Atlantic slope region. *Bull. Mus. Comp. Zool.* **140**, 263–450.

Johnson, R. I. (1972). The Unionidae (Mollusca: Bivalvia) of peninsular Florida. *Bull. Florida State Mus., Biol. Sci.* **16**, 181–249.

Jones, R. O. (1950). Propagation of fresh-water mussels. *Progr. Fish-Cultur.* **12**, 13–25.

Jordan, D. S., Evermann, B. W., and Clark, H. W. (1930). Checklist of the fishes and fish-like vertebrates of North and Middle America north of the northern boundary of Venezuela and Colombia. *Rep. U. S. Comm. Fish. for 1928*, Appendix X, 1–670.

Jorgenson, S. E., and Sharp, R. W. (Ed.) (1971) "Proceedings of a Symposium on Rare & Endangered Mollusks (Naiads) of the U.S." pp. 1–79, U.S. Dep. of the Interior, Fish and Wildl. Serv., Bur. of Sport Fish. and Wildl.

Juday, C. (1908). Some aquatic invertebrates that live under anaerobic conditions. *Trans. Wisconsin Acad. Sci. Arts Lett.* **16**, 10–16.

Kelly, H. M. (1902). A statistical study of the parasites of the Unionidae. *Bull. Ill. State Lab. Natur. Hist.* **5**, 399–418.

Kelly, H. M. (1926). A new host for the aspidogastrid trematode, *Cotylogaster occidentalis. Proc. Iowa Acad. Sci.* **33**, 339.

Kendall, W. C. (1910). American catfishes: habits, culture, and commercial importance. *Rep. U.S. Comm. Fish. for 1910 and Spec. Papers*, pp. 1–39. Separately issued as Bur. Fish. Document No. 733.

Kidder, G. W. (1934). Studies on the ciliates from fresh water mussels. I. The structure and neuromotor system of *Conchophthirius anodontae* Stein, *C. curtus* Engl., and *C. magna* sp. nov. *Biol. Bull.* **66**, 69–90.

Kirtland, J. P. (1840). Fragments of natural history. *Amer. J. Sci. Arts* **39**, 164–168.

Kirtland, J. P. (1851). Remarks on the sexes and habits of some of the acephalous bivalve Mollusca. *Proc. Amer. Ass. Adv. Sci.* **5**, 85–91.

Kofoid, C. A. (1899). On the specific identity of *Cotylaspis insignis* Leidy and *Platyaspis anodontae* Osborn. *Zool. Bull.* **2**, 179–186.

Kofoid, C. A. (1910). The plankton of the Illinois River, 1894–1899, with introductory notes upon the hydrography of the Illinois River and its basin. Part II. Constituent organisms and their seasonal distribution. *Bull. Ill. St. Lab. Nat. Hist.* **8**, 1–360.

Kraemer, L. R. (1970). The mantle flap in three species of *Lampsilis* (Pelecypoda: Unionidae). *Malacologia* **10**, 225–282.

Krumholz, L. A., Bingham, R. L., and Meyer, E. R. (1970). A survey of the commercially valuable mussels of the Wabash and White Rivers of Indiana. *Proc. Indiana Acad. Sci.* **79**, 205–226.

Kuiper, J. G. J., and Wolff, W. J. (1970). The Mollusca of the estuarine region of the rivers Rhine, Meuse and Scheldt in relation to the hydrography of the area. III. The genus *Pisidium. Basteria* **34**, 1–40.

Kunz, G. F. (1894). On pearls, and the utilization and application of the shells in which they are found in the ornamental arts, as shown at the world's Columbian exposition. *Bull. U. S. Fish Comm.* **13**, 439–457.

Kunz, G. F. (1898a). A brief history of the gathering of fresh-water pearls in the United States. *Bull. U. S. Fish Comm.* **17**, 321–330.

Kunz, G. F. (1898b). The fresh-water pearls and pearl fisheries of the United States. *Bull. U. S. Fish Comm.* **17**, 373–426.

Lee, G. F., and Wilson, W. (1969). Use of chemical composition of freshwater clamshells as indicators of paleohydrologic conditions. *Ecology* **50**, 990–997.

Lefevre, G., and Curtis, W. C. (1910a). Experiments in the artificial propagation of fresh-water mussels. *Bull. U. S. Bur. Fish.* **20**, 615–626. Separately issued as Bur. Fish. Document No. 671.

Lefevre, G., and Curtis, W. C. (1910b). Reproduction and parasitism in the Unionidae. *J. Exp. Zool.* **9**, 79–116.

Lefevre, G., and Curtis, W. C. (1911). Metamorphosis without parasitism in the Unionidae. *Science* **33**, 863–865.

Lefevre, G., and Curtis, W. C. (1912). Studies on the reproduction and artificial propagation of fresh-water mussels. *Bull. U. S. Bur. Fish.* **30**, 105–201. Separately issued as Bur. Fish. Document 756.

Leidy, J. (1858). "... observations on entozoa found in the naiades." *Proc. Acad. Natur. Sci. Philadelphia* **9**, 18.

Leidy, J. (1859). Contributions to helminthology. *Proc. Acad. Natur. Sci. Philadelphia* **10**, 110–112.

Leidy, J. (1884). On the reproduction and parasites of *Anodonta fluviatilis*. *Proc. Acad. Nat. Sci. Philadelphia* **35**, 44–46.

Lukacsovics, F. (1966). Hypoxial examination of *Anodonta cygnea* L. on the O_2-consumption of gill tissues and the relation between body dimensions and the respiration of the gill-tissue. *Ann. Inst. Biol. (Tihany) Hung. Acad. Sci.* **33**, 79–94.

Lukacsovics, F., and Labos, E. (1965). Chemo-ecological relationship between some fish species in Lake Balaton and glochidia of *Anodonta cygnea* L. *Ann. Inst. Biol. (Tihany) Hung. Acad. Sci.* **32**, 37–54.

Lukacsovics, F., and Salánki, J. (1964). Effect of substances influencing tissue respiration and of the temperature on the O_2 consumption of the gill tissue in *Unio tumidus*. *Ann. Inst. Biol. (Tihany) Hung. Acad. Sci.* **31**, 55–63.

Lukacsovics, F., and Salánki, J. (1968). Data to the chemical sensitivity of freshwater mussel (*Anodonta cygnea* L.) *Ann. Inst. Biol. (Tihany) Hung. Acad. Sci.* **35**, 25–34.

Mackie, G. L., and Qadri, S. U. (1973). Abundance and diversity of Mollusca in an industrialized portion of the Ottawa River near Ottawa-Hull, Canada. *J. Fish. Res. Bd. Canada* **30**, 167–172.

Mannix, D. P. (1965). Treasure trove of backyard pearls. *True*, November, 48–49, 112–115. Fawcett Publ., Greenwich, Connecticut.

Marshall, R. (1926). Water mites of the Okoboji region. *Univ. Iowa Stud., Stud. Natur. Hist.* **11**, 28–35.

Matteson, M. R. (1948). Life history of *Elliptio complanatus* (Dillwyn, 1817). *Amer. Midl. Natur.* **40**, 690–723.

Matteson, M. R. (1953). Fresh-water mussels used by Illinoian Indians of the Hopewell culture. *Nautilus* **66**, 130–138; **67**, 25–26.

Matteson, M. R. (1955). Studies on the natural history of the Unionidae. *Amer. Midl. Natur.* **53**, 126–145.

McMillan, N. F. (1966). *Margaritifera margaritifera* (L.) in hard water in Scotland. *J. Conchol.* **26**, 69–70.

Meek, S. E., and Clark, H. W. (1912). The mussels of the Big Buffalo Fork of White River, Arkansas. *Rep. Comm. Fish. for 1911 and Spec. Papers*, pp. 1–20. Separately issued as Bur. Fish. Document No. 759.

Meier-Brook, C. (1970). Substate relations in some *Pisidium* species (Eulamellibranchiata: Sphaeriidae). *Malacologia* **9**, 121–125.

Mermilliod, W. (1973). "An investigation for the natural host of the glochidia of *Toxolasma parva*," Undergraduate research paper, Louisiana State University at Baton Rouge.

Merrick, A. D. (1930). Some quantitative determinations of glochidia. *Nautilus* **43**, 89–90.

Miller, C. W., Zuckerman, B. M., and Charig, A. J. (1966). Water translocation of diazinon-C^{14} and parathion-S^{35} off a model cranberry bog and subsequent occurrence in fish and mussels. *Trans. Amer. Fish. Soc.* **95**, 345–349.

Mitchell, R. D. (1955). Anatomy, life history, and evolution of the mites parasitizing fresh-water mussels. *Misc. Publ. Mus. Zool. Univ. Michigan* No. 89, 1–28.

Mitchell, R. D. (1957). On the mites parasitizing *Anodonta* (Unionidae: Mollusca). *J. Parasitol.* **43**, 101–104.

Mitchell, R. D. (1965). Population regulation of a water mite parasitic on unionid mussels. *J. Parasitol.* **51**, 990–996.

Mitchell, R. D., and Wilson, J. L. (1965). New species of water mites (*Unionicola*) from Tennessee unionid mussels. *J. Tenn. Acad. Sci.* **40**, 104–106.

Monk, C. R. (1928). The anatomy and life-history of a fresh-water mollusk of the genus *Sphaerium. J. Morphol. Physiol.* **45**, 473–503.

Monticelli, F. S. (1892). *Cotylogaster michaelis* n.g. n. sp. e revisione degli Aspidobothridae. "Festschrift *Siebenzigsten Geburtstage Rudolf Leuckarts*" 168–214.

Moore, J. P. (1912). The leeches of Minnesota. Part III. Classification of the leeches of Minnesota. *Geol. Natur. Hist. Surv. Minnesota Zool. Ser.* No. 5, 63–143.

Morrison, J. P. E. (1932). A report on the Mollusca of the northeastern Wisconsin lake district. *Trans. Wisconsin Acad. Sci. Arts Lett.* **27**, 359–396.

Morrison, J. P. E. (1955). Some zoogeographic problems among brackish water mollusks. *Annu. Rep. 1954 Amer. Malacolog. Un.* 7–10.

Mullican, H. N., Sinclair, R. M., and Isom, B. G. (1960). Survey of the aquatic biota of the Nolichucky River in the state of Tennessee. Tenn. Stream Pollut. Contr. B. Nashville. pp. 1–28.

Murphy, G. (1942). Relationship of the fresh-water mussel to trout in the Truckee River. *Calif. Fish fame* **28**, 89–102.

Murphy, J. L. (1971). Molluscan remains from four archaeological sites in northeastern Ohio. *Sterkiana* No. 43, 21–25.

Murray, H. D., and Leonard, A. B. (1962). Handbook of unionid mussels in Kansas. *Univ. Kansas Mus. Natur. Hist. Publ.* No. 28, 1–184.

Najarian, H. H. (1955). Notes on aspidogastrid trematodes and hydracarina from some Tennessee mussels. *J. Tenn. Acad. Sci.* **30**, 11–14.

Najarian, H. H. (1961). New aspidogastrid trematode, *Cotylaspis reelfootensis*, from some Tennessee mussels. *J. Parasitol.* **47**, 515–520.

Neel, J. K., and Allen, W. R. (1964). The mussel fauna of the upper Cumberland basin before its impoundment. *Malacologia* **1**, 427–459.

Negus, C. L. (1966). A quantitative study of growth and production of unionid mussels in the river Thames at Reading. *J. Anim. Ecol.* **35**, 513–532.

Nelson, D. J. (1962). Clams as indicators of strontium-90. *Science* **137**, 38–39.

Nelson, D. J. (1964). Biological vectors and reservoirs of strontium-90. *Nature (London)* **203**, 420.

Nelson, D. J. (1967). Microchemical constituents in contemporary and pre-Columbian clamshell. In "Quaternary Paleoecology" (E. J. Cushing and H. E. Wright, Jr., Ed.), pp. 185–204. Yale Univ. Press, New York.

Ortmann, A. E. (1909a). Unionidae from an Indian garbage heap. *Nautilus* **23**, 11–15.

Ortmann, A. E. (1909b). The destruction of the fresh-water fauna in western Pennsylvania. *Proc. Amer. Phil. Soc.* **48**, 90–110.

Ortmann, A. E. (1910). The discharge of the glochidia in the Unionidae. *Nautilus* **24**, 94–95.

Ortmann, A. E. (1911). A monograph of the najades of Pennsylvania. *Mem. Carnegie Mus.* **4**, 279–347.

Ortmann, A. E. (1912). Notes upon the families and genera of the najades. *Ann. Carnegie Mus.* **8**, 222–365.

Ortmann, A. E. (1918). The nayades (freshwater mussels) of the upper Tennessee drainage with notes on synonymy and distribution. *Proc. Amer. Phil. Soc.* **57**, 521–626.

Ortmann, A. E. (1919). A monograph of the naiades of Pennsylvania. Part III. Systematic account of the genera and species. *Mem. Carnegie Mus.* **8**, 1–384.

Ortmann, A. E. (1924). The naiad-fauna of Duck River in Tennesse. *Amer. Midl. Natur.* **9**, 18–62.

Ortmann, A. E. (1925). The naiad-fauna of the Tennessee River system below Walden Gorge. *Amer. Midl. Natur.* **9**, 321–372.

Ortmann, A. E. (1926). The naiades of the Green River drainage in Kentucky. *Ann. Carnegie Mus.* **17**, 167–188.

Ortmann, A. E., and Walker, B. (1912). A new North American naiad. *Nautilus* **25**, 97–100.

Osborn, H. L. (1898a). Observations on the anatomy of a species of *Platyaspis* found parasitic on the Unionidae of Lake Chautauqua. *Zool. Bull.* **2**, 55–67.

Osborn, H. L. (1898b). Observations on the parasitism of *Anodonta plana* Lea by a distomid trematode, at Chautauqua, New York. *Biol. Bull.* **1**, 301–310.

Parker, J. M. (1932). Studies on *Cercaria rhyticerca*, a new rhopalocercous cercaria from *Amblema costata*. M. S. Thesis, Univ. of Illinois, 38 pp.

Parmalee, P. W. (1955). Some ecological aspects of the naiad fauna of Lake Springfield, Illinois. *Nautilus* **69**, 28–34.

Parmalee, P. W. (1967). The fresh-water mussels of Illinois. *Ill. State Mus. Popular Sci. Ser.* **8**, 1–108.

Patrick, R., Cairns, J., Jr., and Roback, S. S. (1967). An ecosystematic study of the fauna and flora of the Savannah River. *Proc. Acad. Natur. Sci. Philadelphia* **118**, 109–407.

Pauley, G. B. (1968a). The pathology of "spongy" disease in freshwater mussels. *Proc. Nat. Shellfish. Ass.* **58**, 13.

Pauley, G. B. (1968b). A disease of the freshwater mussel, *Margaritifera margaritifera*. *J. Invert. Pathol.* **12**, 321–328.

Pauley, G. B. and Becker, C. D. (1968). *Aspidogaster conchicola* in mollusks of the Columbia River system with comments on the hosts' pathological response. *J. Parasitol.* **5**, 917–920.

Pearse, A. S. (1924). The parasites of lake fishes. *Trans. Wisconsin Acad. Sci. Arts Lett.* **21**, 161–194.

Penn, G. H., Jr. (1939). A study of the life cycle of the freshwater mussel, *Anodonta grandis*, in New Orleans. *Nautilus* **52**, 99–101.

Penn, J. H. (1958). Studies on ciliates from mollusks of Iowa. *Proc. Iowa Acad. Sci.* **65**, 517–534.

Pennak, R. W. (1953). "Fresh-water Invertebrates of the United States," pp. 1–769. Ronald Press, New York.

Pilsbry, H. A. (1910). Unionidae of the Panuco river system, Mexico. *Proc. Acad. Natur. Sci. Philadelphia* **61**, 532–539.

Potts, W. T. W. (1954). The inorganic composition of the blood of *Mytilus edulis* and *Anodonta cygnea*. *J. Exp. Biol.* **31**, 376–385.

Prosser, C. L., and Weinstein, S. J. F. (1950). Comparison of blood volume in animals with open and with closed circulatory systems. *Physiol. Zool.* **23**, 113–124.

Rawson, D. S. (1953). The bottom fauna of Great Slave Lake. *J. Fish. Res. B. Can.* **10**, 486–520.

Read, L. B., and Oliver, K. H. (1953). Notes of the ecology of the fresh-water mussels of Dallas County. *Field Lab.* **21**, 75–80.

Reigle, N. (1967). An occurrence of *Anodonta* (Mollusca, Pelecypoda) in deep water. *Amer. Midl. Natur.* **78**, 530–531.

Reuling, F. H. (1919). Acquired immunity to an animal parasite. *J. Infec. Dis.* **24**, 337–346.

Richardson, R. E. (1928). The bottom fauna of the middle Illinois River. *Bull. Ill. State. Lab. Natur. Hist.* **17**, 387–475.

Richardson, W. M., St. Amant, J. A., Botroff, L. J., and Parker, W. L. (1970). Introduction of blue catfish into California. *Calif. Fish Game* **56**, 311–312.

Riggs, C. D., and Webb, G. R. (1956). The mussel population of an area of loamy-sand bottom of Lake Texoma. *Amer. Midl. Natur.* **56**, 197–203.

Roscoe, E. J. (1967). Ethnomalacology and paleoecology of the Round Butte archaeological sites, Deschutes River basin, Oregon. Univ. of Oregon, Bull. No. 6 of the Mus. of Natur. Hist. 1–20.

Roscoe, E. J., and Redelings, S. (1964). The ecology of the fresh-water mussel *Margaritifera margaritifera* (L.). *Sterkiana* No. 16, 19–32.

Salbenblatt, J. A., and Edgar, A. L. (1964). Value activity in fresh-water pelecypods. *Papers Mich. Acad. Sci. Arts Lett.* **49**, 177–186.

Sawyer, R. T. (1972). North American freshwater leeches, exclusive of the Piscicolidae, with a key to all species. *Ill. Biol. Monogr.* No. 46, 1–154.

Scruggs, G. D., Jr. (1960). Status of fresh-water mussel stocks in the Tennessee River. *Spec. Sci. Rep. U. S. Fish Wildl. Serv.* No. 370, 1–41.

Seitner, P. G. (1951). The life history of *Allocreadium ictaluri* Pearse, 1924 (Trematoda: Digenea). *J. Parasitol.* **37**, 223–244.

Sellmer, G. P. (1967). Functional morphology and ecological life history of the gem clam, *Gemma gemma* (Eulamellibranchia: Veneridae). *Malacologia* **5**, 137–223.

Seshaiya, R. V. (1941). Tadpoles as hosts for the glochidia of the fresh-water mussel. *Curr. Sci.* **10**, 535–536.

Shelly, R. M. (1972). In defense of naiades. *Wildl. North Carolina* **36**, 4–8, 26–27.

Shira, A. F. (1913). The mussel fisheries of Caddo Lake and the Cypress and Sulphur rivers of Texas and Louisiana. *U. S. Bur. Fish. Econ. Circ.* No. **6**, 1–20.

Short, Z. F., Palumbo, R. F., Olson, P. R., and Donaldson, J. R. (1969). The uptake of I^{131} by the biota of Fern Lake, Washington, in a laboratory and a field experiment. *Ecology* **50**, 979–989.

Shoup, C. S., Peyton, J. H., and Gentry, G. (1941). A limited biological survey of the Obey River and adjacent streams in Tennessee. *J. Tenn. Acad. Sci.* **16**, 48–76.

Sickel, J. B. (1973). A new record of *Corbicula manilensis* (Philippi) in the southern Atlantic slope region of Georgia. *Nautilus* **87**, 11–12.

Simpson, C. T. (1892). Notes on the Unionidae of Florida and the southeastern states. *Proc. U. S. Nat. Mus.* **15**, 405–436.

Simpson, C. T. (1899). The pearly fresh-water mussels of the United States; their habits, enemies, and diseases; with suggestions for their protection. *Bull. U. S. Fish Comm.* **18**, 279–288.

Simpson, C. T. (1900). Synopsis of the naiades, or pearly fresh-water mussels. *Proc. U. S. Nat. Mus.* **22**, 501–1044.

Sinclair, R. M. (1971). Annotated bibliography on the exotic bivalve *Corbicula* in North America, 1900–1971. *Sterkiana* No. **43**, 11–18.

Smith, H. M. (1899). The mussel fishery and pearl button industry of the Mississippi River. *Bull. U. S. Fish Comm.* **18**, 289–314.

Smith, H. M. (1919). Fresh water mussels. A valuable national resource without sufficient protection. *U. S. Bur. Fish. Econ. Circ.* No. 43, 1–5.

Snyder, N. F. R., and Snyder, H. A. (1969). A comparative study of mollusc predation by Limpkins, Everglade Kites, and Boat-tailed Grackles. *Living Bird, Annu. Rep. Cornell Lab. Ornithol. 8th* 177–223.

Stansbery, D. H. (1964). The mussel (muscle) shoals of the Tennessee River revisited. *Annu. Rep. 1964 Amer. Malacolog. Un.* 25–28.

Stansbery, D. H. (1965a). The molluscan fauna. *In* "The McGraw Site: A Study in Hopewellian Dynamics" (O. H. Prufer, Ed.). *Sci. Publ. Cleveland Mus. Natur. Hist. New Ser.* **3**, 119–124.

Stansbery, D. H. (1965b). The naiad fauna of the Green River at Munfordville, Kentucky. *Annu. Rep. 1965 Amer. Malacolog. Un.* 13–14.

Stansbery, D. H. (1966a). Observations on the habitat distribution of the naiad *Cumberlandia monodonta* (Say, 1829). *Annu. Rep. 1966 Amer. Malacolog. Un.* 29–30.

Stansbery, D. H. (1966b). Utilization of naiads by prehistoric man in the Ohio valley. *Annu. Rep. 1966 Amer. Malacolog. Un.* 41–43.

Stansbery, D. H. (1967). Growth and longevity of naiads from Fishery Bay in western Lake Erie. *Annu. Rep. 1967 Amer. Malacolog. Un.* 10–11.

Stansbery, D. H. (1969). Changes in the naiad fauna of the Cumberland Falls in eastern Kentucky. *Annu. Rep. 1969 Amer. Malacolog. Un.* 16–17.

Stansbery, D. H. (1970a). A study of the growth rate and longevity of the naiad *Amblema plicata* (Say, 1817) in Lake Erie (Bivalvia: Unionidae). *Annu. Rep. 1970 Amer. Malacolog. Un.* 78–79.

Stansbery, D. H. (1970b). Eastern freshwater mollusks. (I) The Mississippi and St. Lawrence river systems. *In* Clarke (1970, pp. 9–21).

Stansbery, D. H. (1971). Rare and endangered freshwater mollusks in eastern United States. *In* Jorgensen and Sharp (1971, pp. 5–18f).

Starrett, W. C. (1971). A survey of the mussels (Unionacea) of the Illinois River: a polluted stream. *Ill. Natur. Hist. Surv. Bull.* **30**, 265–403.

Stearns, R. E. C. (1883). On the shells of the Colorado desert and the region farther east. Part I. The physas of Indio. Part II. *Anodonta californiensis* in a new locality. *Amer. Natur.* **17**, 1014–1020.

Steelman, G. M. (1939). A new macrocerous cercaria. *Trans. Amer. Microsc. Soc.* **58**, 258–263.

Stein, C. B. (1968). Studies in the life history of the naiad, *Amblema plicata* (Say, 1817). *Annu. Rep. 1968 Amer. Malacolog. Un.* 46–47.

Stein, C. B. (1971). Naiad life cycles: their significance in the conservation of the fauna. *In* Jorgensen and Sharp (1971, pp. 19–25).

Sterki, V. (1891a). A byssus in *Unio. Nautilus* **5**, 73–74.

Sterki, V. (1891b). On the byssus of Unionidae. II. *Nautilus* **5**, 90–91.

Stober, Q. J. (1972). Distribution and age of *Margaritifera margaritifera* (L.) in a Madison River (Montana, U.S.A.) mussel bed. *Malacologia* **11**, 343–350.

Strecker, J. K. (1908). The Mollusca of McClennan County, Texas. *Nautilus* **22**, 63–67.

Strecker, J. K. (1931). The distribution of the naiades or pearly fresh-water mussels of Texas. *Spec. Bull. Baylor Univ. Mus.* No. 2, 1–71.

Stromberg, P. C. (1970). Aspidobothrean trematodes from Ohio mussels. *Ohio J. Sci.* **70**, 335–341.

Stunkard, H. W. (1917). Studies on North American Polystomidae, Aspidogastridae, and Paraphistomidae. *Ill. Biol. Monogr.* **3**, 1–114.

Surber, T. (1912). Identification of the glochidia of freshwater mussels. *Rep. U. S. Comm. Fish. for 1912 and Spec. Papers*, pp. 1–10. Separately issued as Bur. Fish. Document No. 771.

Surber, T. (1913). Notes on the natural hosts of fresh-water mussels. *Bull. U. S. Bur. Fish.* **32**, 110–116. Separately issued as Bur. Fish. Document No. 778.

Surber, T. (1915). Identification of the glochidia of fresh-water mussels. *Rep. U. S. Comm. Fish. for 1914*, Appendix V, pp. 1–9. Separately issued as Bur. Fish. Document No. 813.

Taylor, D. W. (1960). Distribution of the freshwater clam *Pisidium ultramontanum*; a zoogeographic inquiry. *Amer. J. Sci.* **258A**, 325–334.

Taylor, D. W. (1966). An eastern American freshwater mussel, *Anodonta*, introduced into Arizona. *Veliger* **8**, 197–198.

Temte, E. F. (1968). "A brief history of the clamming and pearling industry in Prairie du Chien, Wisconsin," pp. 1–37. Graduate seminar paper, Wisconsin State University at La Crosse.

Thomas, G. J. (1959). Self-fertilization and production of young in a sphaeriid clam. *Nautilus* **72**, 131–140.

Thomas, G. J. (1963). Study of a population of sphaeriid clams in a temporary pond. *Nautilus* **77**, 37–43.

Thomas, G. J. (1965). Growth in one species of sphaeriid clam. *Nautilus* **79**, 47–54.

Thut, R. N. (1969). A study of the profundal bottom fauna of Lake Washington. *Ecol. Monogr.* **39**, 79–110.

Tomlinson, J. (1966). The advantages of hermaphroditism and parthenogenesis. *J. Theoret. Biol.* **11**, 54–58.

Trautman, M. B. (1957). "The Fishes of Ohio with Illustrated Keys," pp. 1–683. Ohio State Univ. Press, Columbus, Ohio.

Tucker, M. E. (1927). Morphology of the glochidium and juvenile of the mussel *Anodonta imbecilis*. *Trans. Amer. Microsc. Soc.* **46**, 286–293.

Tucker, M. E. (1928). Studies on the life cycles of two species of fresh-water mussels belonging to the genus *Anodonta*. *Biol. Bull.* **54**, 117–127.

Tuthill, S. J., and Johnson, R. L. (1969). Nonmarine mollusks of the Katalla region, Alaska. *Nautilus* **83**, 44–52.

Tuthill, S. J., and Laird, W. M. (1963–1964). Molluscan fauna of some alkaline lakes and sloughs in southern central North Dakota. *Nautilus* **77**, 47–55, 81–90.

Utterback, W. I. (1915–1916). The naiades of Missouri. *Amer. Midl. Natur.* **4**, 41–53, 97–152, 181–204, 244–273, 311–327, 339–354, 387–400, 432–464.

Utterback, W. I. (1916). Parasitism among Missouri naiades. *Amer. Midl. Natur.* **4**, 518–521.

Utterback, W. I. (1931). Sex behavior among naiades. *Proc. W. V. Acad. Sci.* **5**, 43–45.

Valentine, B. D., and Stansbery, D. H. (1971). An introduction to the naiades of the Lake Texoma region, with notes on the Red River fauna (Mollusca: Unionidae). *Sterkiana* No. 42, 1–40.

Vanatta, E. G. (1910). Unionidae from southeastern Arkansas and N.E. Louisiana. *Nautilus* **23**, 102–104.

van Cleave, H. J. (1940). Ten years of observation on a freshwater mussel population. *Ecology* **21**, 363–370.

van Cleave, H. J., and Williams, C. O. (1943). Maintenance of a trematode, *Aspidogaster conchicola*, outside the body of its natural host. *J. Parasitol.* **29**, 127–130.

van der Schalie, H. (1933). Notes on the brackish water bivalve, *Polymesoda caroliniana* (Bosc). *Occas. Pap. Mus. Zool. Univ. Mich.* No. 258, 1–8.

van der Schalie, H. (1936). The naiad fauna of the St. Joseph River drainage in southwestern Michigan. *Amer. Midl. Natur.* **17**, 523–527.

van der Schalie, H. (1937). A mussel taken from the stomach of *Rana catesbiana* Shaw. *Nautilus* **50**, 104–105.

van der Schalie, H. (1938a). Contributing factors in the depletion of naiades in eastern United States. *Basteria* **3**, 51–57.

van der Schalie, H. (1938b). The naiad fauna of the Huron River, in southeastern Michigan. *Misc. Publ. Mus. Zool. Univ. Mich.* No. 40, 1–83.

van der Schalie, H. (1938c). The naiades (fresh-water mussels) of the Cahaba River in northern Alabama. *Occas. Pap. Mus. Zool. Univ. Mich.* No. 392, 1–29.

van der Schalie, H. (1939a). Additional notes on the naiades (fresh-water mussels) of the lower Tennessee River. *Amer. Midl. Natur.* **22**, 452–457.

van der Schalie, H. (1939b). *Medionidus mcglameriae*, a new naiad from the Tombigbee River, with notes on other naiads of that drainage. *Univ. Mich. Occas. Pap. Mus. Zool.* No. 407, 1–6.

van der Schalie, H. (1945). The value of mussel distribution in tracing stream confluence. *Pap. Mich. Acad. Sci. Arts Lett.* **30**, 355–373.

van der Schalie, H. (1970). Hermaphroditism among North American fresh-water mussels. *Malacologia* **10**, 93–112.

van der Schalie, H., and Parmalee, P. W. (1960). Animal remains from the Etowah site, mound C, Bartow County, Georgia. *Fla. Anthropol.* **13**, 37–54.

van der Schalie, H., and van der Schalie, A. (1950). The mussels of the Mississippi River. *Amer. Midl. Natur.* **44**, 448–466.

Vaughan, T. W. (1892). Mollusks of Dorcheat Bayou and Lake Bisteneau, Louisiana. *Nautilus* **5**, 109–111.

Vertrees, H. H. (1913). "Pearls and Pearling," pp. 1–203. Harding and Fur News Publ., Columbus.

Vickers, G. G. (1940). On the anatomy of *Cercaria macrocerca* from *Sphaerium corneum*. *Quart. J. Microsc. Soc.* **82**, 311–326.

Vidrine, M. F. (1973). Freshwater mussels (Bivalvia: Unionidae) from Evangeline Parish, Louisiana, parasitized by water mites (Acarina: Hydracarina: Unionicolidae) and aspidogastrid trematodes (Trematoda: Aspidogasteridae). *Proc. La. Acad. Sci.* **36**, 53.

Vinyard, W. C. (1955). Epizoophytic algae from mollusks, turtles, and fish in Oklahoma. *Proc. Okla. Acad. Sci.* **34**, 63–65.

von Martens, E. (1890–1910). Land and freshwater Mollusca, pp. 1–706. *In* "Biologia Centrali-Americana" (F. D. Godman and O. Salvin, Eds.). The several portions of the section on Unionidae (pp. 478–540) appeared variously in 1900.

Walker, B. (1915). Habits of *Eupera*. *Nautilus* **29**, 82.

Walker, B. (1918). A synopsis of the classification of fresh-water Mollusca of North America, north of Mexico, and a catalogue of the more recently described species, with notes. *Misc. Publ. Mus. Zool. Univ. Mich.* No. 6, 1–213.

Welsh, J. H. (1961). A female *Lampsilis ovata ventricosa* (Barnes). *Science* **134**, 73, cover.

Wenninger, F. (1921). A preliminary report on the Unionidae of St. Joseph River. *Amer. Midl. Natur.* **7**, 1–13.

Williams, J. C. (1969). Mussel fishery investigations, Tennessee, Ohio and Green Rivers, Final Rep., Kentucky Dep. of Fish and Wildl. Resources and Murray State Univ. Biol. Station, Murray, Kentucky, pp. 1–107.

Wilson, C. B. (1916). Copepod parasites of fresh-water fishes and their economic relations to mussel glochida. *Bull. U. S. Bur. Fish.* **34**, 331–374. Separately issued as Bur. Fish. Document 824.

Wilson, C. B., and Clark, H. W. (1912a). The mussel fauna of the Maumee River. *Rep. U. S. Comm. Fish. for 1911 and Spec. Papers* pp. 1–72. Separately issued as Bur. Fish . Document No. 757.

Wilson, C. B., and Clark, H. W. (1912b). The mussel fauna of the Kankakee basin. *Rep. Comm. Fish. for 1911 and Spec. Papers*, pp. 1–52. Separately issued as Bur. Fish. Document No. 758.

Wilson, C. B., and Clark, H. W. (1912c). Mussel beds of the Cumberland River in 1911. *U. S. Bur. Fish. Econ. Circ.* No. 1, 1–4.

Wilson, C. B., and Clark, H. W. (1914). The mussels of the Cumberland River and its tributaries. *Rep. U. S. Comm. Fish. for 1912 and Spec. Papers.*, pp. 1–63. Separately issued as Bur. Fish. Document No. 781.

Wilson, C. B., and Danglade, E. (1914). The mussel fauna of central and northern Minnesota. *Rep. U. S. Comm. Fish. for 1913*, Appendix V, pp. 1–26. Separately issued as Bur. Fish. Document No. 803.

Wilson, K. A., and Ronald, K. (1967). Parasite fauna of the sea lamprey (*Petromyzon marinus* von Linné) in the Great Lakes region. *Can. J. Zool.* **45**, 1083–1092.

Wolcott, R. H. (1898). New American species of the genus *Atax* (Fab.) Bruz. *Zool. Bull.* **1**, 279–285.

Wolcott, R. H. (1899). On the North American species of the genus *Atax* (Fabr.) Bruz. *Trans. Amer. Microsc. Soc.* **20**, 193–259.

Wolfe, D. A. (1971). Fallout cesium-137 in clams (*Rangia cuneata*) from the Neuse River estuary, North Carolina. *Limnol. Oceanogr.* **16**, 797–805.

Wolff, W. J. (1970). The Mollusca of the estuarine region of the rivers Rhine, Meuse and Scheldt in relation to the hydrography of the area. IV. The genus *Sphaerium. Basteria* **34**, 75–90.

Wurtz, C. B. (1962). Zinc effects on fresh-water mollusks. *Nautilus* **76**, 53–61.

Wurtz, C. B., and Roback, S. S. (1955). The invertebrate fauna of some Gulf coast rivers. *Proc. Acad. Natur. Sci. Philadelphia* **197**, 167–206.

Yokley, P., Jr. (1972). Life history of *Pleurobema cordatum* (Rafinesque 1820) (Bivalvia: Unionacea). *Malacologia* **11**, 351–364.

Yokley, P., Jr. (1973). Freshwater mussel ecology, Kentucky Lake, Tennessee. Project 4–46–R, Tennnessee Game and Fish Comm., Nashville, pp. 1–133.

Young, D. (1911). The implantation of the glochidium on the fish. *Univ. Missouri Bull. Sci. Ser.* **2**, 1–20.

Zabik, M. J., and Bedford, J. W. (1972). The uptake of insecticides by freshwater mussels and the effects of sublethal concentrations of insecticides on these mussels. National Technical Information Service, Springfield, Virginia, Publication PB 214 090, pp. 1–27.

Zetek, J. (1918). The Mollusca of Platt, Champaign, and Vermilion Counties of Illinois. *Trans. Ill. State Acad. Sci.* **11**, 151–182.

Zumoff, C. H. (1973). The reproductive cycle of *Sphaerium simile. Biol. Bull.* **144**, 212–228.

Addenda

Burch's (1973) manual provides an excellent overview of the Nearctic unionacean fauna (see Section II,A).

Conner (1905) identified *Lepomis gibbosus* (Linnaeus) (Centrarchidae) as a glochidial host of *Anodonta cataracta* Say (see Table I).

CHAPTER 9

Snails (Mollusca: Gastropoda)

WILLARD N. HARMAN

I. The Evaluation of Biological Indicator Concepts

A. The Indicator Species

As originally recognized, an indicator species was an organism that, by its presence in a biotope, denoted particular characteristics of that environment that were otherwise difficult to determine (Richardson, 1928). In order for organisms to be useful indicator species they should possess several characteristics: (1) they should be easily recognized by researchers that are not specialists; (2) they should be abundant in their preferred habitats throughout a large geographic region; (3) they should exhibit approximately the same degree of tolerance to a particular phenomena, or be indicative of the same conditions, throughout their range; (4) they should possess a relatively long life span; (5) they should be comparatively sessile, or at least not easily able to avoid temporarily stressed environments by rapid migration. The freshwater Gastropoda satisfy the last two requirements; however, the large majority of them do not fulfill the first three. Like all of the Mollusca, they lack meristic characters that can be objectively utilized in species determination. This has left identification to the subjective opinion of malacologists who often disagree among themselves. As a result, the taxonomy of practically all freshwater families is in need of some revision. It is therefore difficult for the nonspecialist to identify freshwater snails correctly. Practically all freshwater gastropod species with large geographic ranges possess wide environmental tolerances (e.g., Anon. 1969, 1970, 1971a; Harman and Berg, 1971; Horst, 1971; Hunter *et al.*, 1967; Mason *et al.*, 1968; Morrison, 1932; Shoup, 1943; Schwartz and Meredith, 1962). Those species which appear to be stenotopic and have restricted ranges may be limited by their inability to migrate easily from one watershed to the next, rather than by physiological barriers. Data collected concerning the latter species may only indicate the characteristics of the waters of the biotopes in which they are found and not the snails' limits of tolerance to these parameters (Harman and Berg, 1971).

B. Indication by Absence

Several authors (Ingram, 1957; Tarzwell and Gaufin, 1967; Wurtz, 1955) have pointed out that since gastropods associated with polluted environments are also found commonly in clean waters, the absence of "clean water species" is a better indication of environmental conditions than the presence of tolerant ones. Several problems arise in applying this concept: (1) The ecology and physiology of very few freshwater snails is thoroughly known. Species inhabiting clean waters in one region may occur in enriched biotopes in other locations; (2) Many "clean water" gastropods have extremely restricted ranges and may only be utilized as indicators in a few local watersheds; (3) Presence and absence data vary greatly with the experience of the collector, the sampling procedures, and the densities of the populations in the community being sampled; and, (4) The investigator must be knowledgeable about the taxocene concerned in order to recognize the elements that are missing.

C. Indication by Community Structure

Many authors (Gaufin and Tarzwell, 1956; Harrel and Dorris, 1968; Wilhm, 1970b) have considered a high value of species richness (the number of species in an environment) as synonomous with clean waters, and a low richness and/or species diversity as indicative of polluted situations. Several indices have been proposed to simplify the application of this assumption such as the Sequential Comparison Index (Cairns *et al.*, 1968; Cairns and Dickson, 1971), the Biotic Index (Beck, 1969), cluster analyses (e.g., Kaesler *et al.*, 1971), and other more sophisticated diversity indices (Harrel and Dorris, 1968; Wilhm, 1970b; Wilhm and Dorris, 1968) derived from information theory.

Unfortunately serious problems arise between the theoretical and applied concepts of species richness and diversity. There is difficulty defining the boundaries of the biotopes under study and of acquiring statistically adequate samples (Eberhardt, 1969). Most serious is actually defining what "diversity" is. Hurlbert (1971) suggested that species diversity has become a meaningless concept and that the term should be discarded. He asserted that diversity indices, which are necessarily linear in nature, do not represent the actual situation. He propo^es two alternatives: (1) "The proportion of potential inter-individual encounters which is interspecific, assuming every individual in the collection can encounter all other individuals [Eq. 1],

$$\Delta_1 = \left[\frac{N}{N-1}\right]\left[1 - \sum_i \left(\frac{N_i}{N}\right)^2\right] \qquad [1]$$

and "2" the expected number of species in a sample of n individuals selected at random from a collection containing N individuals, S species, and N_i individuals in the ith species [Eq. 2]."

$$E(S_n) = \sum_i \left[1 - \frac{\binom{N-N_i}{n}}{\binom{N}{n}}\right] \qquad [2]$$

More factors affect species richness and diversity in an environment than the stress imposed by some form of pollution (an assumption, so often seen in applied works, is that a low value of richness or diversity indicates a polluted environment). Indeed, interspecific competition in a homogeneous environment often appears to reduce species diversity (Harman, 1972), and there are many naturally homogeneous environments.

D. Application of the Concepts

Whether the researcher uses a particular species as an indicator of environmental conditions, notes the absence of stenotopic species, or utilizes various indices of community structure, the methods used are dependent on the time and funds available and the individual's own competence. Regardless of the techniques utilized, more than a single taxocene must be considered, chemical and physical characteristics of the environment must be observed, and all the resultant data should by synthesized before an evaluation is reached. Overreliance on any one of these parameters can lead to serious misunderstandings of the prevailing environmental conditions.

II. The Freshwater Gastropoda

A. Classification and Origin

There are two subclasses of Gastropoda represented in the fresh waters of the continental United States. One, the Prosobranchia, is represented by the Hydrobiidae, Viviparidae, Pleuroceridae, Valvatidae, and Ampullariidae. Their progenitors invaded fresh water from the sea via estuaries into the coastal rivers. Adaptations necessary for occupation of the freshwater environment were the development of osmoregulation, the suppression of the free-living larval stages (trochophore and veliger) and the development of yolky encapsulated eggs. These species are dioecious (except Valvatidae), possess opercula, have typical molluscan ctenidia, possess comparatively ponderous shells, and are relatively slow-moving organisms. The Hydrobiidae, Viviparidae, and Valvatidae have world-wide, or at least circumboreal distributions, while the Pleuroceridae reach their greatest diversity in the southeastern United States. The Ampullariidae are tropical in distribution.

The other subclass, the Pulmonata, is represented by the Physidae, Lymnaeidae, Planorbidae, and Ancylidae. It is believed that they arose from littoral, marine snaïls due to selection imposed by exposure to the atmosphere (Hunter, 1964). This ancestoral stock evolved into two contemporary orders, the terrestrial Stylommatophora, and the freshwater Basommatophora. The aquatic pulmonates are monoecious, have no opercula, have lost their ctenidia and replaced them with highly vascularized mantle cavities, have relatively lightweight shells, and are comparatively active animals. The Physidae, Lymnaeidae, Planorbidae, and Ancylidae are distributed throughout the world.

B. Taxonomic References

Because one of the most formidable obstacles in the use of freshwater Gastropoda as biological indicators is the difficulty of identifying them, I am including the following list of citations referring to papers, monographs, and major publications useful in determining freshwater snails: Adams (1915), Bailey *et al.* (1931a,b,c), Baker (1902, 1911, 1912, 1928, 1945), Basch (1959a,b,c, 1960, 1962a,b,c, 1963), Beetle (1961), Berry (1943), Bickel (1967, 1968a,b,c, 1970), Branson (1969, undated), Branson and Batch (1971), Chamberlin and Jones (1929), Cheatum and Fullington (1971), Clarke (1973), Clench (1925, 1926, 1962a,b), Clench and Fuller (1965), Clench and Turner (1955, 1956), Clowers (1966), Dazo (1961, 1965), Dundee and Dundee (1969), Getz (1971), Goodrich (1921a,b, 1922, 1923, 1928, 1930, 1931, 1932, 1934a,b,c, 1935a,b, 1936, 1937, 1938, 1939, 1940, 1941, 1942, 1943, 1945), Goodrich and Van der Schalie (1939, 1944), Harman and Berg (1971), Henderson (1924, 1929, 1936a,b), Hickman (1937), Hubendick (1951, 1964), Inaba (1969), LaRocque (1953, 1970), Leonard (1959), Malek (1962), Miles (1958), Morrison (1932), Pilsbry (1934a), Robertson and Blakeslee (1948), Rosewater (1959a,b), Shoup *et al.* (1941), Sinclair (1969), Strecker (1935), Walker (1918, 1925), Walter (1969, 1972), Winslow (1918, 1926), and Wurtz (1949). Little of substance was found concerning the freshwater Gastropoda of the west coast. Taylor (in litt.) stated that "There are no useful references for determination of freshwater snails in western North America." Additional publications concerning Pleistocene mollusks that have been found valuable in determining contemporary species are La Rocque (1960), Pilsbry (1934b), Roy (1964), and Taylor (1960, 1966, 1970). A few citations referring to species that may easily be introduced into our southern states are also included: Anon. (1968), Ferguson and Richards (1963), Van der Schalie (1948).

In order to further assist the reader, the first mention of a species occurring in the United States will be accompanied by the author and date of description. If not otherwise noted, contemporary descriptions may be found in Harman and Berg (1971).

C. Life Cycles

Because eggs, immature snails, and adults react differently to the stresses imposed by various pollutants it is important to know something of the life cycles of these organisms. In the temperate regions the freshwater Gastropoda, particularly the smaller species, tend to have annual life cycles (a total life span of one year). There may be one reproductive period in the spring or in the fall, or two or more reproductive periods over the duration of a summer, with the original cohort replaced or supplemented. Some snails reproduce continuously during the entire summer period. Some species overwinter as small immatures, reaching maturity the following summer. Others overwinter at relatively large sizes and mature early in the spring, while still others overwinter in the mature state. Some of the larger species normally possess a biennial life cycle, following similar reproductive patterns as the annuals. If one assumes that the varices on the shells of specimens indicate a reduction of growth during the winter period, it is evident that a fair amount of individuals normally having biennial life cycles live for a period encompassing three summers. Intraspecific variation in the life cycles of various populations is pronounced and can be attributed to both genetic and environmental causes (DeWitt, 1954; Eisenberg, 1970, Hunter, 1964).

1. Pulmonata

Clampitt (1970) and DeWitt (1955), working with *Physa gyrina* Say 1821, Clampitt, with *P. integra* Haldeman 1841; and Herrmann (1972), with *P. heterostropha* Say 1817, noted that these snails had annual life cycles with early spring and summer cohorts, the first maturing before winter, the second maturing in spring. Vlasblom (1971) reported that *Aplexa hypnorum* Linnaeus 1746 had an annual life cycle with one cohort developing during one summer and maturing the next. Hunter (1964) observed several reproductive patterns in *P. fontinalis* (Linnaeus). The planorbids, *Helisoma anceps* Menke 1830 and *Planorbis albus* Müller, were observed, respectively, by Herrmann (1972) and Hunter (1964). The former exhibited a biennial life cycle, with young produced in June and July; the latter, an annual life cycle, with oviposition in July and August. Working with the family Lymnaeidae, Walton and Jones (1926) noted that *Limnaea* (=*Lymnaea*) *truncatula* (Müller) produced several cohorts during the summer. McCraw (1961) noted oviposition and hatching in the fall of *L. humilis* Say 1822, with the young and adults overwintering. The adults resumed oviposition in the spring and then died. Hunter (1964) observed a biennial life cycle in *Lymnaea stagnalis* Linnaeus 1758, which oviposited in May and June, the young supplementing the previous years population. The ancylids *Ferrissia rivularis* (Say) 1819 and *Ancylus fluviatilis* Müller were observed by Burkey (1971) and Hunter (1964), respectively. *Ferrissia rivularis* had an annual life cycle and reproduced all summer, while *A. fluviatilis* reproduced in only April and May.

2. Prosobranchia

The life cycles of the freshwater Prosobranchia appear to possess most of the variations typical of the Pulmonata. Most smaller species in temperate regions have an annual life cycle and breed once a year (Hunter, 1964). *Valvata tricarinata* (Say, 1817) oviposits in late spring and continues egg-laying into late summer (Heard, 1963). *Pomatiopsis cincinnatiensis* (Lea, 1840) (Baker, 1928) reaches maturity before winter, ovipositing in early spring (Van der Schalie and Getz, 1962). *Bithynia tentaculata* (Linnaeus 1758) young are first observed in June and July, at which time adults from the proceeding year reach the end of their life span (Pinel-Alloul and Magnin, 1971).

The larger species tend to have biennial and triennial life cycles with some individuals living four summers. *Viviparus contectoides* (Binney 1805) [=*georgianus* (Lea 1834)] breeds in the spring, giving birth to living young (Van Cleave and Lederer, 1932). A similar situation has been observed in *Campeloma rufum* (Haldeman 1841) (concept of Baker, 1928; Van Cleave and Altringer, 1937) and in the genus *Lioplax* (Van Cleave and Chambers, 1935). Winterkill of *Viviparus japonicus* Von Martens 1860 (LaRocque, 1953) has been observed in Lake Erie by Wolfort and Hiltunen (1968).

Pleurocera acuta Rafinesque 1831 and *Goniobasis livescens* (Menke 1830) mate in the fall and oviposit the following spring, living as long as four years (Dazo, 1965). Van Cleave (1932) observed oviposition of *P. acuta* from spring into summer. Extensive decimation of populations of *P. canaliculatum undulatum* (Say 1829) (concept of Goodrich, 1940) has been observed in the winter by Magruder (1934).

III. Freshwater Gastropod Ecology

The physiological ecology of freshwater Gastropoda has been reviewed by Hunter (1964). Various other aspects of ecology have been touched upon in Fretter (1968), Fretter and Graham (1962), Purchon (1968) and Wilbur and Yonge (1966). Therefore, I will concentrate only on those parameters normally measured by the limnologist or field biologist. One problem that must be kept in mind is that these data often are not readily comparable because of different analytical procedures and because the time and techniques of sampling vary considerably.

A. Chemistry

1. Hydrogen Ion Concentration

Harman and Berg (1971) have tabulated the pH ranges experienced by 41 species of freshwater snails in more than 570 biotopes in central New York State. The hydrogen ion concentrations of the waters sampled varied from 6.7 to 9.0. The majority of pulmonates collected occurred from a pH of slightly above 7.0 to about 8.4. Most of the prosobranchs showed somewhat narrower ranges, from about 7.4 to approximately 8.3 (See Table I). The most common species occurred over the widest pH ranges. The rarest species were collected over narrow ranges of hydrogen ion concentration. It cannot be ascertained whether these organisms are uncommon because

TABLE I

pH Values at Which Central New York Gastropoda Were Collected[a]

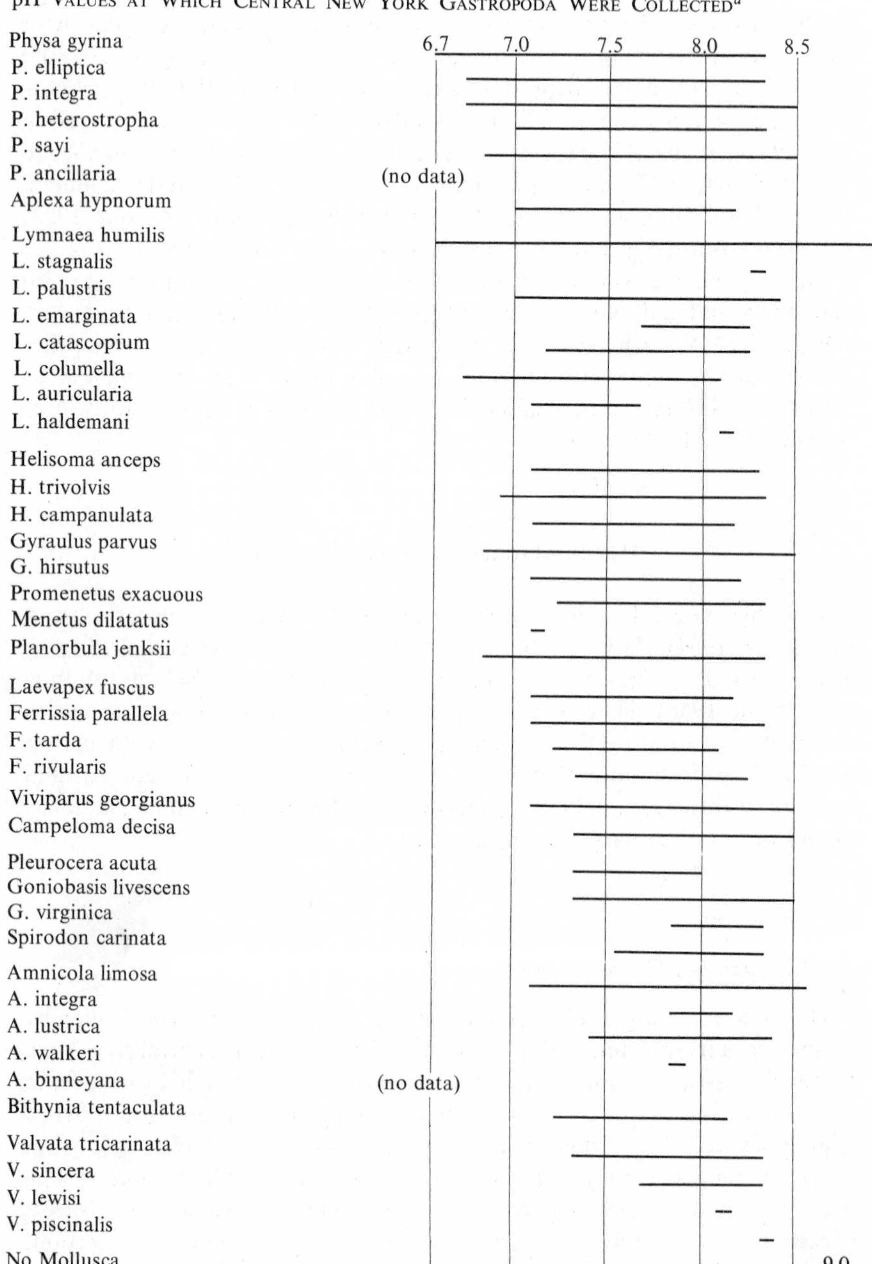

[a] From Harman and Berg, 1971.

of stenotopic tendencies or that they appear to be stenotopic because the small number of biotopes in which they were collected do not reflect the full range of pH values encountered in the watersheds concerned.

Morrison (1932) collected data on several species in Wisconsin lakes where the pH of the local waters ranged from 5.1 to 8.3. The results indicate that the pH tolerances of Wisconsin snails differ considerably from the pH tolerances of the same species of snails collected in New York by Harman and Berg. For example, *Physa sayi* Tappan 1839 occurred over a pH range of 5.68 to 7.96 in Wisconsin and a range of 6.9 to 8.5 in New York; *Helisoma anceps* in Wisconsin, 6.03 to 8.02, but in New York, 7.1 to 8.4; *Ferrissia parallela* (Haldeman 1841), in Wisconsin, 6.05 to 8.51, but in New York, 7.3 to 8.1. These data indicate that these species can tolerate the ranges of pH experienced in both geographic regions and that the data simply reflect the characteristics of the local waters or that populations of gastropods from one region have become adapted to the mean hydrogen ion concentration of the waters in that region resulting in great intraspecific variation in reference to this parameter.

Extraordinarily high pH values in natural waters do not appear detrimental to some species of freshwater pulmonates. In Moe Pond, at the State University of New York's College at Oneonta Biological Field Station at Cooperstown, New York, the pH in 1970–1971 commonly varied between 6.9 and 8.5 (Herrmann, 1972). During three weeks in June and July 1972 the pH there fluctuated around 10 (maximum 10.3). Populations of *P. heterostropha*, *H. anceps*, and *F. parallela* from the pond have not exhibited any deleterious effects (Harman, unpublished).

Very low pH values occurring in natural waters may be lethal to freshwater snails. Gastropods are not found in Scandanavian Lakes that maintain pH values between 4.4 and 5.2 (Økland, 1969b). Schwartz and Meredith (1962) cite several sources that indicate that hydrogen ion concentrations below 5.5 affect oxygen levels to the extent that fish and "other aquatic organisms" may be eliminated. However, Bell (in litt.) recorded *C. decisum* living in waters having a pH value as low as 3.9. References pertaining to artificial manipulation of pH values will be discussed in Section IV,A,1.

A serious problem concerning the validity of these data results from the normal variation of pH in natural waters. For example, on July 30–31, 1969, in Canadarago Lake, New York, Fuhs (1972) collected pH readings from various parts of the lake that ranged from 6.2 to 11.0. Another problem is the ability to record pH values that the organisms are actually exposed to. Yongue and Cairns (1971) submerged polyurethane sponges in a pond in North Carolina. pH readings taken after the establishment of microbial communities in the sponges showed that water from them had a pH of 6.8 while that of the surrounding liquid was 10.5. Normal survey techniques could result in greatly misleading data in either of these cases.

2. Alkalinity

As with hydrogen ion concentration, Harman and Berg (1971) have tabulated the alkalinity (as $CaCO_3$ in ppm) in which 41 species of central New York freshwater gastropods were collected (Table II). The normal range of alkalinity was from about 20 ppm to about 180 ppm with a few stations extremely higher, extending the total range up to a maximum of 665 ppm in a polluted situation. As with pH values, the more common organisms exhibited the widest ranges; the uncommon, the narrowest. Again pulmonates appeared more eurytopic than the prosobranchs. There were no definitive results that showed that any species was eliminated by low alkalinity. However, in several localities the vigor of populations, as determined by adult size and population density, was apparently affected. Several authors—for examples, Harrison *et al.* (1970) and Houp (1970)—have noted that population size and vigor vary directly with alkalinity. Harrel and Dorris (1968) recognized a direct relationship between high alkalinity and high productivity. In 1943 Shoup studied the distribution of freshwater snails in Tennessee streams and attained an almost linear relation between the distributions of 47 species and stream alkalinity.

Horst (1971) studied McCargo Lake in New York State, where alkalinity values ranged from 64 to 300 ppm during the course of the study. Populations of *P. sayi*, *Gyraulus parvus* (Say 1817), *Promenetus exacuous* (Say 1817), *Helisoma trivolvis* (Say 1821), *V. georgianus*, *Amnicola limosa* (Say 1817), and *V. tricarinata* are established there.

A factor directly related to alkalinity, although often not measured in limnological surveys, is the presence of calcium. Calcium availability often limits primary productivity in aquatic biotopes and therefore may indirectly affect snails by reducing food and cover. Several authors, e.g., Boycott (1936), Hunter (1964), and Macan (1950), have asserted that a closer correlation exists between the presence of calcium and selected species than between them and alkalinity. However, Hunter *et al.* (1967) have shown that utilization of calcium may be very different in various populations of the same species. According to Hunter (1964), waters with more than 25 mg/liter of calcium can support all molluscan species in a geographic region, waters with 10 to 25 mg/liter support about 55%, waters with 5 to 10 mg/liter support about 40%, and waters with less than 3 mg/liter support less than 5%.

3. Total Carbon Dioxide

As with the other parameters, Harman and Berg (1971) reported on the ranges of total carbon dioxide in ppm experienced by 41 species of central New York Gastropoda. The range encountered in more than 570 biotopes was between 0 and 43 ppm (Table III). Snails occurred throughout the entire range. As before, common species exhibited the wider ranges, rare

TABLE II

Alkalinity in $CaCO_3$ (ppm) at Which Central New York Gastropoda Were Collected[a]

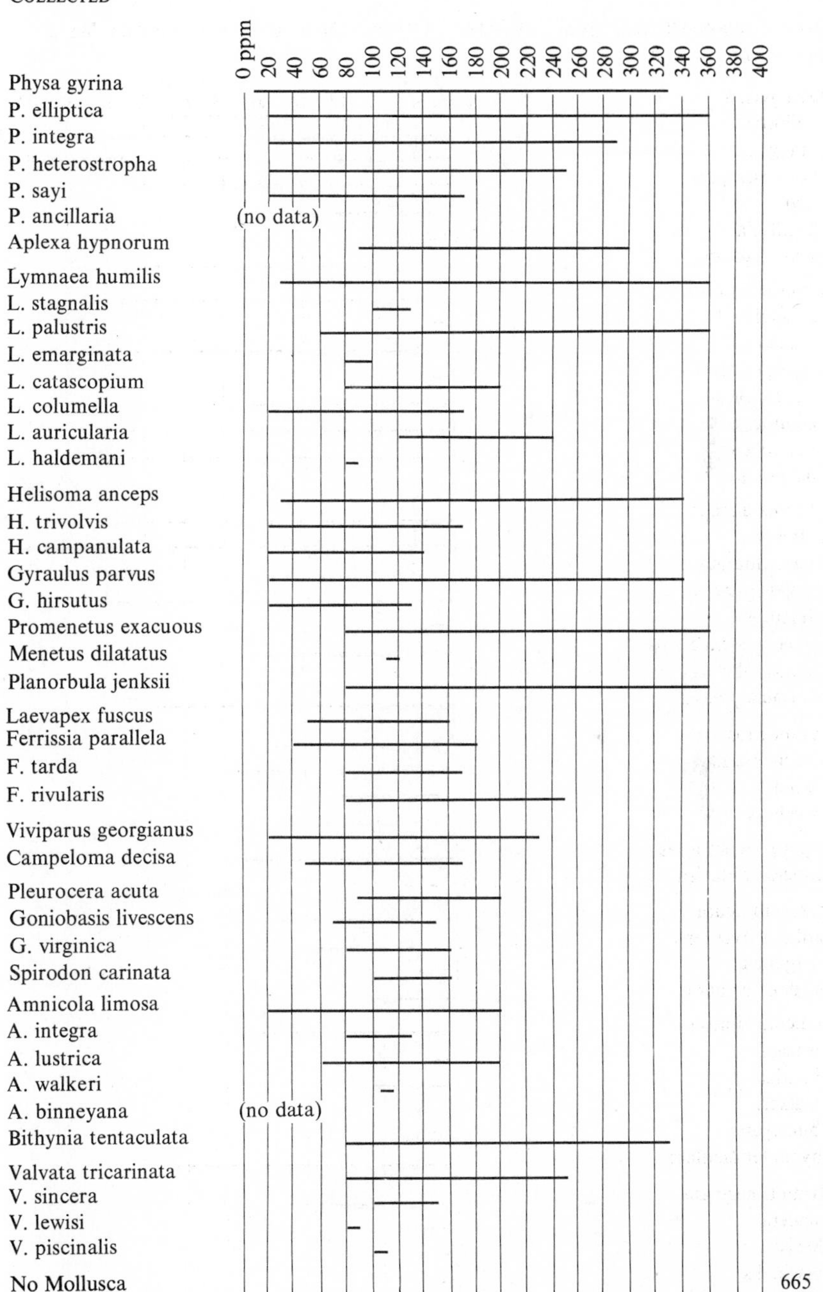

[a]From Harman and Berg, 1971.

TABLE III

TOTAL CARBON DIOXIDE (ppm) AT WHICH CENTRAL NEW YORK GASTROPODA WERE COLLECTED[a]

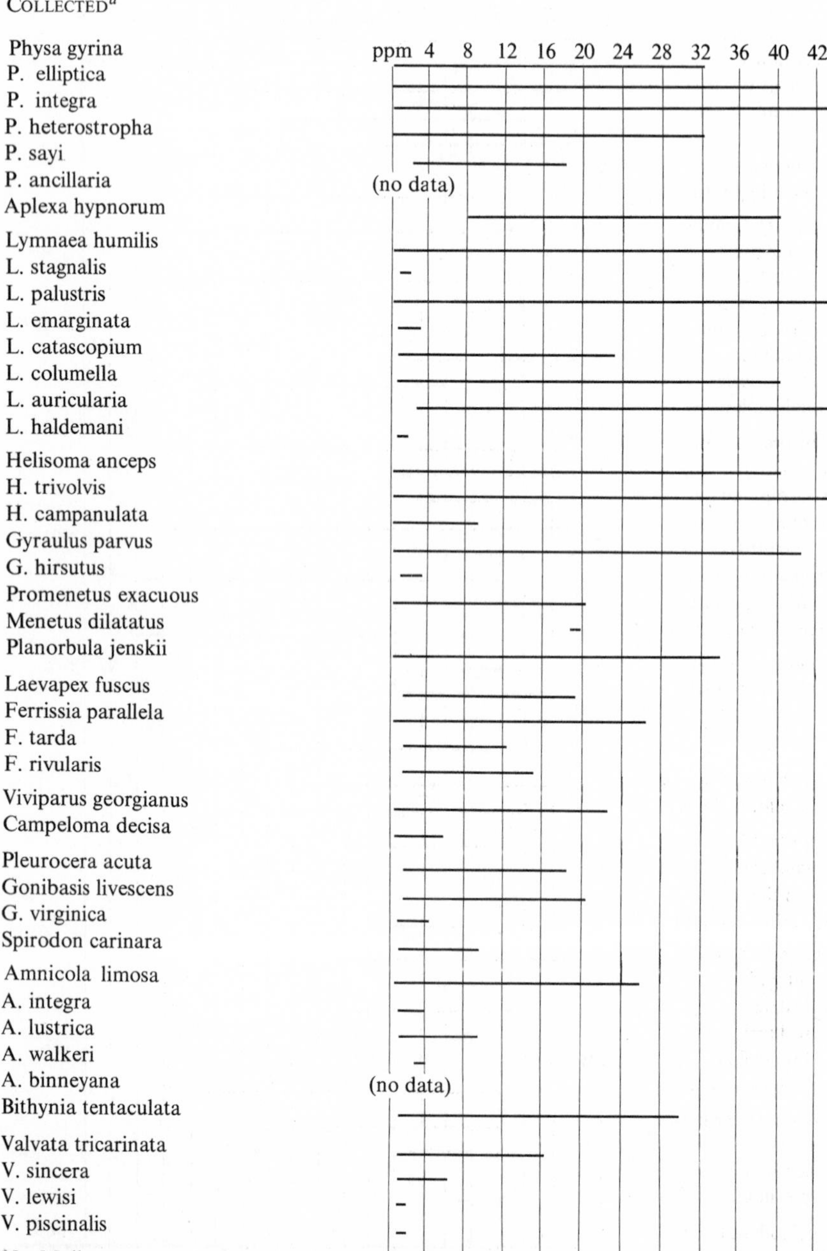

[a]From Harman and Berg, 1971.

species, the narrower. With the exception of the water-dependent Ancylidae, the Pulmonata generally appeared strongly eurytopic, the Prosobranchia stenotopic. Horst (1971) recorded total carbon dioxide in McCargo Lake from 0 to 54 ppm where populations of *P. sayi*, *G. parvus*, *P. exacuous*, *H. trivolvis*, *V. georgianus*, *A. limosa*, and *V. tricarinata* are present. There is no evidence that any species is limited by total carbon dioxide in solution.

The apparent stenotopic tendencies of the Prosobranchia, compared to those of the Pulmonata, regarding hydrogen ion concentration, alkalinity, and total carbon dioxide in central New York is, in the opinion of the author, a reflection of the chemical attributes of the biotopes that these subclasses inhabit. The monoecious, light-weight, active, often semi-aquatic pulmonates disperse widely into ephemeral aquatic environments such as roadside ditches, marshes, swamps, temporary streams, small ponds, and other similar biotopes where the parameters discussed fluctuate extremely. The dioecious, usually ponderous, sluggish, water-dependent prosobranchs, because of their poor dispersal abilities, remain in the larger, more homeostatic lakes and rivers and never experience the broad ranges of these values that the pulmonates do. Therefore, these data do not necessarily imply that pulmonates are more eurytopic concerning these factors than are the prosobranchs. This is not to imply that southeastern prosobranchs, existing in rivers undisturbed by Pleistocene glaciation, have not become more specialized, and therefore more stenotopic than their northern counterparts.

4. *Dissolved Oxygen*

Gastropods need oxygen in order to carry on normal metabolic activities. Therefore, it is obvious that dissolved oxygen must be limiting in some biotopes. The hypolimnions of eutrophic lakes, by definition devoid of oxygen, do not support snails. The Pulmonata, because they are able to utilize atmospheric air for respiration, can exist in anaerobic waters for extended periods of time. However, their eggs must be in contact with dissolved oxygen for development to occur (Richard, 1965). Harman and Berg (1971) have listed 24 species of pulmonates, four of which were found in anaerobic waters (*P. integra*, *Lymnaea palustris* Müller 1774, *L. auricularia* (Linneaus 1758), and *H. trivolvis*) (Table IV). Horst (1971) collected *P. sayi*, *G. parvus*, *P. exacuous*, and *H. trivolvis* in waters with dissolved oxygen values as low as 0.4 ppm. Turner (1972) recorded *P. heterostropha* from waters lacking oxygen. Gaufan and Tarzwell (1952) also collected *P. integra* in practically anaerobic waters.

Boycott (1936) stated that prosobranchs are more limited than pulmonates to waters containing dissolved oxygen because they must utilize oxygen from the water in respiration. However, some prosobranchs have been

TABLE IV

CONCENTRATIONS OF DISSOLVED OXYGEN (ppm or mg/liter) AT WHICH CENTRAL NEW YORK GASTROPODA WERE COLLECTED[a]

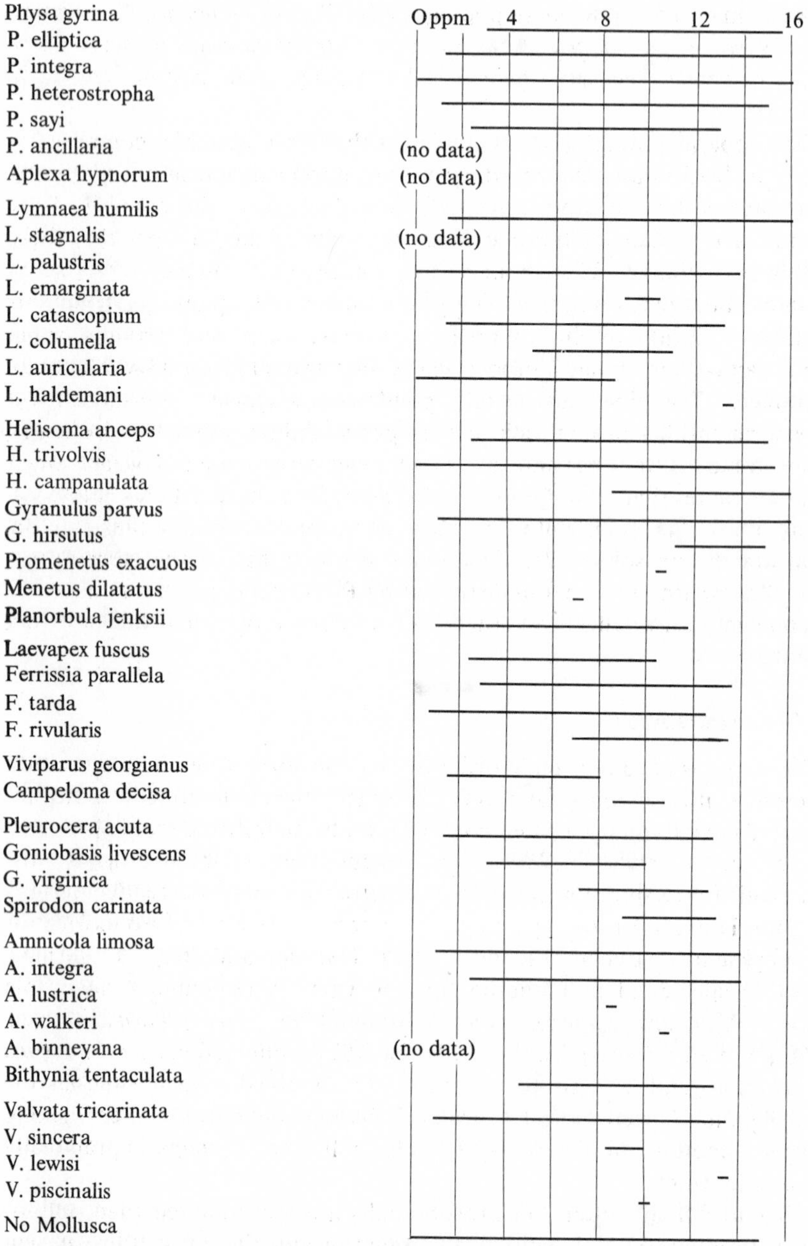

Species	0 ppm 4 8 12 16
Physa gyrina	
P. elliptica	
P. integra	
P. heterostropha	
P. sayi	
P. ancillaria	(no data)
Aplexa hypnorum	(no data)
Lymnaea humilis	
L. stagnalis	(no data)
L. palustris	
L. emarginata	
L. catascopium	
L. columella	
L. auricularia	
L. haldemani	
Helisoma anceps	
H. trivolvis	
H. campanulata	
Gyranulus parvus	
G. hirsutus	
Promenetus exacuous	
Menetus dilatatus	
Planorbula jenksii	
Laevapex fuscus	
Ferrissia parallela	
F. tarda	
F. rivularis	
Viviparus georgianus	
Campeloma decisa	
Pleurocera acuta	
Goniobasis livescens	
G. virginica	
Spirodon carinata	
Amnicola limosa	
A. integra	
A. lustrica	
A. walkeri	
A. binneyana	(no data)
Bithynia tentaculata	
Valvata tricarinata	
V. sincera	
V. lewisi	
V. piscinalis	
No Mollusca	

[a]From Harman and Berg, 1971.

collected in waters comparatively low in dissolved oxygen. Harman and Berg (1971) found *V. georgianus, A. limosa* and *V. tricarinata* in waters with less than 2 ppm oxygen. Horst (1971) found *A. limosa*, *V. tricarinata*, and *V. georgianus* in McCargo Lake where a low surface reading of 0.4 ppm was recorded. Richardson (1925) collected *Campeloma subsolidium* (Anthony 1844) (concept of Clench, 1962) in water with 0.5 ppm oxygen. Sutter and Moore (1922) found *Goniobasis virginica* (Gmelin 1791) and *Campeloma decisum* (Say 1817) in anerobic waters.

The larger prosobranchs are seldom found in ephemeral environments. However, it may be incorrect to attribute their absence in these biotopes to periodic shortages of oxygen. Much more limiting may be the inability of these organisms to immigrate into these isolated environments and the occasional shortages of water occurring there (Harman and Berg, 1971).

Berg and Ockelmann (1959) have shown that two types of respiratory activity occur in freshwater snails. One group tends to maintain oxygen consumption, despite a lowering of dissolved oxygen, until a critical threshold is reached. The other exhibits a slow reduction in consumption as the dissolved oxygen decreases. However, other requirements, such as food availability and temperature considerations, may override these phenomena so that it may not be possible to demonstrate a correlation between respiratory physiology and the distribution of snails in nature.

B. Physical Factors

1. Temperature

In natural waters, temperature seldom eliminates species living within their normal geographic ranges. Members of all four families of Nearctic pulmonates, and two families (Hydrobiidae and Pleuroceridae) of prosobranches have been observed actively foraging in ice-covered waters (Harman, unpublished). McDonald (1969) cites sources which indicate that *L. stagnalis* may be supercooled to −3.5°C without ill effects. Cheatum (1934) froze *L. stagnalis*, *L. pulustris*, *L. emarginata* (Say 1821), *L. megasoma* Say 1824 (concept of Baker, 1911), *H. antrosum percarinatum* (Walker 1909) (concept of Baker, 1928), *H. campanulatum smithi* (Baker 1912) *H. trivolvis*, *P. parkeri* "Currier" DeCamp 1881 (concept of LaRocque, 1953), and *P. sayi* in cakes of ice. All withstood this treatment after having 84 hours to become acclimated to these low temperatures. McDonald (1969) states the lymnaeids "... are more sensitive to warm temperatures, usually not surviving continuous exposure to temperatures greater than 30°C." However, she noted that *L. peregra* (Müller) has been found in warm water springs at 45°C. Harman and Berg (1971) have reared northeastern species of Lymnaeidae, Physidae, and Planorbidae through several generations at about 35°C.

The indirect effects of temperature may be biological or chemical in nature. This parameter has profound effects on primary productivity, in both the amount and the quality of the biomass produced. In temperatures below 25°C diatoms are the most abundant algae; between 25° and 35°C green algae predominate; at higher temperatures blue-green algae compose the largest number of species (Cairns, 1970). Since the quality of food appears important to snails (Butler *et al.*, 1969; Calow, 1970; Eisenberg, 1970), and their omnivorous habits allow for little active selection (Clampitt, 1970), the types of food available in the greatest quantity may have profound effects on the vigor of the populations concerned. Individuals of *L. stagnalis* living in 16 m of water in Seneca Lake, New York, average around 27 mm in height while those living at the surface in Little York Lake, New York attain sizes of 45 mm (Harman and Berg, 1971). Reproductive patterns (Section II,C) are usually different between populations of one species where they occur in habitats with varying thermal regimes.

The interactions between temperature, dissolved oxygen, alkalinity, and other limnological factors have been discussed by many authors. For a comprehensive treatment see Hutchinson (1957).

2. Substrate Characteristics

Several authors, in attempting to determine factors affecting species presence and distribution, have found little or no correlation with chemical factors (Harman, 1972; Harman and Berg, 1971; Harry *et al.*, 1957; de Meillon *et al.*, 1958; Schutte and Frank, 1964). The physical factors affecting the characteristics of local biotopes such as parent rock and soils, angle of repose, exposure, fetch, current, nutrient sources, and productivity all affect benthic substrates, whether organic or inorganic.

The observation that a particular freshwater gastropod species is frequently encountered on a particular substrate has been recognized by naturalists for many years. Baker's (1928) classic work on the snails of Wisconsin includes notes on the types of substrates that many species are normally found on. Sparks (1961), Zimmerman (1960) and other authors have listed associations of freshwater species according to "water types," classifications that are strongly related to the same factors that determine the characteristics of substrates, the microhabitats on which gastropods live. Goodrich (1921) noted that several species of *Goniobasis* were isolated from each other by waters with different substrate characteristics. Foin (1971) noted the same conditions in respect to the distribution of *Oxytrema proxima* (Say 1825) (concept of Tryon, 1873). Houp (1970) recognized that the distribution of *P. acuta* was closely correlated with stream substrates in central Kentucky. Vlasblom (1971) observed preferences for particular substrate types by *A. hypnorum* in the Netherlands. In Ohio

Bickel (1965) noted that *P. integra* was always associated with certain macrophytes and submerged logs. Harman and Berg (1971), working in central New York, observed that snail species in that region occurred on particular substrates or combinations thereof, and contended that substrate conditions were the principal factors involved in the potential establishment of species in various habitats. Harman (1972) showed that definite relationships exist between mollusk distribution and substrate patterns in three New York lakes. He listed four associations of snails found on various substrate types and correlated the numbers of species found in 348 locations, including several biotope types, with the types of substrates in each and found that $r = 0.79$, a highly significant correlation.

IV. Pollution Ecology of Freshwater Gastropoda

A. Chemical Pollutants

The toxicity of most substances is influenced by such factors as temperature, turbidity, hydrogen ion concentration, dissolved oxygen, carbon dioxide, and water hardness (Anon., 1971b). For these reasons, laboratory conditions under which quantitative material is gathered must be fully presented to provide a valid assessment of the effects of chemicals on organisms. In the interest of space, detailed procedures have not been mentioned in this review. The original sources should be obtained for precise evaluation of the data.

It should be noted that dilutions of chemicals which cause no detrimental effects on organisms in short-term laboratory tests may cause serious long-term effects in nature. Many pollutants accumulate for periods of time in sediments. Due to varying conditions, such as changes of temperature or pH, these compounds may return into solution in lethal amounts. Low concentrations of chemicals may affect the fitness of the organisms stressed by changing behavior patterns, feeding rates, and by increasing oxygen consumption (Cairns, 1968a).

1. *Industrial Wastes*

Acid drainage from regions where anthracite and bituminous coal, gold, barite, manganese, lead, and zinc were mined affected more than 48,000 miles of rivers in the United States as early as 1962 (Cairns *et al.*, 1971). This amounted to total acid loads of about 3.5 million tons per year resulting in waters that normally possess pH values much lower than 6.0 (Herricks and Cairns, 1972). Since hydrogen ion levels below 5.5 are deleterious to most benthic organisms (Schwartz and Meredith, 1962), these streams are probably devoid of gastropods.

Økland (1969a; 1972, in litt.) has evidence that the burning of fossil fuels, resulting in increasing amounts of sulfur in the atmosphere, has resulted in a lowering of pH values in Scandanavian waters. In many lakes trout, which are apparently more tolerant of low hydrogen ion values than snails, have already been decimated. The same environmental conditions have recently become apparent in our New England States where increased acidity and sulfur content of inland waters can be traced to input of acids from the atmosphere via precipitation (Johnson et al., 1972).

Although some are necessary for normal metabolism, excess amounts of heavy metals are toxic to organisms. Snails are particularly sensitive to zinc, copper, mercury, and silver (Wurtz, 1962). The Ystwyta River in Wales was devoid of mollusks 35 years after the termination of lead mining in that region and was still carrying up to 0.7 ppm Zn (Wurtz, 1962). Wurtz also reported that one year after water was pumped from a base-metal (Zn, Pb, Cu) mine in preparation for renewing operations in New Brunswick, Canada, the molluscan fauna was still absent for 12 miles below the outfall into the Northwest Minamichi River. At the time of observation five species of mollusks were present upstream from the source of pollution.

Heavy metal toxicity is affected by hardness and pH (Arthur and Leonard, 1970). Wurtz (1962) conducted several experiments (using TL_m's expressed by the concentration of a chemical needed to eradicate 50% of a laboratory population during a predetermined period of time) with *P. heterostropha* using $ZnSO_4 \cdot 7H_2O$ (22.6% Zn) in soft (20 ppm total hardness) and hard (100 ppm total hardness) waters. The pH was maintained above 7.0 in both cases. Adult snails (12 to 15 mm shell height) exhibited 96-hour TL_m's of 3.16 ppm Zn in hard water and 1.11 ppm Zn in soft water, both at a temperature of 70 ± 2°F. Young (3 to 6 mm shell length) *P. heterostropha* were also tested. At 68°F 96-hour TL_m's were 1.70 ppm Zn in hard water, 0.434 ppm Zn in soft water. *Helisoma campanulatum* was bioassayed in a like manner. Ninety-six hour TL_m's at 73°F were 1.27 ppm Zn in both hard and soft water. He concluded that *H. campanulatum*, having hemoglobin in their circulatory systems, were more tolerant of Zn than *P. heterostropha* that utilize hemocyanin. Cairns and Scheier (1958) and Patrick *et al* (1968) worked with *P. heterostropha* and Zn^{2+} as $ZnCl_2$. Ninety-six hour TL_m's in soft water (40 ppm total hardness) at 20°C were between 0.79 and 1.27 ppm. In hard water (165 ppm total hardness) at 20°C, the 96-hour TL_m was 2.66 to 5.57.

Bioassays using copper sulfate $CuSO_4 \cdot 5H_2O$ (25.5% Cu) and *P. heterostropha* were also undertaken by Wurtz (1962). Adult—96 hour TL_m's in hard water (100 ppm total hardness) were 0.069 ppm Cu at 70 ± 2°F. Young snails in water comparable to Wurtz's experiments with zinc sulfate had 96-hour TL_m's of 0.05 ppm Cu in hard waters and 0.034 ppm Cu in soft water (20 ppm total hardness). Arthur and Leonard (1970) determined 96-hour TL_m's values for *C. decisum* as 1.7 mg/liter Cu and *P. integra* as 0.039

mg/liter Cu. The highest total copper concentration at which no effect on these species was noted after six weeks exposure was between 8.0 and 14.8 μg/liter at 15° C.

Extremely high pH values may also be detrimental to gastropod populations. In 1967 at Carbo, Virginia, a fly ash pond collapsed releasing 130 million gallons of slurry into the Clinch River via Dumps' Creek. This constituted 40% of the rivers normal flow (Cairns *et al.*, 1972). The slurry was primarily $Ca(OH)_2$ having a pH between 12.0 and 12.7. This spilled a slug, composed of about 90% hydroxide alkalinity and 10% carbonate alkalinity, that eliminated snails 11.7 miles below the outfall (fish were killed for 90 miles downstream). This area had not recovered two years after the incident.

Spills of petroleum products are everyday occurrences in our urbanized environments. However, very little has been written regarding the tolerance of gastropods to these stresses. Cairns *et al.* (1971) reported a spill of ethyl benzene and creosote at Salem, Virginia, into the Roanoke River. Between 400 to 600 gallons entered the stream in a 1- to 2-hour period with a resultant estimated concentration of 1000 ppm. The benthic community was affected up to seven miles below the spill. However "... some snails ..." survived this exposure. Bugbee and Walter (1972) observed the effects of a gasoline spill on Grace Cooledge Creek, South Dakota, during May 1970. Six months after the spill Trichoptera and Ephemeroptera were still absent and the sediments possessed a high organic carbon content. Although samples were apparently too small to judge accurately the effects on snails, *Physa* sp., *Helisoma* sp., and *Ferrissia* sp. were collected at and below the spill.

Cairns and Scheier (1957, 1962) studied the effects of naphthenic acids on *P. heterostropha*. These compounds occur in petroleum that is comparatively poor in paraffins and are used in insecticides, paper production, and in rubber and mineral oil processing (Cairns and Scheier, 1962). The experiments were conducted in soft and hard waters (see sources for chemical composition of water) and at 20° and 30°C. The TL_m for *P. heterostropha* at 20°C, in soft water was 6.6 ppm; at 30°C in soft water, 15.6 ppm; in hard water at 20°C, 12.6 ppm; in hard water at 30°C, 12.2 ppm (Patrick *et al.*, 1968). *Physa heterostropha* was also used in experiments to determine the effect of various low oxygen levels on tolerance to naphthenic acids (Cairns and Scheier, 1957). For these organisms, 96-hour TL_m's in waters containing 5 to 9 ppm oxygen were 6.6 to 7.5 ppm; under a two-hour low oxygen level of 2 ppm per 24 hours the 96-hour TL_m was 2.0 ppm, indicating less tolerance under the additional stress of low oxygen conditions. However, when 2.0 ppm tests were conducted with a "race" of *P. heterostropha* from an environment with naturally low dissolved oxygen there was no difference from the test run at the 5 to 9 ppm dissolved oxygen levels. This would indicate the intraspecific variation in physiology expected of pulmonate snails and reminds one that overgeneralization of such data, collected on a single population of a eurytopic species, is risky.

Before 1965 alkyl benzene sulfonate (ABS) was used extensively in the production of detergents. Cairns *et al.* (1964) and Patrick *et al.* (1968) studied the toxicity of this compound on *P. heterostropha*. The temperature was kept at 20°C and dissolved oxygen maintained between 5 to 9 ppm (for the specific composition of the water, see the original paper). It was determined that the toxicity in hard and soft waters was identical. The 96-hour TL_m was between 34.2 and 35.8 ppm. This approximates the results of Surber and Thacher (1963) for *Goniobasis* sp. They used detergent containing 27% ABS. *Goniobasis* sp. existed two weeks at 10 ppm ABS with no apparent effects but was eliminated at 32 ppm over the same time period. At 16 ppm between 40 and 80% of the population was decimated. Since 1965 detergents have been used in which biodegradable linear alkylate sulfonate (LAS) has replaced ABS. Arthur (1970) studied the toxicity of LAS on *C. decisum* and *P. integra* at 15°C. Ninety-six-hour TL_m's for *C. decisum* and *P. integra* were 27 mg/liter and 9 mg/liter, respectively. The maximum acceptable LAS concentrations for *C. decisum* for six weeks were 0.4 to 1.0 mg/liter. Trisodium nitrilotriacetate (Na_3NTA) has been proposed as a replacement for polyphosphates in detergents (Flannagan, 1971). Flannagan studied the toxicity of the compound in *Physa* sp., *L. stagnalis* and *H. trivolvis*. It is found that Na_3 NTA was not directly toxic to these snails at concentrations up to 500 mg/liter. Mortality was greater in soft water than in hard because of the buffering capacity of the latter. Snails were not affected until pH values exceeded 9.2, and then pH was considered the direct cause of their elimination.

Chloride wastes from Olin Chemical Corporations' soda ash operation in the North Fork of the Holston River over the last 75 years has apparently eliminated *Io fluvialis* (Say 1825) (Goodrich, 1940), *Oxytrema unicale* (Haldeman 1841) (Stansbery, 1972), and *Leptoxis sùbglobsa* (Say 1825) (Stansbery, 1972). All of these species occur above the pollution source. Other industrial wastes have been checked for toxicity to *P. heterostropha* (Patrick *et al.*, 1968). Ninety-six-hour TL_m's were recorded at 20 $\pm$2°C with a dissolved oxygen of 5 to 9 ppm. Ammonium (NH_3–H calculated was 90.0 ppm; CN^-, 0.432 ppm; Cr^{2+} (when introduced as $K_2Cr_2O_7$), 17.3 ppm; Cr^{2+} (when introduced as $K_2Cr_2O_4$), 16.8 ppm; KCL, 940.0 ppm; and phenol, 94.0 ppm.

2. *Pesticides*

In tropical climates, where schistosomiasis is a serious health problem, many chemicals have been used as molluscicides. Because these, or closely related compounds, may be inadvertantly or unscrupulously introduced into freshwater environments in the United States, a few papers should be mentioned to introduce the reader to this rich literature (Boyce *et al.*, 1967; Chu *et al.*, 1968; Jobin and Unrau, 1967; Ritchie and Berrios-Duran, 1969; Ritchie *et al.*, 1969; Seiffer and Schof, 1967). An excellent review of the chemical control of freshwater snails is included in Muirhead-Thomson's "Pesticides and Freshwater Fauna (1971)."

In the northern United States, and particularly the Lake States, schistosomes causing cercerial dermatitis (swimmer's itch) are serious pests. These trematodes utilize several species of Lymnaeidae, Physidae, and Planorbidae as intermediate hosts. Copper sulfate and hydrated lime are used for mollusk control, and therefore disease control, in these areas (Mackenthunan and Ingram, 1967). Solutions of the chemicals mentioned are applied at the rate of two pounds per 100 ft^2 for control (Levy and Folstad, 1969). In Minnesota, where no permit is required to apply many kinds of molluscicides or aquatic herbicides, 158,023 pounds of copper sulfate were used to control snails and vegetation in 1968 (Levy and Folstad, 1969). The same authors claim that during that year 90,000 pounds of $CuSO_4$ were used for snail control in Michigan, and in Wisconsin 57,503 pounds were used as molluscicides and herbicides. In small, lentic waters these treatments are very effective because the desired concentrations of Cu^{2+} (*ca* 32 ppm) are not only lethal to snails but also eliminate their food and cover.

Because of rampant cultural eutrophication throughout the United States, increasing use of herbicides must be taking its toll of freshwater Gastropoda by elimination of plants and direct toxicity to snails. The active ingredients usually used are listed by Johnson (1965) and Ferguson (1971). The effects of these chemicals on freshwater snails are practically unknown. Organotins and organoleads are very toxic to snails and their eggs (Muirhead-Thompson, 1971), as are the bivalent herbicides such as Acrolein (2-propenal), Paraquat (1,1′-dimethyl 1-4, 4′-bipyridynium), and Diquat (1-1′-ethylene-2-2′-dipyridylium). Harp and Campbell (1964) studied the effects of Silvex [2-(2,4,5-trichlorophenoxy)propionic acid] on benthic organisms. With a concentration of 4.6 ppm of active ingredient, which exceeded recommended dosage, *H. trivolvis*, *P. gyrina*, and *P. sayi* appeared unaffected. Walker (1971) reported that after six weeks exposure to 0.5 ppm of disodium endothall the biomass of snails in a 1-hectare pond had decreased from 1.920 gm/m^2 to 0.196 gm/m^2. Zischkale (1952) tested the effects of several herbicides on *Physa* sp. and *Helisoma* sp. at a pH of 7.2 and a temperature between 27° and 29°C. The results were as follows: Using Esteron 44 (the isopropylester of 2,4-dichlorophenoxyacetic acid), the maximum tolerance for *Physa* sp. was 5.0 ppm and the minimum lethal dose 5.5 ppm. The maximum tolerance for *Helisoma* sp. was 10.0 ppm. The use of sodium arsenite (sodium metaarsenite) in a solution containing 6.5 pounds per gallon resulted in a maximum tolerance of *Physa* sp. at 4.0 ppm and a minimum lethal dose of 4.5 ppm. *Helisoma* sp. had a maximum tolerance of 3.0 ppm and a minimum lethal dose of 3.5 ppm with the same compound. Hanson (1952) treated 15 acres of lentic water with Chlordane (1,2,4,5,6,7, 8,8-octochloro-2,3,3a,4,7,7a-hexahydro-4,7-methanomdene) at a dosage of one pound per acre. Snails remained abundant and did not appear affected. Fredeen and Duffy (1970) noted TDE [2,2-bis (*p*-chlorophenyl)-1, 1-dichloraethane] residues in *Campeloma* sp. Seventeen miles upstream from the application point, 0.002 ppm were found, 10 miles downstream, 0.101 ppm were present, 45 miles downstream no residues were present. Pimentel (1971) collated data collected by several investigators. Atrazine (2-chloro-4-ethylamino-6-isopropylamino-*s*-triazine) in the environment at concentrations of 0.5 to 2.0 ppm was correlated with a 4-fold increase in the snail populations. Neburon (1-*y*-butyl-3-3,4-dichlorophenyl-1-methylurea) concentrations of 1 to 10 ppm were accompanied by a doubling of the snail populations. Snails survived 380 ppm of Picloram (4-amino-3,5,6-trichloropicolinic acid), but at 530 ppm 100% mortality occurred. After one treatment of Simazine [2-chloro-4,6-bis(ethyl amino)-*s*-triazine] at 0.5 to 10 ppm about 50% of the snails were eliminated. One week after treatment of 2,4-D (2,4-dichlorophenoxyacetic acid or its sodium salt or amine) snail populations were reduced by 50%. Concentrations of the herbicide in the environment were between 1 to 4 ppm.

A potential source of toxic pollution is bis (tri-*n*-butyltin) oxide formulated in rubber. The active toxicant is slowly released into the water. This material has been used in antifouling paints, and is highly toxic to mollusks and their eggs (Berrios-Duran and Ritchie, 1968). If vessels plying our inland waterways utilize these protective coatings, many gastropods may be exposed.

Insecticides have received tremendous publicity through their apparent pollution of the entire ecosystem. The appalling lack of information concerning the effects of these compounds on freshwater mollusks is exemplified by a 62-page chapter entitled "Impact of Pesticides on Aquatic Invertebrates in Nature" in Muirhead-Thomson's excellent volume (1971). The word "mollusk" is not even mentioned.

DDT (dichlorodiphenyltrichloroethane) is accumulated in the tissues of freshwater snails (Pimental, 1971; Dindal, 1970). Dindal assumed the accumulation of Cl^{36} DDT in waterfowl tissues was enhanced by their feeding on snails. Keenleyside (1967) noted that snails replaced insects eradicated by the forest spraying of DDT in New Brunswick and became an important component of the diet of young salmon. The implication is that gastropods are comparatively resistant to DDT. Another chlorinated hydrocarbon, dieldrin (1,2,3,4,10,10-hexachloro-*exo*-6,7-epoxy-1,4,4a,5,6,

7,8,8a-octahydro-1,4-*endo-exo*-5,8-dimethanonapthalene) was reported by Wallace and Brady (1971) as a chronic pollutant from a woolen mill in South Carolina. This compound enters the stream concerned at concentrations of ca. 200 ppb. The dominant gastropod above the outfall is *Goniobasis* sp., below the only snails found in abundance were *Physa* sp. These snails contained 33 to 62 ppm dieldrin in their tissues. Pimental (1971) cited sources indicating that freshwater snails were unaffected by concentrations of 0.1 ppm of toxaphene (a mixture of chlorinated camphenes). In a marsh that was treated with two pounds per acre (105 ppm in H_2O) snails were eradicated in 10 days. Grzenda *et al.* (1964) noted that pleurocerids apparently were not affected by spills of toxaphene and BHC (a gamma isomer of 1,2,3, 4,5,6-hexachlorocyclohexane) that drastically altered invertebrate macrofauna and fish populations. Two populations of *P. gyrina*, one from Belzoni, Mississippi, and another from nearby State College, were exposed to toxaphene and endrin (1,2,3,4,10,10-hexachloro-6,7-epoxy-1,4,4a,5,6,7,8,8a-octahydro-1,4-*endo-endo* 5,8-dimethanonaphthalene). The population from Belzoni, collected from a pesticide contaminated ditch, showed a higher tolerance to both insecticides than the population from the cleaner waters from near State College (Naqvi and Ferguson, 1968). DN-111 (2 cyclohexyl-4,6-dinitrophenol, dicyclohexylamine salt) at field applications of 1 to 10 ppm reduced snail populations, and DNOC (sodium salt of 4,6-dinitro-*o*-cresol) at 1 ppm eliminated all snails (Pimentel, 1971).

B. Thermal Pollution

Water utilized as coolant in industrial processes elevates the temperatures of the aquatic environments that it enters, often with serious ecological consequences. Death of organisms can occur because temperatures exceed their tolerance limits, because oxygen consumption is increased (Akerlund, 1969; Boycott, 1936; Cheatum, 1934; Hunter, 1964) as oxygen levels decrease, and because resistance to toxins is decreased (Cairns, 1970) while the toxicity of some compounds, such as zinc, increases (Jensen *et al.*, 1969). Indirect effects, such as changes in growth, disruption of reproductive activities (Cairns 1970), changes in behavior patterns (Jensen *et al.*, 1969), variation in the quality of food and growth of toxic organisms (Cairns, 1970), and competitive replacement, also occur. High temperatures may also act as barriers to stenothermal organisms (Cairns, 1971).

The Martin's Creek stream electric generating station on the Delaware River elevated the August 1959 temperatures from about 80° F to more than 105° F (Coutant, 1962). Oxygen levels remained relatively normal in the area because of physical oxygenation over a cobble substrate. *Physa* sp., *Ferrissia* sp., and *Gyraulus* sp. occurred throughout the year above, and about one mile below, the outfall. Where the heated water entered the river, only *Physa* sp. persisted, and then only during the cooler months.

Water of subnormal temperatures may also have deleterious effects on ecosystems. Lehmkuhl (1972) attributed a reduction in benthos as far as 70 miles below an impoundment in the Saskatchewan River to low summer temperatures. The hypolimnion drains of Cannonsville and Pepacten Reservoirs in southeastern New York normally maintain temperatures lower than 15°C throughout the year in the streams below them (Harman *et al.*, unpublished). This results in extremely low food supplies, very slow growth, and inhibition of reproduction. Only isolated specimens of *Physa* spp. have been found in these environments.

C. Impoundment, Canalization, and Siltation

Lentic waters are an extremely ephemeral geological phenomenon. The most stable, although dynamic, aquatic biotopes are the major rivers. As would be expected, our most distinctive, and probably stenotopic, Gastropoda occur in these environments. Unfortunately, many of our rivers are becoming chains of impoundments utilized for flood control, hydroelectric power, navigation, irrigation, urban water supplies, and recreation. The construction of reservoirs in tropical climates has increased the habitats of pulmonate vectors of schistosomiasis, causing serious health problems. However, these same impoundments can eliminate gastropods specialized for existence in the erosional areas of rivers. In these artificial lentic biotopes, biological and physical phenomena contribute to a lowering of dissolved oxygen and an increase of carbon dioxide in solution, accompanied with a reduction or loss of molluscan species (Stansbery, 1970a,b; 1971). In the Tennessee drainage basin the pleurocerid genera *Io*, *Lithasia*, *Pleurocera*, *Eurycaelon*, *Anculosa*, and *Spirodon* are greatly reduced in species numbers or are extinct (Stansbery, 1970a; 1971). Sinclair (undated) claimed that *Pleurocera canaculata* (Say 1821) (concept of Goodrich, 1937) and *Lithasia verrucosa* (Rafinesque 1820) (concept of Goodrich, 1941) are the most tolerant pleurocerids occurring in the Ohio and Tennessee Rivers. *Lithasia geniculata* Haldeman 1840 (Goodrich, 1934a, 1941) and *L.g. fuliginosa* (Lea 1847) (Goodrich, 1934a, 1941) are now the most ubiquitous pleurocerids in the Duck River, Tennessee. Five associated "forms" have been deletriously effected by the construction of impoundments (Sinclair, undated).

In reservoirs in the upper Delaware River in New York even the pulmonates are essentially eliminated. In these impoundments, which are used as water supplies for New York City (Oglesby *et al.*, 1972), the water levels fluctuate as much as 20 m annually, and daily changes of 2 m have been recorded (Lord, personal communication). These same procedures have been used for snail control in tropical reservoirs (Jobin, 1970; Palmer *et al.*, 1969). Large amounts of precipitation in 1972 in New York State resulted in ephemerally large flows of water out of these reservoirs (compared to the usual restricted flow) which flushed snails away in the downstream areas. This technique has also been utilized as a snail control procedure (Jobin and Ippen, 1964). The effects of impoundments on the thermal regime of rivers has already been mentioned (see Section IV,B).

One of the most serious affects of the construction of reservoirs is the resultant destabilization of substrates and increased siltation that occurs. Canalization, for the purpose of more efficient drainage and irrigation (and associated with the construction of highways), usually has the same effects and has also been used for snail control (Palmer *et al.*, 1969). Silt in suspension has an abrasive action on molluscan shells that erodes the periostracum, allowing carbonic acids quickly to erode the $CaCO_3$ layers underneath. It also reduces light penetration, reducing primary productivity, and decreasing the dissolved oxygen levels (Cairns, 1968b; Chutter, 1969). When this silt is deposited, it fills the interstices between rocks, eliminating much of the available surface for the growth of organisms and eliminating many benthic species. In the case of gastropods, their eggs do not develop properly when covered with silt (Chutter, 1969). Siltation also increases the possibility of adsorption and absorption of various toxic chemicals (Cairns, 1968b). Tarzwell and Gaufin (1967) considered areas of shifting sands and silt "aquatic deserts" as a result of these stresses.

Harman *et al.* (unpublished) have studied the distribution of mollusks in the Delaware watershed north of the Beaverkill in southeastern N.Y. More than three-fourths of that watershed has been deleterously affected. The major streams exhibit vast reaches practically devoid of mollusks, because of impoundment and canalization. Isolated specimens of *Physa* spp., *H. anceps*, and *G. parvus* are the only snails found in these biotopes.

C. Cultural Eutrophication

The terms cultural eutrophication, organic pollution, or artificial enrichment are impossible to define quantitatively. First, one must make sure confusion does not exist between artificial enrichment and ephemeral seasonal characteristics of environments [e.g., a springbook after leaf fall (Hynes, 1966)], natural succession in lotic environments (Harrel and Dorris, 1968; Pennak, 1971), or the effects of rich soils and allochthonous vegetation (Hynes, 1966; 1970). Second, there appears to be no easily obtained chemical, physical, or biological parameter that can be quantitatively measured to indicate exclusively that organic pollution exists. In an excellent survey of the Ohio River Basin, Mason *et al.* (1968) measured dissolved oxygen, pH, alkalinity, and turbidity. The means and ranges of all these parameters have been calculated where snails were present and where they were not. There are no significant differences. If one assumes snails were absent in polluted biotopes, these data indicate that these factors have little or no relevance in characterizing polluted environments.

Low levels of dissolved oxygen and "septic" conditions are normally associated with cultural eutrophication, and often limit organisms in these situations. Records of Gastropoda exposed to low oxygen levels under apparently natural conditions were discussed in Section III,A. Gaufin and Tarzwell (1966) observed *P. integra* in septic waters devoid of oxygen. Ingram (1957) collated the results of several researchers: *Valvata tricarinata* was found in water with 1.4 ppm oxygen; *C. subsolidium*, 0.5 ppm; *H. trivolvis*, 2.8 ppm with *C. integrum*; and *P. integra*, 0.2 ppm. Burdick (1940) collected *H. trivolvis* and *P. gyrina* in a "polluted" creek. Ingram (1957) cites sources that indicate that *P. heterostropha*, *G. virginica* and *C. decisum* were found in septic environments. In Georgia (Anon., 1969; 1970; 1971) *Physa* sp., *Ferrissia* sp., and *Promenetus* sp. were often found in "polluted, moderately polluted, and grossly polluted," situations. Mason *et al.* (1970) considered the pulmonate families Physidae, Lymnaeidae, Planorbidae, and Ancylidae "pollution associated" in the Klamath River, Oregon. Wurtz (in Ingram, 1957) considered 19 species of freshwater snails ". . . able to survive at least some degree of pollution." Although most were pulmonates, *C. integrum* and *C. decisum* (Viviparidae) and *B. tentaculata* (Hydrobiidae) were included. Whipple, Fair, and Whipple (*in* Ingram, 1957) added *V. tricarinata* (Valvatidae) to that list. Mason *et al.* (1968) considered *Ferrissia* spp., *Lymnaea* spp., *Gyraulus* spp., *Pleurocera* spp., and *Goniobasis* spp. extremely eurytopic organisms in the Ohio River Basin. It is apparent that various authors have, at one place or another, considered representatives of every widespread family of freshwater Gastropoda tolerant of at least moderately polluted waters.

It may be of more value to observe changes in gastropod faunas over time in increasingly eutrophying environments, in order to ascertain the response of snails to organic enrichment. Hall *et al.* (1970), under controlled conditions, determined that mild enrichment [2.72 kg of 10:1:1 (urea:superphosphate:potassium chloride) per week per 650 m^3 of water] enhanced vigor and growth of benthic populations. Carr and Hiltunen (1965) observed a sixfold increase in gastropod biomass in polluted waters of Lake Erie between 1930 and 1961. *Bithynia tentaculata* comprised 51% of the total.

It is well known that highly eutrophic waters tend to become homogeneous and therefore decrease in diversity, or species richness (e.g., Jonasson, 1969). Harman and Forney (1970) noted a decrease in diversity of gastropods in Oneida Lake, New York between 1917 and 1967. An average of 100 cm^2

sample of the bottom of the lake in 1917 would have yielded 9 specimens of *A. lustrica*, 3 of *L. catascopium*, 3 of *G. parvus*, and one each of *P. integra*, *P. sayi*, *P. exacuous*, and *V. tricarinata*. In 1967 the same average sample yielded four specimens of *B. tentaculata*. Howmiller *et al.* (1971) observed a decrease in gastropod diversity at 10 stations in Green Bay, Lake Michigan, between 1952 and 1969. In 1952 *Campeloma* sp., *Helisoma* sp., *V. tricarinata* and *Viviparus* sp. were found. In 1969 only two stations supported living snails, determined as *Amnicola* sp. Harman (1968a,b) noted correlations between the distributions of *B. tentaculata*, *P. acuta*, *G. livescens*, and *G. virginica*, the migratory capabilities of these species, and areas of cultural eutrophication in central New York. It appears that *B. tentaculata*, by direct competition, has all but eliminated *P. acuta* and *G. virginica* (large river species) from the Oswego watershed. *Goniobasis livescens* still exists in oligotrophic waters in association with *B. tentaculata*, and in headwater streams, enriched or not, where *B. tentaculata* is not present.

V. Freshwater Gastropoda as Biological Indicators

Aurele LaRocque, his students, and other paleomalacologists have made excellent studies of the evolution and characteristics of fossil lakes by using gastropod shells found in the sediments as indicators (e.g., LaRocque, 1960; Pilsbry, 1934b; Roy, 1964; Taylor, 1960; 1966).

Certain heavy metals, pesticides such as DDT (Section IV,A,2), and radioactive materials (Hiyama and Shimizu, 1969; Nelson, 1962, 1963; Price, 1965; Van der Borght and Van Puymbroeck, 1970; Wilhm, 1970a) are concentrated in the flesh and shells of mollusks. Analysis of gastropods living in biotopes subjected to these materials can indicate quantitatively the amounts of those types of pollutants occurring there.

If habitats appear optimal for gastropods and the local species possess the migratory abilities to invade the area, yet they are not found, the chemical and physical characteristics of the environment should be determined. Low pH values (Sections III,A,1 and IV,A,1), the presence of heavy metals (Section IV,A,1), pesticides that gastropods are sensitive to (Section IV,A,2), low oxygen levels (Section III,A,4), or limiting thermal regimes (Section IV, B) may be present.

If the chemical and physical characteristics of a body of water appear conducive to a high diversity of organisms, yet only one or two species are present, the stress on that community may be caused by organic pollution. In the eastern Great Lakes drainage basin the introduced prosobranch, *B. tentaculata*, is often the most abundant snail in enriched biotopes (Section IV,D,4). Throughout the United States, the eurytopic snails which are most resistant to these factors, and even septic conditions (Section IV,D) are pulmonates, particularly the families Physidae and Planorbidae. The most resistant species are *P. heterostropha*, *P. integra*, *P. gyrina*, *G. parvus*, *H. anceps*, and *H. trivolvis*, but almost every common species has been found in polluted environments (Section IV,D).

References

Adams, C. C. (1915). The variations and ecological distribution of the snails in the genus *Io*. *Proc. Nat. Acad. Sci. U.S.* **12**, 7–184.

Akerlund, G. (1969). Oxygen consumption of the ampullariid snail *Marisa cornuarietis* L. in relation to body weight and temperature. *Oikos* **20**, 529–533.

Anonymous (1968). A guide for the identification of the snail intermediate hosts of schistosomiasis in the Americas. *Pan Amer. Health Organization Sci. Publ.* (*WHO*) **168**, 1–122.

Anonymous (1969). Dalton Water Quality Survey. Georgia Water Quality Contr. Bd., 47 Trinity Avenue, S.W., Atlanta, Georgia 30334, (Access to this and the two following publications was provided by the Georgia Water Quality Contr. Bd., Water Quality Contr. Sect., Environ. Protect. Div., Georgia Dept. of Natur. Resources.)

Anonymous (1970). Coosa River Basin Study. Georgia Water Quality Contr. Bd., 47 Trinity Avenue, S.W., Atlanta, Georgia 30334.

Anonymous (1971a). Chattahoochee River Basin Study. Georgia Water Quality Contr. Bd., 47 Trinity Avenue, S.W., Atlanta, Georgia 30334.

Anonymous (1971b). Water Quality Criteria Data Book. Volume 3. Effects of Chemicals on Aquatic Life. Water Quality Office, Environ. Protect. Agency, Washington, D.C.

Arthur, J. W. (1970). Chronic effects of linear alkylate sulfonate detergent on *Gammarus pseudolimnaeus*, *Campeloma decisum* and *Physa integra*. *Water Res.* **4**, 251–257.

Arthur, J. W., and Leonard, E. N. (1970). Effects of copper on *Gammarus pseudolimnaeus*, *Physa integra* and *Campeloma decisum* in soft water. *J. Fish. Res. Bd. Can.* **27**, 1277–1283.

Bailey, J., Pearl, R., and Winsor, C. P. (1931a). Variation in *Goniobiasis virginica* and *Anculosa carinata* under natural conditions. I. *Biol. Gen.* **8**, 607–630.

Bailey, J., Pearl, R., and Winsor, C. P. (1931b). Variation in *Goniobiasis virginica* and *Anculosa carinata* under natural conditions. II. *Biol. Gen.* **9**, 48–69.

Bailey, J., Pearl, R., and Winsor, C. P. (1931c). Variation in *Goniobiasis virginica* and *Anculosa carinata* under natural conditions. III. *Biol. Gen.* **9**, 301–336.

Baker, F. C. (1902). The Mollusca of the Chicago area, Part 2. *Bull. Natur. His. Surv. Chicago Acad. Sci.* **3**, 131–418.

Baker, F. C. (1911). The Lymnaeidae of North and Middle America recent and fossil. *Chicago Acad. Sci. Spec. Publ.* **3**, 539, Pl. 1–57.

Baker, F. C. (1912). A new *Planorbis* from Michigan. *Nautilus* **25**, 118–120.

Baker, F. C. (1928). The fresh water Mollusca of Wisconsin, Part 1. Gastropoda. *Wis. Geol. Natur. Hist. Surv. Bull.* **70(1)**, 1–507, Pl. 1–28.

Baker, F. C. (1945). "The Molluscan Family Planorbidae." Univ. Illinois Press, Urbana, Illinois.

Basch, P. F. (1959a). The anatomy of *Laevapex fuscus*, A freshwater limpet (Gastropoda: Pulmonata). *Mus. Zool. Univ. Mich. Misc. Publ.* **108**, 1–56.

Basch, P. F. (1959b). Studies on Ancylidae. *Amer. Malacol. Un. Bull.* **25**, (Reprint) no pagination.

Basch, P. F. (1959c). *Gundlachia* in Michigan. *Amer. Malacol. Un. Bull.* **25**, (Reprint) no pagination.

Basch, P. F. (1960). Anatomy of *Rhodacmea cahawbensis* Walker, 1917, a river limpet from Alabama. *Nautilus* **73**, 88–95.

Basch, P. F. (1962a). Radulae of North American Ancylid snails I. Subfamily Rhodacmeinae. *Nautilus* **75**, 97–101.

Basch, P. F. (1962b). Radulae of North American Ancylid snails II. Subfamily Neoplanorbinae. *Nautilus* **75**, 145–148.

Basch, P. F. (1962c). Radulae of North American freshwater limpet snails. III. *Ferrissia* and *Laevapex*. *Nautilus* **76**, 28–33.

Basch, P. F. (1963). A review of the recent freshwater limpet snails of North America (Mollusca: Pulmonata). *Mus. Comp. Zool. Harvard Univ. Occas. Pap. Mollusks* **2**, 385–412.

Beck, W. M. (1969). Stream monitoring. Biological parameters. *Fla. Eng. Ind. Exp. Sta. Bull.* **135**, 1–168.

Beetle, D. E. (1961). A checklist of Wyoming recent Mollusca. *Sterkiana* **3**, 1–9.

Bell, H. L. (in litt. 1972). From a letter received 16 June 1972.

Berg, K., and Ockelmann, K. W. (1959). The respiration of freshwater snails. *J. Exp. Biol.* **36**, 690–708.

Berrios-Duran, L. A., and Ritchie, L. S. (1968). Molluscicidal activity of bis(tri-*n*-butyltin) oxide formulated in rubber. *Bull. W. H. O.* **39**, 310–311.

Berry, E. G. (1943). The Amnicolidae of Michigan: Distribution, ecology, and taxonomy. *Mus. Zool. Univ. Mich. Misc. Publ.* **57**, 1–68, Pl. 1–9.

Bickel, D. (1965). The role of aquatic plants and submerged structures in the ecology of a freshwater pulmonate snail *Physa integra* Hald. *Sterkiana* **18**, 17–20.

Bickel, D. (1967). Preliminary checklist of recent and pleistocene molluscs of Kentucky. *Sterkiana* **28**, 7–20.

Bickel, D. (1968a). *Goniobasis semicarinata* and *G. indianensis* in Blue River, Indiana. *Nautilus* **81**, 133–138.

Bickel, D. (1968b). *Goniobasis curreyana lyoni*, a pleurocerid snail of West-Central Kentucky. *Nautilus* **82**, 13–18.

Bickel, D. (1968c). Checklist of the Mollusca of Tennessee. *Sterkiana* **31**, 15–39.

Bickel, D. (1970). Pleistocene non-marine mollusca of the Gatineau Valley and Ottawa areas of Quebec and Ontario, Canada. *Sterkiana* **38**, 1–50.

Boyce, C. B. C., Jones, T. W., and Vantongeren, W. A. (1967). The molluscicidal activity of N-tritylmorpholine. *Bull. W. H. O.* **37**, 1–11.

Boycott, A. E. (1936). The habitats of freshwater mollusks in Britain. *J. Anim. Ecol.* **5**, 116–186.

Branson, B. A. (1969). Distribution notes on western and southwestern snails. *Sterkiana* **36**, 21.

Branson, B. A. (undated). Checklist and distribution of Kentucky aquatic gastropods. *Kentucky Fish. Bull.* **54**, 1–20, Pl. 15.

Branson, B. A., and Batch, D. L. (1971). Annotated distribution records for Kentucky Mollusca. *Sterkiana* **43**, 1–9.

Bugbee, S. L., and Walter, C. M. (1972). The effects of gasoline on the biota of a mountain stream. *Annu. Meeting Midwest Benthol. Soc., 20th, March 29–31*. Iowa State Univ., Ames, Iowa.

Burdick, G. E. (1940). VI. Studies on the invertebrate Fish Food in certain lakes, bays, streams, and ponds of the Lake Ontario Watershed pp. 147–166. *In State N.Y. Conserv. Dept., Biol. Surv. Lake Ontario Watershed. Suppl. 29th Annu. Rep., 1939* **16**, 1–261.

Burkey, A. J. (1971). Biomass turnover, respiration and interpopulation variation in the stream limpet *Ferrissia rivularis*. *Ecol. Monogr.* **41**, 235–251.

Butler, J. M., Ferguson, F. F., and Oliver-Gonzalez, J. (1969). A system for mass producing the snail *Marisa cornuarietis*, for use as a biological control agent of Schistosomiasis. *Caribbean J. Sci.* **9**, 135–139.

Cairns, J. (1968a). The need for regional water resources management. *Trans. Kansas Acad. Sci.* **41**, 480–490.

Cairns, J. (1968b). Suspended solids standards for the protection of aquatic organisms. *Purdue Univ. Eng. Bull. Part I.* **129**, 16–27.

Cairns, J. (1970). Ecological management problems caused by heated waste water discharge into the aquatic environment. *Water Resources Bull.* **6**, 868–878.

Cairns, J. (1971). Thermal pollution—a cause for concern. *J. Water Pollut. Contr. Fed.* **43**, 55–66.

Cairns, J., and Dickson, K. L. (1971). A simple method for the biological assessment of the effects of waste discharges on aquatic bottom-dwelling organisms. *J. Water Pollut. Contr. Fed.* **43**, 755–772.

Cairns, J., and Scheier, A. (1957). The effects of periodic low oxygen upon the toxicity of various chemicals to aquatic organisms. *Purdue Univ. Eng. Bull.* **96**, 165–176.

Cairns, J., and Scheier, A. (1958). The effects of temperature and hardness of water upon the toxicity of zinc to the pond snail, *Physa heterostropha* (Say). *Notulae Natur. Acad. Natur. Sci. Philadelphia* **308**, 1–11.

Cairns, J., and Scheier, A. (1962). The effects of temperature and water hardness upon the toxicity of Napthenic acids to the common Bluegill sunfish *Lepomis macrochiras* Raf, and the Pond snail, *Physa heterostropha* Say. *Notulae Natur. Acad. Natur. Sci. Philadelphia* **353**, 1–12.

Cairns, J., Scheier, A., and Hess, N. E. (1964). The effects of alkyl benzene sulfonate on aquatic organisms. *Ind. Water Wastes* **9**, no pagination, 7 pp.

Cairns, J., Albaugh, D. W., Busey, F., and Chanay, M. D. (1968). The sequential comparison index—a simplified method for non-biologists to estimate relative differences in biological diversity in stream pollution studies. *J. Water Pollut. Contr. Fed.* **40**, 1607–1613.

Cairns, J., Crossman, J. S., Dickson, K. L., and Herricks, E. E. (1971). The Recovery of Damaged Streams. *Ass. Southeast. Biol. Bull.* **18**, 49–106.

Cairns, J., Dickson, K. L., and Crossman, J. S. (1972). The response of aquatic communities to spills of hazardous materials. *Proc. 1972 Nat. Conf. Hazardous Mater. Spills* 179–197.

Calow, P. (1970). Studies on the natural diet of *Lymnaea peregra obtusa* (Kobelt) and its possible ecological implications. *Proc. Malocol. Soc. London*, **39**, 203–215.

Carr, J. F., Hiltunen, J. K. (1965). Changes in the bottom fauna of Western Lake Erie from 1930–1961. *Limnol. Oceanogr.* **10**, 551–569.

Chamberlin, R. V., and Jones, D. T. (1929). A descriptive catalogue of the Mollusca of Utah. *Bull. Univ. Utah* **19**, 1–203.

Cheatum, E. P. (1934). Limnological investigations on respiration, annual migratory cycle, and other related phenomena in fresh-water pulmonate snails. *Trans. Amer. Microsc. Soc.* **53**, 348–407.

Cheatum, E. P., and Fullington, R. W. (1971). The aquatic and land Mollusca of Texas, Bulletin 1. Supplement: Keys to the families of the recent land and freshwater snails of Texas. *Dallas Mus. of Natur. Hist.*

Chu, K. Y., Massoud, J., and Arfaa, F. (1968). Comparative studies of the molluscicidal effect of cuprous chloride and copper sulfate in Iran. *Bull. W.H.O.* **39**, 320–326.

Chutter, F. M. (1969). The effects of silt and sand on the invertebrate fauna of streams and rivers. *Hydrobiologia* **34**, 57–76.

Clampitt, P. T. (1970). Comparative ecology of the snails *Physa gyrina* and *Physa integra* (Basommatophora: Physidae). *Malacologia* **10**, 113–148.

Clarke, A. H. (1973). The Freshwater mollusks of the Canadian interior Basin. *Malacologia* **13**, 1–510.

Clench, W. J. (1925). Notes on the genus *Physa* with descriptions of three new subspecies. *Univ. Michigan Occas. Pap. Mus. Zool.* **161**, 1–10.

Clench, W. J. (1926). Some notes and a list of shells of Rio, Kentucky. *Nautilus* **40**, 7–12, 65–67.

Clench, W. J. (1962a). A catalogue of the Viviparidae of North America with notes on the distribution of *Viviparus georgianus* Lea. *Mus. Comp. Zool. Harvard Univ. Occas. Pap. Mollusks* **2**, 261–287.

Clench, W. J. (1962b). Three new species of *Physa. Univ. Michigan, Occas. Pap. Mus. Zool.* **168**, 1–6.

Clench, W. J., and Fuller, S. L. H. (1965). The genus *Viviparus* (Viviparidae) in North America. *Mus. Comp. Zool. Harvard Univ. Occas. Pap. Mollusks* **2**, 385–412.

Clench, W. J., and Turner, R. D. (1955). The North American Genus *Lioplax* in the Family Viviparidae. *Mus. Comp. Zool. Harvard Univ. Occas. Pap. Mollusks* **2**, 1–20.

Clench, W. J., and Turner, R. D. (1956). Freshwater mollusks of Alabama, Georgia, and Florida. *Bull. Fla. State Mus.* **1**, 96–238.

Clowers, S. (1966). Pleistocene mollusca of the box marsh deposit, Admaston Township, Renfrew County, Ontario, Canada. *Sterkiana* **22**, 31–59.

Coutant, C. C. (1962). The effect of heated water effluent upon the macroinvertebrate riffle fauna of the Delaware River. *Proc. Penn. Acad. Sci.* **36**, 58–71.

Dazo, B. C. (1961). Some studies on the genus *Io. Proc. Amer. Malacol. Un., 28th Annu. Meeting* p. 6 (Abstract)

Dazo, B. C. (1965). The morphology and natural history of *Pleurocera acuta* and *Goniobiasis livescens* (Gastropoda: Cerithiacea: Pleuroceridae). *Malacologia* **3**, 1–80.

deMeillon, B., Frank, G. H., and Allansen, B. R. (1958). Some aspects of snail ecology in South Africa. *Bull. W.H.O.* **18**, 771–783.

DeWitt, R. M. (1954). Reproduction, embryonic development, and growth in the pond snail *Physa gyrina* Say. *Trans. Amer. Microsc. Soc.* **73**, 125–137.

DeWitt, R. M. (1955). The ecology and life history of the pond snail *Physa gyrina. Ecology* **36**, 40–44.

Dindal, D. L. (1970). Accumulation and excretion of Cl^{36} DDT in Mallard and Lesser Scaup Ducks. *J. Wildl. Management* **34**, 74–92.

Dundee, D. S., and Dundee, H. A. (1969). Notes concerning two Texas molluscs, *Cochliopa texana* Pilsbry and *Lyrodes cheatumi* Pilsbry. (Mollusca: Hydrobiidae). *Trans. Amer. Microsc. Soc.* **88**, 206–210.

Eberhardt, L. L. (1969). Some aspects of species diversity models. *Ecology* **50**, 503–505.

Eisenberg, R. M. (1970). The role of food in the regulation of the pond snail, *Lymnaea elodes. Ecology* **51**, 680–84.

Ferguson, F. F. (1971). Some current controls for aquatic weeds inimical to public health. *E. Afr. Med. J.* **48**, 456–459.

Ferguson, F. F., and Richards, C. S. (1963). Fresh-water Mollusks of Puerto Rico and the U.S. Virgin Islands. *Trans. Amer. Microsc. Soc.* **82**, 391–395.

Flannagan, J. F. (1971). Toxicity evaluation of trisodium nitrilotriacetate to selected aquatic invertebrates and amphibians. *Fish. Res. Bd. Can. Tech. Rep. No. 258*, 1–15 (mimeo).

Foin, T. C. (1971). The distribution pattern of the freshwater prosobranch gastropod *Oxytrema proxima* (Say). *J. Elisha Mitchell Sci. Soc.* **87**, 1–10.

Fredeen, F. J. H., and Duffy, R. (1970). Insecticide residues in some components of the St. Lawrence River ecosystem following applications of DDD. *Pestic. Monit. J.* **3**, 219–226.

Fretter, V. (1968). "Studies in the Structure, Physiology, and Ecology of Mollusks." Academic Press, New York.

Fuhs, W. G. (1972). Canadarago Lake eutrophication study. Lake and tributary surveys 1968–1970. Methodology and data. N.Y. State Dept. of Environ. Contr., Environ. Quality, Res. and Develop. Unit.

Gaufin, A. R., and Tarzwell, C. M. (1956). Aquatic macro-invertebrate communities as indicators of organic pollution in Lytle Creek. *Sewage Ind. Wastes* **28**, 906–924.

Getz, L. L. (1971). On the occurrence of *Pomatiopsis cincinnatiensis* in Wisconsin. *Sterkiana* **43**, 20.

Goodrich, C. (1921a). River barriers to aquatic animals. *Nautilus* **35**, 1–4.

Goodrich, C. (1921b). Three new species of Pleuroceridae. *Univ. Mich. Occas. Pap. Mus. Zool.* **91**, 1–5.

Goodrich, C. (1922). The Anculosae of the Alabama River drainage. *Mus. Zool. Univ. Mich. Misc. Publ.* **7**, 1–57.

Goodrich, C. (1923). Variations in *Goniobiasis edgariana* Lea. *Nautilus* **36**, 115–119.

Goodrich, C. (1928a). The group of *Goniobasis catenaria. Nautilus* **42**, 28–32.

Goodrich, C. (1928b). *Strephobasis*: A section of *Pleuocera. Univ. Mich. Occas. Pap. Mus. Zool.* **192**, 1–18.

Goodrich, C. (1930). *Goniobasis* of the vicinity of Muscle Shoals. *Univ. Mich. Occas. Pap. Mus. Zool.* **209**, 1–25.

Goodrich, C. (1931). The Pleurocerid genus *Eurycaelon. Univ. Mich. Occas. Pap. Mus. Zool.* **223**, 1–9.

Goodrich, C. (1932). The Mollusca of Michigan. *Univ. Mich., Mich. Handb. Ser.* **5**, 1–120.

Goodrich, C. (1934a). Studies of the Gastropod family Pleuroceridae I. *Univ. Mich. Occas. Pap. Mus. Zool.* **284**, 1–19.

Goodrich, C. (1934b). Studies of the Gastropod family Pleuroceridae II. *Univ. Mich. Occas. Pap. Mus. Zool.* **295**, 1–6.

Goodrich, C. (1934c). Studies of the Gastropod family Pleuroceridae III. *Univ. Mich. Occas. Pap. Mus. Zool.* **300**, 1–11.

Goodrich, C. (1935a). Studies of the Gastropod family Pleuroceridae IV. *Univ. Mich. Occas. Pap. Mus. Zool.* **311**, 1–11.

Goodrich, C. (1935b). Studies of the Gastropod family Pleuroceridae V. *Univ. Mich. Occas. Pap. Mus. Zool.* **318**, 1–12.

Goodrich, C. (1936). *Goniobasis* of the Coosa River, Alabama. *Mus. Zool. Univ. Mich. Misc. Publ.* **31**, 1–60.

Goodrich, C. (1937). Studies of the Gastropod Family Pleuroceridae VI. *Univ. Mich. Occas. Pap. Mus. Zool.* **347**, 1–12.

Goodrich, C. (1938). Studies of the Gastropod family Pleuroceridae VII. *Univ. Mich. Occas. Pap. Mus. Zool.* **376**, 1–12.

Goodrich, C. (1939). Pleuroceridae of the St. Lawrence River Basin. *Univ. Mich. Occas. Pap. Mus. Zool.* **404**, 1–4.

Goodrich, C. (1940). The Pleuroceridae of the Ohio River drainage system. *Univ. Mich. Occas. Pap. Mus. Zool.* **417**, 1–21.

Goodrich, C. (1941). Studies of the Gastropod family Pleuroceridae VIII. *Univ. Mich. Occas. Pap. Mus. Zool.* **447**, 1–13.

Goodrich, C. (1942). The Pleuroceridae of the Atlantic coastal plain. *Univ. Mich. Occas. Pap. Mus. Zool.* **456**, 1–6.

Goodrich, C. (1943). The Walker-Beecher paper of 1876 on the Mollusca of the Ann Arbor area. *Mus. Zool. Univ. Mich. Occas. Pap.* **475**, 1–26.

Goodrich, C. (1945). *Goniobasis livescens* of Michigan. *Mus. Zool. Univ. Mich. Misc. Publ.* **64**, 1–33.

Goodrich, C., and van der Schalie, H. (1939). Aquatic Mollusks of the Upper Peninsula of Michigan. *Mus. Zool. Univ. Mich. Misc. Publ.* **43**, 1–56.

Goodrich, C., and van der Schalie, H. (1944). A revision of the Mollusca of Indiana. *Amer. Midl. Natur.* **32**, 257–326.

Grzenda, A. R., Lauer, G. J., and Nicholson, H. P. (1964). Water pollution by insecticides in an agricultural river basin. II. The zooplankton, bottom fauna, and fish. *Limnol. Oceanogr.* **9**, 318–323.

Hall, D. J., Cooper, W. E., and Werner, E. E. (1970). An experimental approach to the production dynamics and structure of freshwater animal communities. *Limnol. Oceanogr.* **15**, 839–928.

Hanson, W. R. (1952). Effects of some herbicides and insecticides on biota of North Dakota marshes. *J. Wildl. Manage.* **16**, 299–308.

Harman, W. N. (1968a). Replacement of Pleurocerids by *Bithynia* in polluted waters of central New York, *Nautilus* **81**, 77–83.

Harman, W. N. (1968b). Interspecific competition between *Bithynia* and Pleuroceridae. *Nautilus* **82**, 72–73.

Harman, W. N. (1972). Benthic substrates: their effect on fresh-water Mollusca. *Ecology* **53**, 271–277.

Harman, W. N., and Berg, C. O. (1971). The freshwater Gastropoda of central New York with illustrated keys to the genera and species. *Search: Cornell Univ. Agr. Exp. Sta. Entomol. (Ithaca)* **1(4)**, 1–68.

Harman, W. N., and Forney, J. L. (1970). Changes in the molluscan community in Oneida Lake, N. Y. between 1917 and 1967. *Limnol. Oceanogr.* **15**, 454–460.

Harp, G. L., and Campbell, R. S. (1964). Effects of the herbicide silvex on benthos of a farm pond. *J. Wildl. Manage.* **28**, 308–317.

Harrel, R. C., and Dorris, T. C. (1968). Stream order, morphometry, physico-chemical conditions, and community structure of benthic macroinvertebrates in an intermittent stream system. *Amer. Midl. Natur.* **80**, 220–251.

Harrison, A. D., Williams, N. V., and Grieg, G. (1970). Studies on the effects of calcium bicarbonate concentrations on the biology of *Biomphalaria pfeifferi* (Krauss) (Gastropoda: Pulmonata). *Hydrobiologia* **36**, 317–327.

Harry, H. H., Cumbie, B. G., and deJesus, J. M. (1957). Studies on the quality of freshwaters of Puerto Rico relative to the occurrence of *Australorbis glabratus* (Say). *Amer. J. Trop. Med. Hyg.* **6**, 313–322.

Heard, W. H. (1963). Reproductive features of *Valvata. Nautilus* **77**, 64–68.

Henderson, J. (1924). Mollusca of Colorado, Utah, Montana, Idaho, and Wyoming. *Univ. Colo. Stud.* **13**, 65–223.

Henderson, J. (1929). Non-marine mollusca of Oregon and Washington. *Univ. Colo. Stud.* **17**, 47–190.

Henderson, J. (1936a). Mollusca of Colorado, Utah, Montana, Idaho and Wyoming—supplement. *Univ. Colo. Stud.* **23**, 81–145.

Henderson, J. (1936b). The non marine mollusca of Oregon and Washington supplement. *Univ. Colo. Stud.* **23**, 251–280.

Herricks, E., and Cairns, J. (1972). The recovery of stream macrobenthic communities from the effects of acid mine drainage. *Symp. Coal Mine Drainage Res. Inst., 4th April 25–27* pp. 370–398.

Herrmann, S. H. (1972). The population dynamics of two species of freshwater gastropods, *Physa heterostropha* Say and *Helisoma anceps* (Menke). Master's thesis, State Univ. N.Y., Oneonta.

Hickman, M. E. (1937). A contribution to the mollusca of East Tennessee. Bound Thesis, Univ. of Tennessee Library, Knoxville, Tennessee.

Hiyama, Y., and Shimizu, M. (1969). Uptake of radioactive nucleotides by aquatic organisms: the application of the expotential model. *In* "Environmental Contamination by Radioactive Materials." Int. At. Energy Ag., Vienna.

Horst, T. J. (1971). Ecology of *Amnicola limosa* and selected gastropod species of McCargo Lake. Master's Thesis, State Univ. N.Y., Brockport.

Houp, K. H. (1970). Population dynamics of *Pleurocera acuta* in a central Kentucky limestone stream. *Amer. Midl. Natur.* **83**, 81–88.

Howmiller, R. P., and Beeton, A. M. (1971). Biological evaluation of environmental quality. Green Bay, Lake Michigan. *J. Water Pollut. Contr. Fed.* **43**, 123–133.

Hübendick, B. (1951). Recent Lymnaeidae, their variation, morphology, taxonomy, nomenclature and distribution. *Kung L. Svenska Vetenska Psakadem. Handlingar* **3**, 1–223.

Hübendick, B. (1964). Studies on Ancyclidae, The Subgroups. *Meddelanden Fran Goteborgs Musei Zoologiska Avendelining*, 137. (*Goteborgs Kungl. Vetenskaps-Och Vitterhets-Samhalles Handlingar. Sjatte Foljden. Ser. B.*) **9**, 1–72.

Hunter, W. R. (1964). Physiological aspects of ecology in non-marine molluscs. *In* "Physiology of Mollusca" (K. M. Wilbur and C. M. Yonge, (eds.), Vol. 1, pp. 83–116. Academic Press, New York.

Hunter, W. R., Apley, M. L., Burky, A. J., and Meadows, R. T. (1967). Interpopulation variations in calcium metabolism in the stream limpet, *Ferrissia rivularis* (Say). *Science* **155**, 338–340.

Hurlbert, S. H. (1971). The nonconcept of species diversity: a critique and alternative parameters. *Ecology* **52**, 577–586.

Hutchinson, G. E. (1957). "A Treatise on Limnology," Vol. 1, Geography, physics, and chemistry. Wiley, New York.

Hynes, H. B. N. (1966). "The Biology of Polluted Waters," Liverpool Univ. Press, Liverpool.

Hynes, H. B. N. (1970). "The Ecology of Running Waters." Univ. Toronto Press, Toronto.

Inaba, A. (1969). Cytotaxonomic studies of Lymnaeid snails. *Malacologia* **7**, 143–168.

Ingram, W. M. (1957). Use and value of biological indicators of pollution: Fresh water clams and snails. *In Trans. Sem. Biol. Probl. Water Poll. Cincinnati, Ohio* C. M. Tarzwell, (ed.), pp. 94–135. Robert A. Taft Sanitary Eng. Center.

Jensen, L. D., Davis, R. M., Brooks, A. S., and Meyers, C. D. (1969). The effects of elevated temperature upon aquatic invertebrates. Edison Res. Project No. 49, Rep. No. 4, 1–252. (Avaliable from Edison Electric Inst., 750 3rd Avenue, New York, N. Y. 10017.)

Jobin, W. R. (1970). Control of *Biomphalaria glabrata* in a small reservoir by fluctuation of the water level. *Amer. J. Tropical Med. Hyg.* **19**, 1049–1054.

Jobin, W. R., and Ippen, A. T. (1964). Ecological design of irrigation canals for snail control. *Science* **145**, 1324–1326.

Jobin, W. R., and Unrau, G. O. (1967). Chemical control of *Australorbis glabratus*. *Publ. Health Rep.* **82**, 63–71.

Johnson, M. G. (1965). Control of aquatic plants in farm ponds in Ontario. U.S. Fish and Wildlife Service. **27**, 23–30.

Johnson, N. M., Reynolds, R. C., and Likens, G. E. (1972). Atmospheric sulfur: its effect on the chemical weathering of New England. *Science* **177**, 514–516.

Jonasson, P. M. (1969). Bottom fauna and eutrophication, 274–305. *In* "Eutrophication: Causes, Consequences, Correctives." Printing and Publ. Office, Nat. Acad. Sci. Washington, D.C.

Kaelser, R. L., Cairns, J., and Bates, J. M. (1971). Cluster analysis of non-insect macro-invertebrates of the Upper Potomac River. *Hydrobiologia* **37**, 173–181.

Keenleyside, M. H. A. (1967). Effects of forest spraying with DDT in New Brunswick on food of young Atlantic salmon. *J. Fish. Res. Bd. Can.* **24**, 107–822.

LaRocque, A. (1953). Catalogue of the Recent Mollusca of Canada. *Bull. Nat. Mus. Can.* **129**, 1–406.

LaRocque, A. (1960). Molluscan faunas of the Flagstaff formation of central Utah. *Geol. Soc. Amer. Mem.* **78**, 1–100.

LaRocque, A. (1970). Un manuscript Inedit de l'abbe Provancher sur les Mollusques du Canada. *Sterkiana* **37**, 1–23; **41**, 1–33.

Lehmkuhl, D. M. (1972). Change in thermal regime as a cause of reduction of benthic fauna downstream of a reservoir. *J. Fish. Res. B. Can.* **29**, 1329–1332.

Leonard, A. N. (1959). Handbook of Gastropods in Kansas. *Univ. Kansas Mus. Natur. Hist. Misc. Publ.* **20**, 1–224.

Levy, G. F., and Folstad, J. W. (1969). Swimmers itch. *Environment* **11**, 14–21.

Lord, P. H. Personal communication, Box # 142, West Oneonta, New York 13861.

Macan, T. T. (1950). Ecology of fresh-water Mollusca in the English Lake District. *J. Anim. Ecol.* **19**, 124–146.

Mackenthunan, K. M., and Ingram, W. M. (1967). Biological associated problems in freshwater environments. Their identification, investigation, and control. U.S. Dept. of the Interior, Fed. Water Pollut. Contr. Administration.

Magruder, S. R. (1934). Notes on the life history of *Pleurocera canaliculatum undulatum* (Say). *Nautilus*, **48**, 26–28.

Malek, E. A. (1962). "Laboratory Guide and Notes for Medical Malacology." Burgess Publ., Minnesota.

Mason, W. T., Lewis, P. A., and Anderson, J. B. (1968). Macroinvertebrate collections and water quality monitoring in the Ohio River Basin 1963–1967. Cooperative rep. Off. of Tech. Progr., Ohio Basin Regions and Anal. Quality Contr. Lab. Water Quality Office. Environ. Protect. Ag., Cincinnati, Ohio 45202.

Mason, W. T., Anderson, J. B., Kreis, R. D., and Johnson, W. C. (1970). Artificial substrate sampling of macroinvertebrates in a polluted reach of the Klamath River. *Ore. J. Water Pollut. Contr. Fed.* **42(8)**, Part II, R315–R328.

McCraw, B. M. (1961). Life history and growth of the snail *Lymnaea humilis* Say. *Trans. Amer. Microsc. Soc.* **70**, 16–27.

McDonald, S. (1969). *Lymnaea stagnalis*. *Sterkiana* **36**, 1–17.

Miles, C. D. (1958). The family Succineidae (Gastropoda: Pulmonata) in Kansas. *Univ. Kansas Sci. Bull.* **38**, 1499–1543.

Morrison, J. P. E. (1932). A report on the mollusca of the northeastern Wisconsin Lake district. *Trans. Wisconsin Acad. Sci. Arts Lett.* **27**, 359–396.

Muirhead-Thomson, R. C. (1971). "Pesticides and Freshwater Fauna". Academic Press, New York.

Naqvi, M., and Ferguson, D. E. (1968). Pesticide tolerances of selected freshwater invertebrates. *J. Miss. Acad. Sci.* **14**, 121–127.

Nelson, D. J. (1962). Clams as Indicators of Strontium-90. *Science* **137**, 38–39.

Nelson, D. J. (1963). The Strontium and Calcium relationships in Clinch and Tennessee River mollusks. *In* "Radioecology" (V. Schultz and A. W. Klement, ed.), pp. 203–211. *Reinhold, New York and Amer. Inst. Biol. Sci., Washington, D.C.*

Oglesby, R. T., Carlson, C. A., and McCann, J. A. (1972). "River Ecology and Man". Academic Press, New York.

Økland, J. (1969a). Om forsuring av vassdrag ot betydningen av surhetsgraden (pH) for fiskens naeringsdyr i ferskvann. *Fauna* **22**, 140–147.

Økland, J. (1969b). Distribution and ecology of the freshwater snails of Norway. *Malacologia*, **9**, 143–151.

Økland, J. (in litt., 1972). From a letter received 29 June 1972.

Palmer, J. R., Colon, A. Z., Ferguson, F. F., and Jobin, W. R. (1969). The control of schistosomiasis in Patillas, Puerto Rico. *Publ. Health. Rep.* **84**, 1003–1008.

Patrick, R., Cairns, J., and Scheier, A. (1969). The relative sensitivity of diatoms, snails, and fish to twenty common constituents of industrial wastes. *Progr. Fish Cult.*, **July**, 137–140.

Pennak, R. W. (1971). Toward a classification of lotic habitats. *Hydrobiologica* **38**, 321–334.

Pilsbry, H. A. (1934a). Review of the Planorbidae of Florida, with notes on other members of the family. *Proc. Acad. Natur. Sci. Philadelphia* **86**, 29–66.

Pilsbry, H. A. (1934b). Mollusks of the Fresh-water Pliocene beds of the Kettleman Hills and neighboring oil fields, California. *Proc. Acad. Natur. Sci. Philadelphia* **86**, 541–570.

Pimental, D. (1971). Ecological effects of pesticides on non-target species. Executive Office of the President, Off. of Sci. and Technol.

Pinel-Alloul, B., and Magnin, E. (1971). Cycle vital et croissance de *Bithynia tentaculata* L. (Mollusca; Gastropoda; Prosobranchia) du Lac St. Louis pres de Montreal. *Can. J. Zool.* **49**, 759–766.

Price, T. J. (1965). Accumulation of radionuclides and the effects of radiation on mollusks. *In* "Seminar on biological problems in water pollution," pp. 202–210. Taft Sanitary Eng. Center, Cincinnati, Ohio.

Purchon, R. D. (1968). "The Biology of the Mollusca". Pergamon, Oxford.

Richard, A. G. (1965). The development rate and oxygen consumption of snail eggs at various temperatures. *Z. Naturforsch. Zob* 347–349.

Richardson, R. D. (1925). Changes in the small bottom fauna of Peoria Lake, 1920 to 1922. *Bull. Ill. State Natur. Hist. Surv.* **15**, 327–389.

Ritchie, L. S., and Berrios-Duran, L. A. (1969). Chemical stability of molluscicidal compounds in water. *Bull. W. H. O.* **40**, 471–473.

Ritchie, L. S., Berrios-Duran, L. A., and Sierra, R. (1969). A field screening test on a slow-release formulation of sodium pentachlorophenate for molluscicidal use. *Bull. W. H. O.* **40**, 474–476.

Robertson, I. C. S., and Blakeslee, C. L. (1948). The Mollusca of the Niagara Frontier Region. *Bull. Buffalo Soc. Natur. Sci.* **19**, 1–191.

Rosewater, J. (1959a). Calvin Goodrich: A bibliography and catalogue of his species. *Mus. Comp. Zool. Harvard Univ. Occas. Pap. Mollusks*, **2**, 189–208.

Rosewater, J. (1959b). Mollusca of the Salt River, Kentucky. *Nautilus* **73**, 57–63.

Roy, E. C. (1964). Pleistocene non-marine mollusca of Northwestern Wisconsin. *Sterkiana* **15**, 5–75.

Schutte, C. H. J., and Frank, G. H. (1964). Observations on the distribution of freshwater mollusca and chemistry of the natural waters in the south-eastern Transvaal and adjacent northern Swaziland. *Bull. W. H. O.* **30**, 389–400.

Schwartz, J., and Meredith, W. G. (1962). Mollusks of the Cheat River watershed of West Virginia and Pennsylvania, with comments on present distributions. *Ohio J. Sci.* **62**, 203–207.

Seiffer, E. A., and Schoof, H. F. (1967). Tests of 15 experimental molluscicides against *Australorbis glabratus. Publ. Health Rep.* **82**, 833–839.

Shoup, C. S., Peyton, J. H., and Gentry, G. (1941). A limited biological survey of the Obey River and adjacent streams in Tennessee. *Rep. Reelfoot Lake Biol. Sta.* **5**, 48–76. (also *J. Tenn. Acad. Sci.* **16**, 48–76.)

Shoup, C. S. (1943). Distribution of freshwater gastropods in relation to total alkalinity of streams. *Nautilus* **56**, 130–134.

Sinclair, R. M. (1969). The pleurocerid fauna of the Tennesse River. *Annul. Rep. Amer. Malacol. Un.* **36**, 45–47.

Sinclair, R. M. (undated). Endangered Pleurocerid snail populations, a study in adaptation (mimeo).

Sparks, B. W. (1961). 10. The ecological interpretation of quaternary non-marine mollusca. *Proc. Linnean Soc. London. 172nd session, 1959–60* 71–80.

Stansbery, D. H. (1970a). 2. Eastern freshwater mollusks (I) the Mississippi and St. Lawrence River systems. *Malacologia* **10**, 9–21.

Stansbery, D. H. (1970b). Dams and the extinction of aquatic life. Lecture: U.S. At. Energy Comm., 10 December 1970. (mimeo).

Stansbery, D. H. (1971). Rare and endangered freshwater mollusks in the eastern United States, *Proc. Symp. Rare Endangered Mollusks (Naiads) U.S.* pp. 5–18.

Stansbery, D. H. (1972). The Mollusk Fauna of the North Fork Holston River at Saltville, Virginia. *Bull. Amer. Malacol. Un. February,* 45–46.

Strecker, J. K. (1935). Land freshwater snails of Texas. *Trans. Texas Acad. Sci.* **17**, 4–44.

Surber, E. W., and Thatcher, T. O. (1963). Laboratory studies of the effects of alkyl benzine sulfonate (ABS) on aquatic invertebrates. *Trans. Amer. Fish. Soc.* **92**, 152–160.

Sutter, R., and Moore, E. (1972). Stream Pollution Studies. State of N. Y. Conservation Comm. Albany.

Tarzwell, C. M., and Gaufin, A. R. (1967). Some important biological effects of pollution often disregarded in stream surveys, *In* "Biology of Water pollution" L. E. Keup, W. M. Ingram, and K. M. Mackenthunan (eds.), pp. 21–31. A collection of selected papers on stream pollution, waste water, and water treatment. Fed. Water Pollut. Contr. Administration, Cincinnati, Ohio, CWA-3.

Taylor, D. W. (1960). Late Cenozoic Molluscan faunas from the high plains. *U.S. Geol. Surv. Prof. Pap.* **337**, 1–94.

Taylor, D. W. (1966). Summary of North American Blancan nonmarine mollusks. *Malacologia* **4**, 1–172.

Taylor, D. W. (1970). West American freshwater mollusca. I. Bibliography of Pleistocene and Recent species. *San Diego Natur. Hist. Mus. Mem.* **4**, 1–73.

Taylor, D. W. (in litt. 1972). From a letter received 18 August 1972.

Tryon, G. W. (1873). Land and freshwater shells of North America. Part 4 Strepomatidae. *Smithsonian Misc. Collect.* **253**, 287, fig. 555.

van Cleave, J. (1932). Studies on snails of the genus *Pleurocera* I. The eggs and egg laying habits. *Nautilus* **46**, 29–34.

van Cleave, J., and Altringer, A. (1937). Studies on the life cycle of *Campeloma rufum*, a freshwater snail. Amer. Natur. **71**, 167–184.

van Cleave, J., and Chambers, R. (1935). Studies on the life history of a snail of the genus *Lioplax. Amer. Midl. Natur.* **16**, 913–920.

van Cleave, H. J., and Lederer, L. G. (1932). Studies on the life cycle of the snail *Viviparus contectiodes. J. Morphol.* **53**, 499–522.

Van der Broght, O., and Van Puymbroeck, (1970). Initial uptake distribution and loss of soluble Ru-106 in marine and freshwater organisms in laboratory conditions. *Health Phys.* **19**, 801–811.

van der Schalie, H. (1948). The land and freshwater snails of Puerto Rico. *Mus. Zool. Univ. Mich. Misc. Publ.* **70**, 1–134, 14 pl.

van der Schalie, H. and Getz, L. L. (1962). Distribution and natural history of the snail *Pomatiopsis cincinnatiensis* (Lea). *Amer. Midl. Natur.* **68**, 203–231.

Vlasblom, A. G. (1971). Furhter investigations into the life cycle and soil dependence of the water snail *Aplexa hypnorum. Basteria* **35**, 95–108.

Walker, B. (1918). A Synopsis of the classification of the fresh-water Mollusca of North America, north of Mexico, and a catalogue of the more recently described species, with notes. *Univ. Mich. Misc. Publ.* **6**, 1–213.

Walker, B. (1925). New species of North American Ancylidae and Lancidae. *Univ. Mich. Occas. Pap. Mus. Zool.* **165**, 1–10.

Walker, C. R. (1971). The toxicological effects of herbicides and weed control on fish and other organisms in the aquatic ecosystem. *Proc. Eur. Weed Res. Council, 3rd Int. Symp. Aquatic Weeds* 119–127.

Wallace, J. B., and Brady, E. U. (1971). Residue levels of dieldrin in aquatic invertebrates and effect of prolonged exposure on populations. *Pestic. Monit. J.* **5**, 295–300.

Walter, H. J. (1969). Illustrated biomorphology of the "*Angulata*" lake form of the Basommatophoran snail *Lymnaea catascopium* Say. *Malacol. Rev.* **2**, 1–102.

Walter, H. J. (1972). Suggested additions and changes to Billy G. Isom's list of aquatic molluscs of Tennessee Valley area (unpublished).

Walton, C. L., and Jones, W. N. (1926). Further observations on the life-history of *Limnaea truncatula. Parasitology* **18**, 144–147.

Wilber, K. M., and Yonge, C. M. (1966). "Physiology of Mollusca," Vol. 2. Academic Press, New York.

Wilhm, J. L. (1970a). Transfer of radioisotopes between detritus and benthic macroinvertebrates in laboratory microecosystems. *Health Phys.* **18**, 277–284.

Wilhm, J. L. (1970b). Range of diversity index in benthic macroinvertebrate populations. *J. Water Pollut. Contr. Fed.* **42(5)**, R221-R224.

Wilhm, J. L., and Dorris, T. C. (1968). Biological parameters for water quality criteria. *Bioscience* **18**, 477–481.

Winslow, M. L. (1918). *Pleurobema clava* (Lam.) and *Planorbis dilatatus buchanensis* (Lea) in Michigan. *Univ. Mich. Occas. Pap. Mus. Zool.* **51**, 1–7.

Winslow, M. L. (1926). The varieties of *Planorbis campanulatus* (Say). *Univ. Mich. Occas. Pap. Mus. Zool.* **180**, 1–11.

Wolfert, D. R., and Hiltunen, J. K. (1968). Distribution and abundance of the Japanese snail, *Viviparus japonicus*, and associated macrobenthos in Sandusky Bay, Ohio. *Ohio J. Sci.* **68**, 32–40.

Wurtz, C. B. (1949). *Physa heterostropha* (Say). *Nautilus* **63**, 2–7.

Wurtz, C. B. (1955). Stream biota and stream pollution. *Sewage Ind. Wastes* **27**, 1270–1278.

Wurtz, C. B. (1962). Zinc effects on freshwater mollusks. *Nautilus* **76**, 53–61.

Yongue, W. H., and Cairns, J. (1971). Micro-habitat pH differences from those of the surrounding water. *Hydrobiologia* **38**, 453–461.

Zimmerman, J. (1960). Pleistocene molluscan faunas of the Newell Lake deposit, Logan County, Ohio. *Ohio J. Sci.* **60**, 13–39.

Zischkale, M. (1952). Effects of rotenone and some common herbicides on fish-food organisms. *Field Lab.* **20**, 18–24.

Chapter 10

Insects (Arthropoda: Insecta)

SELWYN S. ROBACK

I. Introduction

Few people now doubt the importance of the insects as part of the overall fauna of the aquatic biotope. They occupy almost all conceivable habitats, have a complete range of food habits, and possess a great variety of adaptations to aquatic respiration. The questions I hope to shed some light on are (1) What parts of the faunas of undamaged and damaged situations do the insects compose, and (2) What are the proportions of the individual families of aquatic insects under these conditions?

An undamaged stream is one which supports a diverse and balanced fauna and flora, with all trophic levels proportionally represented and no obvious population imbalance. Patrick (1949) refers to this as a "healthy" stream. In making this evaluation, the physical structure and available habitats must be considered, as well as background studies of similar stream sections whose sources and amounts of contamination are known. The latter are specially important, if not essential. I realize there is, to the engineer and perhaps to others, a lack of precision in this definition. The problem is that our knowledge of the biological systems involved is itself imprecise, and our definitions cannot be better than our knowledge.

In theory one could say that an undamaged station is a stream or lake section which is occupied by aquatic organisms up to its full potential for supporting such life. Such a definition, though difficult from a practical view, would cover situations such as naturally saline water, hot springs, highly oligotrophic streams, etc., where the undamaged condition (due to unusual water quality or naturally low nutrients) is a restricted fauna and flora. The definition given in the preceding paragraph is essentially a special

TABLE I
THE FAUNAS OF 13 UNDAMAGED STATIONS[a]

Station	Protozoa		Invertebrates other than insects		Insects		Fish		Total
	sp	%	sp	%	sp	%	sp	%	sp
A	51	29	24	13	74	42	29	16	178
B	84	54	24	16	31	19	17	11	156
C	35	29	10	8	58	47	20	16	123
D	65	36	23	13	71	39	21	12	180
E	40	33	19	16	43	36	18	15	120
F	30	22	27	19	50	36	32	23	139
G	57	35	23	14	60	37	21	13	161
H	67	40	40	24	46	27	16	9	169
I	32	30	10	10	54	52	9	9	105
J	29	33	6	7	43	49	10	11	88
K	63	44	24	17	44	31	13	9	144
L	64	39	17	10	58	36	24	15	163
M	59	48	8	6	49	40	8	6	124
Mean	52	36	20	13	52	38	18	13	142
Range	29–84	22–54	6–27	6–24	31–74	19–52	8–32	6–23	88–180

[a]The number of species (sp.) and percent of the total undamaged fauna (%) of each station are given in each column.

TABLE II
THE INSECT FAUNAS OF 13 UNDAMAGED STATIONS[a]

Station	Odonata		Ephemeroptera		Plecoptera		Hemiptera		Megaloptera		Coleoptera		Trichoptera		Diptera		Total
	sp	%	sp	%	sp	%	sp	%	sp	%	sp	%	sp	%	sp	%	sp
A	15	21	13	18	2	3	7	10	2	3	14	20	7	10	11	15	71
B	11	36	6	19	0	0	2	6	0	0	3	10	2	6	7	23	31
C	9	16	12	21	5	8	5	8	3	5	8	14	4	7	12	21	58
D	7	10	15	21	6	9	5	7	2	3	10	14	8	11	17	25	70
E	10	23	4	9	0	0	7	16	2	5	6	14	3	7	11	26	43
F	4	8	9	18	2	4	3	6	1	2	10	20	5	10	16	32	50
G	8	13	7	12	2	3	7	12	3	5	17	28	6	10	10	17	60
H	7	15	5	11	0	0	5	11	1	2	14	30	1	2	13	28	46
I	8	15	8	15	7	13	3	6	2	4	6	11	7	13	13	24	54
J	4	9	5	12	6	14	4	9	2	5	14	33	2	5	6	14	43
K	7	16	4	9	1	2	7	16	1	2	9	20	5	11	10	23	44
L	11	19	7	12	4	7	2	3	2	3	15	26	7	12	10	17	58
M	3	6	7	14	1	2	5	10	0	0	11	22	8	16	14	29	49
Mean	8	16	8	15	3	5	5	9	2	3	11	20	5	9	12	23	52
Range	3–15	6–36	4–15	9–21	0–7	0–14	2–7	3–16	0–3	0–5	3–17	10–33	1–8	2–16	6–17	14–32	31–71

[a]These are the same stations cited in Table I. The numbers of species (sp.) and percent of the insect faunas represented (%) are given for each order. The Lepidoptera are omitted.

case of the above, more comprehensive, definition, and is easier to apply to the average body of water.

Any downward change from that state constitutes "biological damage," and does not necessarily bear any relation to use, economics, or aesthetics. In severely damaged areas, the fauna may be uniformly depressed (as by industrial pollution) or unbalanced with some species completely dominant (as by organic waste).

I have used the words "damaged" and "undamaged" rather than "polluted" and "unpolluted". For the present, I find the terms "polluted" and "pollution" much too ambiguous and emotionally charged. Pollution can mean organic sewage wastes, organic or inorganic industrial waste, detergents, heat, silt, insecticide runoff, excessive runoff due to urbanization, channelization, dams, radioactivity, or any combination of these. It is therefore obvious that one person's pollution can be another person's necessary price for progress. The same is true of such biologically meaningless terms such as "impairment," "relative worth of the biological environment," "economic benefit," etc. These words, though used in legal documents and arguments, are really value-loaded political terms. When one speaks of use, whose use? The change that would render a body of water "polluted" to a fisherman would not be noticed by a motor boat enthusiast. The same is true of the other terms—relative worth to whom? Whose economic impairment or benefit? Although stream channelization brings undoubted returns to the farmer in the form of more arable land, less flooding, and better swamp drainage—to the fisherman, ecologist, or conservationist the stream in

TABLE III
THE FAUNAS OF 10 DAMAGED STATIONS[a]

Station	Protozoa		Invertebrates other than insects		Insects		Fish		Total
	sp.	%	sp.	%	sp.	%	sp.	%	sp.
N	36	36	9	9	33	33	22	22	100
O	30	73	11	27	0	0	0	0	41
P	37	69	6	11	8	15	3	5	54
Q	31	69	12	27	2	4	0	0	45
R	19	39	2	4	27	55	1	2	49
S	31	50	15	24	0	0	16	26	62
T	26	53	12	25	10	20	1	2	49
U	46	60	1	1	23	30	7	9	77
V	57	64	23	26	5	6	4	4	89
W	30	38	5	6	43	55	0	0	78
Mean	34	55	10	16	15	22	5	7	64
Range	19–57	36–73	1–23	1–27	0–43	0–55	0–22	0–26	41–100

[a]The numbers of species (sp.) and percents of the total fauna (%) of each station are given in each column.

question is irreparably damaged. In discussing pollution it is easy to sink in a morass of undefinable verbiage (see Doudoroff and Warren, 1957).

As shown in Table I, the insects constituted from 19 to 52% (mean 38%) of the total fauna of 13 undamaged stations, and the mean of the total fauna of these stations was 142 species. When one considers ten severely damaged stations (Table III), it can be seen that the insects constituted 0 to 55% of the total fauna (mean 22%). The mean number of species in the total fauna was down to 64. It can also be seen that the Protozoa and other invertebrates, although lower in numbers of species under damaged conditions, constituted higher percentage of the total fauna than they did under undamaged conditions. The insects and fish were more severely affected by the contaminants involved, which include industrial, strip mining, and sewage effluents.

When the insect faunas are considered by families (Tables II and IV) it can be seen that under undamaged conditions the means ranged from 3% (Megaloptera) to 23% (Diptera). Under damaged conditions the Odonata and Diptera, while lower in numbers of species, increased their percentages, while the Coleoptera and Trichoptera, although also lower in numbers of species, remained at the same percentage levels.

In all stations considered in Tables II through IV, the Lepidoptera and Neuroptera were either absent or so rarely present that they were not considered.

A. Quantitative Measurements

Quantitative measurements in a stream or river are still only a reasonable approximation. It is, as far as I am concerned, impossible to obtain accurate and reproducible quantitative measurements in a river comparable to those obtained in lakes or pond bottoms. See Wurtz (1960).

B. Indicator Organisms

The concept of indicator organisms is a beguiling one and has a long history in pollution biology. It promises a quick, easy way of making judgements. Unfortunately, there is a large gap between the promise and reality. As far as the insects are concerned, I am convinced that the concept has little validity, and that the presence or absence of any species of insect in a stream indicates (as far as damage is concerned) no more or less than the bald fact of its presence or absence.

The majority of insect species found in the average eutrophic body of water (the dominant type we have to deal with) can tolerate a broad enough range of water chemistry and physical conditions to render them useless, singly, as indicators of the degree of (or lack of) damage to any body of water.

TABLE IV
THE INSECT FAUNAS OF 10 DAMAGED STATIONS[a]

Station	Odonata		Ephemeroptera		Plecoptera		Hemiptera		Megaloptera		Coleoptera		Trichoptera		Diptera		Total
	sp.	%	sp.	%	sp.	%	sp.	%	sp.	%	sp.	%	sp.	%	sp.	%	sp.
N	6	18	7	21	1	3	1	3	0	0	5	16	6	18	7	21	33
P	0	0	0	0	0	0	1	13	0	0	5	63	0	0	2	24	8
Q	2	100	0	0	0	0	0	0	0	0	0	0	0	0	0	0	2
R	6	22	2	7	1	4	4	15	1	4	2	7	1	4	10	37	27
T	5	50	0	0	0	0	0	0	1	10	0	0	1	10	3	30	10
U	7	30	5	22	0	0	1	4	0	0	3	13	4	18	3	13	23
V	2	40	0	0	0	0	0	0	0	0	1	20	1	20	1	20	5
W	6	14	3	7	0	0	5	12	3	7	10	23	3	7	13	30	43
X	0	0	1	20	0	0	0	0	0	0	3	60	0	0	1	20	5
Y	1	14	1	14	0	0	0	0	0	0	0	0	1	14	4	58	7
Mean	4	29	2	9	<1	1	1	5	<1	2	3	20	2	9	4	25	16
Range	0–7	0–100	0–7	0–22	0–1	0–4	0–5	0–15	0–3	0–10	0–10	0–63	0–6	0–20	0–13	0–58	2–43

[a]Eight of these stations are the same stations cited in Table III. The numbers of species (sp.) and percents of the insect faunas (%) represented are given for each order.

In addition to water chemistry, the presence or absence of a species can be determined by such factors as: (1) its presence or absence in the species pool available for colonizing the area studied; (2) the season in which the collection is made; (3) flow conditions at the time of study; and (4) chance.

It is probably more valid to speak of indicator assemblages of species (insects and other groups) or indicator communities, but at present our knowledge is not sufficient to define these units in the higher invertebrates. The paper by Cairns, Lanza and Parker (1972) is a good approach to this problem.

C. Chemical and Biological Data

The data offered here (Table I–XIV) are from stream surveys conducted from 1951 to date by the Limnology Department, Academy of Natural Sciences of Philadelphia, under the direction of Dr. Ruth Patrick, to whom I am deeply indebted. The chemical tests were performed by Miss Faerie Lynn Carter, Miss Yvonne Swabey, Miss Nancy Hess, Mr. Nick Nitti, and Mr. Robert Haug. The species determinations were performed by Mr. T. Dolan IV, Dr. Arden Gaufin, Dr. J. Hanson, Dr. C. Hodge IV, Dr. J. Lattin, Mr. J. Lutz, Mr. J. Richardson, Dr. M. Sanderson, Dr. M. J. Westfall Jr., and myself. All errors of omission or commission are my responsibility.

The superscript number after each species name gives the number of records involved in the set of chemical data given for that species. It also gives some indication of how common the particular species is. Overall, the data from 110 survey stations were used to make up the tables.

The odonate data are from Roback and Westfall (1967). The data from Patrick *et al.* (1967) and Roback and Richardson (1969) are incorporated into the other tables. Except for the Odonata, 15 chemical parameters are given. These are

Alkalinity (Alk), ppm (methyl orange)	Nitrate nitrogen (NO_3), ppm
Chloride (Cl), ppm	Nitrite nitrogen (NO_2), ppm
Dissolved oxygen (DO), ppm	Phosphate (PO_4), ppm
Iron (Fe), ppm	Sulfate (SO_4), ppm
Total hardness, ppm	Turbidity, ppm
Calcium (Ca), ppm	Biochemical oxygen demand (BOD), ppm
Magnesium (Mg), ppm	pH
Ammonia nitrogen (NH_3), ppm	

The tests were done in accordance with the edition of "Standard Methods" current at the time the tests were done.

In interpreting the tables of chemical data, it should be remembered that the stations surveyed are not a random sampling of North American bodies of water, but that the selection was biased toward those bodies of water associated with industrial, urban, and suburban development—and consequently no mountain streams or lakes are represented. Genera such as *Rhyacophila* and *Epeorus* are therefore not to be found in the species lists.

D. Extreme Tolerance Lists

The ten extreme tolerance lists (pp. 363–370) were extracted from the Tables V through XIII in order to enable the reader to see at a glance those species or genera which can tolerate the extremes of some of the chemical parameters given. Naturally most of these parameters are not isolated, but it was not feasible, without computer analysis, to explore all of these interrelationships. For example, all stations with high Cl were also high in total hardness, but the reverse was not always true. List No. 4 gives the taxa associated with the former situation; list No. 8, those with the latter. Brackish situations generally have high Mg and SO_4, but this is not true where the Cl comes from other than a marine source. Hopefully, with computers we will be able to explore the relationship of aquatic insects to more meaningful combinations of chemical parameters.

E. Reference Material

The references for this chapter are grouped under general keys (comprehensive works with keys to all or most of the aquatic insects); general ecology and "pollution" biology (works on sampling or the ecology and relationships to "pollution" of aquatic insects, generally not restricted to a single order); and references on the systematics and ecology of the individual orders of aquatic insects. The references are not intended to be exhaustive (this could constitute a volume in itself) but are, rather, a core of references listing those works which I consider to be essential to the average worker in aquatic pollution problems. Anyone wishing to delve further into any specific area or order should consult the references of these works. Those books with especially extensive references are marked with asterisks.

II. Odonata

A. Life History

The eggs of odonates are deposited in many ways—dropped into water, attached to objects in water, deposited in gelatinous strips or masses, or inserted in soft plant tissue underwater and in twigs above water. The eggs hatch in roughly 13 to 35 days, with temperature being very important in the duration of the egg stage. The nymphs go through from about 10 to over 20 instars. The duration of the nymphal stage may be as short as 36 days and as long as 5 years. Species which live in ponds and temporary pools develop more quickly than those in cooler, permanent bodies of water. Temperature is important, but not the only factor in determining the duration of nymphal development. To emerge, the nymph crawls out of the water and attaches itself to a suitable substrate. The adult emerges through a longitudinal split in the thorax or head.

B. Habitats

Nymphs of the odonates are found in a great variety of habitats. These, and the forms chiefly found in them, are as follows:

1. Roots along the edge of a stream—Aeschnidae, *Hagenius*
2. Emergent vegetation—Most Zygoptera, *Macromia*, *Hagenius*, *Sympetrum*, *Tramea*
3. Mats of *Elodea*, *Myriophyllum*—Mostly Coenagrionidae
4. Mud bottom surface, deposits of plant debris—*Cordulegaster*, *Libellula*
5. Burrowing in mud—Most Gomphidae
6. Surface of rocks and wood—*Argia*, *Neurocordulia*

C. Food

The nymphs of all odonates are predaceous, feeding on Protozoa, other aquatic insects and, in some rare cases, even fish. Prey is captured by means of the modified extensile labium, characteristic of the odonate nymphs. This labium grasps the prey and then retracts, bringing the prey to the mandibles and maxillae.

In early instars prey is detected primarily by the antennae, but in the later instars visual detection tends to predominate, although a combination of the two is used in most cases. The prey can be detected at distances up to 20 cm. The clambering forms, Aeschnidae and Zygoptera, actively pursue their prey, while the bottom crawlers and burrowers wait in ambush for the prey to pass near.

D. Respiration

In the Anisoptera respiration is accomplished by means of modified rectum, the inner surface of which is foliate, extensively supplied with tracheae, and enlarged to form a basketlike structure. Water is taken into this basket and then expelled. This pumping action is also used as a means of "jet" locomotion.

Respiration in the Zygoptera is more complex. The three caudal lamellae with their tracheae serve as part, but not all, of the respiratory mechanism.

Under ideal conditions some nymphs can survive without them and regenerate those lost. At low oxygen tensions this is not true, and increased mortality results from their loss. In addition, respiration in the Zygoptera can take place in the rectum, over the body surface, and possibly at the wing sheaths. Many species can survive out of water, in moist air, for several weeks.

E. Chemical Parameters

As can be seen from Table XIV the Odonata are not, as a whole, a sensitive group. Except for pH and high BOD, odonates have a few-to-moderate number of species tolerant of the extremes of the chemical parameters listed.

With regard to number of species found under the extremes of water quality, the damselflies and dragonflies are fairly even, although from my experience the damselfly populations are, in almost all cases, considerably larger.

Above pH 8.5 only one damselfly species was found. Of the 15 odonate species found above an alkalinity of 210 ppm, ten were damselflies. In the case of chloride greater than 1000 ppm, eight damselfly and seven dragonfly species were found; in brackish situations with a dissolved oxygen less than 4 ppm, a 1:1 species relationship existed between the two suborders.

In hard waters, the dragonfly nymphs predominated 6:3; and where sulfate was greater than 400 ppm, 3:1.

Where BOD. was greater than 5.9 ppm, the dragonfly to damselfly species ratio was 9:6; and only one species of damselfly, *Ischnura verticalis*, was found where the BOD was greater than 10 ppm.

As mentioned above, the damselflies (*Ischnura* sp., for example) are successful in consistently maintaining good population sizes under chemical extremes. *Ischnura* was found in six of the ten categories listed, with *Argia* and *Enallagma*, 4 times each (see extreme tolerance lists, Section I,B). This is especially true under conditions of high organic loading, when *Ischnura* and the other damselflies often dominated the insect populations present. The dragonflies are more diverse in their tolerances—with no dragonfly genus being found in more than three categories.

III. Ephemeroptera

A. Life History

The mayflies are, for all or part of their life cycle, undoubtedly the most truly aquatic of all the insects that have adapted to the aquatic environment. As the German name for the order, *Eintagsfliegen*, indicates, the adult portion of the life cycle is very short, generally two to three days, and serves

TABLE V

	pH	Alkalinity	Cl	DO
Calopterygidae				
Calopteryx dimidiata Burm.[5]	5.6–6.9	2–23	2–34	7–11
Calopteryx maculata (Beauv.)[14]	6.0–7.7	2–40	1–69	4–10
Hetaerina americana (Fabr.)[20]	6.8–8.4	20–206	3–4430	4–12
Hetaerina titia (Drury)[13]	6.4–8.0	13–21	1–7	6–8
Lestidae				
Archilestes grandis (Ramb.)[2]	7.0–7.2	83–95	14–16	4–6
Coenagrionidae				
Argia apicalis (Say)[35]	6.8–8.2	20–220	1–86	5–9
Argia moesta (Hagen)[49]	6.4–8.3	9–220	1–5570	6–10
Argia sedula (Hagen)[31]	6.7–8.0	202–220	3–4430	6–10
Argia tibialis (Ramb.)[27]	6.3–7.8	10–177	1–965	6–12
Argia translata Hagen[4]	6.7–7.9	18–220	2–71	7–10
Argia violacea (Hagen)[9]	6.0–8.4	2–64	<1–5570	8–10
Enallagma antennatum (Say)[2]	6.5–6.9	16–36	3–6	6–8
Enallagma basidens Calvert[2]	7.8–7.9	206–220	25–69	—
Enallagma carunculatum Morse[10]	7.4–8.5	48–107	20–23	8–12
Enallagma civile (Hagen)[16]	7.2–8.2	24–110	6–5570	6–10
Enallagma divagans Selys[9]	6.0–7.6	2–27	2–59	6–10
Enallagma exsulans (Hagens)[19]	6.8–8.4	22–238	3–146	4–10
Enallagma pallidum Root[2]	6.7	21	10	7
Enallagma signatum (Hagen)[30]	5.6–8.5	10–222	2–1040	4–12
Enallagma traviatum Selys[1]	8.4–8.6	216–222	675–1040	8–9
Enallagma vesperum Calvert[2]	6.9–7.8	23–220	11–25	11–11
Enallagma weewa Byers[8]	6.4–7.8	9–80	3–27	7–9
Ischnura posita (Hagen)[10]	6.8–7.8	15–244	3–2500	1–9
Ischnura verticalis (Say)[28]	6.8–8.5	18–201	7–95	4–12
Protoneuridae				
Neoneura aaroni Calvert[2]	7.8–8.2	180–184	80–85	6–7
Cordulegasteridae				
Cordulegaster maculatus Selys[4]	6.4–7.3	10–23	1–12	7–9
Gomphidae				
Progomphus obscurus (Ramb.)[11]	6.9–8.4	24–213	2–83	6–11
Hagenius brevistylus Selys[14]	5.6–8.4	9–118	1–20	4–12
Ophiogomphus rupinsulensis (Walsh)[5]	7.6–8.3	81–103	9–20	8–12
Erpetogomphus designatus Hagen[20]	6.8–8.2	21–206	1–5570	5–10
Dromogomphus spinosus Selys[32]	6.6–8.4	15–238	1–5570	6–10
Dromogomphus spoliatus Hagen[6]	7.6–8.1	58–208	14–83	5–8
Lanthus albistylus (Hagen)[2]	7.5–8.0	66–180	11–32	8–9
Gomphus dilatatus Rambur[4]	6.8–8.1	20–182	5–83	6–7
Gomphus externus Hagen[4]	7.6–7.9	119–208	20–71	6–8
Gomphus fraternus (Say)[2]	7.5–8.2	9–107	18–56	7–9
Gomphus hybridus Wllmsn.[4]	7.8–8.3	93–103	13–146	6–11
Gomphus lineatifrons Calv.[4]	8.2	87–104	18–20	8–9
Gomphus vastus Walsh[18]	7.3–8.4	23–109	2–5570	6–10
Gomphus exilis Selys[14]	7.7–8.2	84–104	13–20	6–11
Gomphus lividus Selys[22]	6.4–8.4	9–116	1–2580	7–10
Gomphus spicatus? Hagen[4]	8.2	81–98	18–19	9

[a]All results are in parts per million (ppm) except pH. Superscript numbers following species names indicate the number of records comprising the following ranges.

Ranges of Chemical Analysis of Waters in Which Species of the Order Odonata Were Collected[a]

Total hardness	Ca	Mg	NO_3	SO_4	BOD	Turbidity
8–70	3–12	1–10	0.13–2.81	1.9–45.2	0.1–2.2	12–34
6–323	2–84	1–27	0.05–4.92	2.5–247.6	0.6–7.9	50–> 72,000
17–4020	4–1570	2–41	0.03–0.51	2.9–289.7	0.6–6.7	2–> 72,000
4–232	2–61	< 1–19	0.04–0.34	19.9–45.4	0.6–2.0	7–548
135–136	35–36	11–12	0.68–2.66	49.3–51.0	0.6–4.0	22–25
15–336	3–84	< 1–33	0.03–4.92	2.5–247.8	0.4–7.9	4–> 72,000
8–4980	2–1955	< 1–32	0.06–0.94	2.5–172.8	0.1–6.0	2–> 72,000
15–4020	3–1570	1–34	0.04–1.39	2.5–116.6	0.4–5.4	2–241
4–405	2–90	< 1–78	0.04–0.51	2.3–289.7	0.2–6.7	8–165
12–208	2–58	1–33	0.07–1.52	2.5–160.2	1.1–3.0	16–> 72,000
36–4980	12–1955	< 1–32	0.05–2.81	<2.0–116.4	0.1–2.9	2–32
27–29	5–7	3–4	0.01	21.6–49.0	—	21–35
208–227	30–47	27–33	<0.01	29.7–44.0	2.0	66–71
41–177	40–51	9–12	0.03–0.05	21.6–28.8	0.7–2.8	4–19
65–4980	16–1955	3–32	0.01–2.66	16.8–160.2	0.6–4.9	2–78
12–171	2–48	1–12	0.07–2.81	2.5–45.4	1.1–3.0	7–140
21–373	4–101	< 1–34	0.02–1.58	2.9–213.8	0.7–6.0	4–241
21	6	1	0.04	3.5	0.6–0.7	15
8–560	3–97	1–77	0.03–1.58	1.9–160.2	0.4–4.9	4–140
435–560	88–97	52–77	—	112.0–127.0	0.5–1.7	230–250
37–208	8–30	4–33	0.15–0.21	10.8–29.7	0.4–2.0	230–250
11–192	3–7	1–2	0.03–0.47	2.5–143.3	0.1–0.9	10–161
13–900	3–88	1–180	0.05–0.87	2.8–480.0	0.4–29.0	8–99
19–405	4–90	2–41	<0.01–4.92	8.3–289.7	0.8–18.4	6–> 72,000
222–226	59–61	18–18	0.03–0.05	23.5–31.0	—	90–100
7–29	2–10	1–3	0.05–0.19	2.4–5.1	0.2–2.2	6–18
17–320	3–61	2–43	0.06–0.22	4.8–17.8	0.3–1.8	18–161
4–176	2–42	< 1–17	0.05–0.44	1.9–72.2	0.6–4.6	5–90
128–171	36–40	13–17	<0.01–0.17	10.5–31.0	0.3	26–54
10–4980	3–1955	1–35	<0.01–0.76	16.7–116.4	0.3–5.4	2–>72,000
17–4980	3–1955	1–34	0.05–0.75	4.8–172.8	0.6–6.0	2–660
70–245	21–64	4–21	0.04–0.53	6.9–32.5	0.2	49–140
80–570	23–170	6–34	0.25–0.94	22.9–350.0	1.1–1.7	6–25
29–220	10–58	1–18	0.08–0.22	5.1–20.5	—	7–161
180–232	54–61	11–19	0.04–0.79	20.1–48.6	0.9–4.4	71–>72,000
159–176	39–50	12–17	0.01–0.24	21.4–45.2	2.0	8–46
91–143	32–42	8–18	0.23–0.56	33.3–66.9	2.8–6.0	26–227
159–176	39–42	13–17	0.01–0.24	21.4–28.0	1.8	25–46
14–4980	3–1955	1–32	0.01–0.75	2.7–116.4	0.6–6.0	2–647
101–176	27–42	6–17	0.01–0.75	21.4–66.9	0.3–4.6	10–227
6–2600	2–1050	1–39	0.01–1.90	2.4–450.0	0.3–5.4	2–73
155–164	36–40	14–18	<0.01–0.14	16.8–31.7	1.9	26–78

(*continued*)

TABLE V (*continued*)

	pH	Alkalinity	Cl	DO
Gomphus amnicola Walsh[4]	7.6–7.9	118–142	42–71	7–8
Gomphus ivae Wlmsn.[2]	5.6–6.4	9–10	2–3	8–9
Gomphus laurae Wlmsn.[7]	5.6–6.9	9–33	2–5	6–9
Gomphus plagiatus Selys[39]	6.3–7.6	9–46	1–185	5–10
Gomphus spiniceps (Walsh.)[20]	7.2–8.2	67–114	6–127	4–12
Aeshnidae				
Basiaeschna janata (Say)[12]	6.0–8.2	2–110	5–5570	4–10
Boyeria vinosa (Say)[41]	5.6–8.1	10–238	1–5570	5–10
Anax junius (Drury)[9]	6.8–8.5	20–197	14–95	4–12
Nasiaeshna pentacantha Ramb.[11]	6.7–8.3	20–206	1–71	5–12
Aeshna u. umbrosa Walk.[7]	7.0–7.5	12–95	5–32	4–10
Macromiidae				
Didymops transversa (Say)[8]	6.0–8.0	2–211	15–88	5–10
Corduliidae				
Neurocordulia alabamensis Hodges[3]	5.6–6.4	9–10	2–5	8–9
Neurocordulia molesta Walsh[41]	6.3–8.1	10–184	1–86	4–8
Neurocordulia obsoleta (Say)[3]	7.2–8.1	24–103	2–19	8–11
Neurocordulia virginiensis Davis[11]	6.4–8.3	9–109	1–146	7–10
Epicordulia princeps (Hagen)[7]	6.8–8.0	20–220	<1–25	5–9
Tetragoneuria cynosura (Say)[5]	6.0–6.9	2–17	5–34	6–10
Libellulidae				
Perithemis tenera (Say)[11]	6.8–8.4	17–132	6–20	6–10
Libellula luctuosa Burm.[5]	7.2–8.1	34–177	7–95	4–12
Libellula pulchella Drury[2]	7.2	34–58	7–19	8–10
Libellula vibrans Fabr.[2]	6.8	18–20	5–7	7
Plathemis lydia (Drury)[16]	4.8–8.4	3–197	2–63	6–11
Sympetrum vicinum (Hagen)[2]	7.0–7.2	83–95	14–16	4–6
Erythemis simplicicollis (Say)[2]	7.3	39	5–2500	8
Pachydiplax longipennis Burm.[4]	6.7–7.8	16–46	3–12	7–11

only for reproduction. The eggs are laid directly in the water, either singly or in packets, where they sink and adhere to rocks or vegetation. Hatching time varies from one to two weeks, the time depending on the temperature. The numphal stage lasts from five or six weeks to over two years, with about one year the average. The number of instars seems to be variable, with not too many life cycles fully known. Those reported range from 20 to 30 in stars.

B. Habitats

Mayfly nymphs can be found in all kinds of bodies of water, from temporary ponds through all types of streams to large lakes. Burks (1953) gives a list of the water types in Illinois and the characteristic mayflies of each. Within any body of water, mayfly nymphs are found on rocks, in the bottom,

TABLE V (*continued*)

Total hardness	Ca	Mg	NO_3	SO_4	BOD	Turbidity
180–216	53–58	11–17	0.22–0.79	34.0–48.6	0.9–4.4	>72,000
8–11	3	1	0.13–0.20	1.9–3.2	0.1–0.4	13–34
8–31	3–10	1–3	0.09–0.20	3.2–23.0	0.1–2.2	12–39
6–176	2–50	1–14	0.05–0.38	2.2–46.4	0.5–2.2	5–548
92–285	22–94	4–17	<0.01–0.65	4.8–72.2	0.3–4.6	0–672
29–4980	10–1955	1–32	0.08–2.80	5.1–116.4	0.3–4.0	2–54
7–4980	2–1955	1–34	0.02–1.91	1.9–350.0	0.4–2.7	2–241
82–405	15–90	11–41	0.02–1.33	16.1–289.7	0.3–6.7	16–118
10–387	3–87	1–41	0.11–0.25	2.8–260.2	0.4–5.1	7–261
23–135	6–36	2–13	0.10–2.66	2.8–112.5	0.2–4.0	6–25
70–232	12–62	8–20	0.04–2.81	12.1–55.1	1.4–6.3	18–87
8–38	3–14	<1–1	0.13–0.20	1.9–3.2	0.1–0.4	4–34
7–227	2–61	<1–18	0.03–3.10	2.3–32.5	0.3–4.4	8–>72,000
18–170	4–40	2–17	0.07–0.17	6.6–30.2	0.3–2.1	24–93
6–176	2–50	1–12	0.05–0.75	2.4–55.1	0.6–6.0	5–70
29–245	10–64	1–33	0.05–0.42	<2.0–53.2	<0.1–6.0	12–140
18–82	4–14	<1–11	0.09–2.81	2.9–45.2	0.8–3.0	4–165
16–173	4–37	1–20	0.11–0.91	8.3–118.5	0.7–3.3	15–53
115–405	30–90	9–41	0.33–0.95	31.3–289.7	2.1–6.7	11–85
126–214	30–49	13–23	0.63–0.95	112.5–174.8	2.1–4.8	11–84
19–29	4–10	1–2	0.19–0.21	5.1–8.3	1.2–2.2	12–53
12–382	2–81	1–35	0.03–1.91	2.5–422.0	0.3–4.9	4–140
135	34–36	11–12	0.68–2.66	49.3–51.1	0.6–3.9	22–25
37–900	14–60	<1–180	0.05	2.9–480.0	0.4	4–8
14–65	3–15	1–7	0.04–0.33	3.6–66.1	0.7–0.8	15–119

and on vegetation. The following is a brief list of these habitats and their prime mayfly occupants, by genus. Many can be found on range of habitats.

1. Soft mud bottom of lakes and side of rivers—*Hexagenia*, *Pentagenia*, and *Ephoron*
2. Burrowing in the banks of moderate to large rivers—*Tortopus*
3. Under rocks in moderate to slow streams—*Stenonema*, *Heptagenia*, *Ephemerella*, *Potamanthus*, and *Baetis*
4. Bottom sand and gravel in riffles—*Ephemera*
5. On stones, wood, vegetation, trailing plants in very fast flow—*Isonychia*, *Leptophlebia*, *Baetis*, and *Neocloeon*
6. Clean bottom sand in small streams—*Dolania*
7. On rocks, wood, and vegetation in slow silty areas—*Caenis*, *Tricorythodes*, *Ephemerella*, *Leptohyphes*, and *Stenonema*
8. Bottom and vegetation in ponds and backwaters—*Callibaetis* and *Caenis*

As the divisions between the habitats are not sharp, any one of them may grade into several others along its margins. The same can be said of sizes and types of rivers and lakes.

C. Food

Mayflies are perhaps the prime grazers in the aquatic food web. Most are algal feeders; some are scavengers, feeding both on algae and vegetable detritus; and a few (e.g., *Isonychia*) appear to be predators. Some Neotropical mayfly nymphs (*Campylocia*, for example) appear to be obligate predators—at least they can inflict a painful bite, as I can personally attest. The nymph of *Asthenopus* bores into rotten logs, and has powerful triangular mandibles to go along with the habit.

D. Respiration

Mayfly respiration is accomplished by means of abdominal gills. These are typically present laterally on abdominal segments one through seven. These gills may vary considerably in shape and tracheation. In the bottom-dwelling *Hexagenia* the gills are biramous, extend dorsally, and are finely fringed. Other mayflies which live in silty situations (*Caenis*, *Leptohyphes*, and *Tricorythodes*) have gills which lie flat on the dorsum of the abdomen, with the second gill becoming operculate to protect the finely fringed gills below. This both protests the gills from the silt and increases the gill surface. In many Leptophlebiidae which live in ponds or other low dissolved oxygen situations, the gills are finely fringed. In *Callibaetis*, which also typically lives in pondlike situations, the gill surface is increased by enlargement and folding. In most mayflies the gills are simple plates, sometimes with accessory ventral gill tufts.

E. Chemical Parameters

The mayflies have generally been considered to be very sensitive to pollutants—in most cases meaning organic loading. Table XIV and the preceding habitat lists show that, as a generality, this is not true. At least two mayflies can tolerate dissolved oxygen less than 4 ppm. Ten species can tolerate a BOD greater than 5.9 ppm; and one (*Baetis*) BOD greater than 10 ppm. Three species can tolerate brackish water situations; six species high alkalinity (MO); and six species, high SO_4.

In addition to the hard chemical data, my own observation over years of work is that the mayflies are not as sensitive as is generally believed. I have occasionally found *Stenonema*, for example, in situations certainly inimical to sensitive species—oil covered rocks in a highly damaged (organic and chemical) river.

It will be noted in Table VI that the *Stenonema* determinations are only to group level, and *Baetis* and its relatives only to species (sp.) or subspecies (spp.). As far as I am concerned, consistently reliable nymphal determinations are not yet possible in these genera. This is most unfortunate, as these genera and their close relatives are widely distributed in North American waters. It is also unfortunate that data on heavy metal and insecticide tolerance are not more readily available.

IV. Plecoptera

A. Life History

Eggs of the Plecoptera are deposited either in flight, by the female touching the tip of her abdomen to the water, or after the female crawls into the water's edge. The eggs are in the form of a ball, which disintegrates when submerged, and hatching takes placed in two or three weeks. Periods of up to three months are known in some of the larger Setipalpia (Systellognatha). The nymphs (the few that have been reared from the egg) go through 22 to 23 instars. Most stoneflies have a one year life cycle, but a few have two to three year cycles.

Temperature appears to play a large part in nymphal development: Warmer water accelerates development, while cooler water does the reverse. Emergence takes place over the entire year, with the genera *Capnia* and *Nemoura* emerging in the winter to early spring. The majority of genera emerge during the spring and summer.

B. Habitats

Most stonefly nymphs are found on rocks on stream bottoms or between the bottom rubble. In streams without rocks, I have found *Perlesta* and *Paragnetina* on logs and boards. *Pteronarcys* occurs in bottom detritus and among trailing roots along stream margins. In some rivers I have found *Taeniopteryx* on trailing tree leaves, and in mats of *Myriophyllum* and algae in very fast water.

C. Food

The stonefly nymphs are quite variable in their food habits. The genus *Pteronarcys* feeds mostly on plant material, with some occasional insect parts. The genera *Arcynopteryx* and *Isoperla* feed primarily on animal matter, with some plant debris mixed in. The genera *Acroneuria* and *Paragnetina* appear to be primarily carnivorous, feeding on chironomids, blackflies, and mayfly nymphs. The Capniidae and Taeniopterygidae are primarily herbivorous.

TABLE VI

Ephemeroptera	Fe	pH	Alkalinity	Cl	DO	Total hardness
Siphlonuridae						
Siphlonurus nr. *marshalli Trav.*[1]	0.01	6.9	15	3	8	13
Isonychia sp.[26]	<0.01–2.89	5.5–8.8	4–130	1–56	4–14	7–216
Oligoneuridae						
Homeoneuria dolani Edm. Bern., Trav.[1]	0.01	7.0	22	1	8	13
Heptagenia spp.[28]	<0.01–0.89	5.5–8.3	4–124	1–40	5–11	8–178
Stenonema (*bipunctatum* gp.) spp.[15]	<0.01–0.61	7.7–8.5	36–130	11–56	8–14	149–287
Stenonema (*interpunctatum* gp.) spp.[25]	<0.01–2.89	5.6–8.4	5–205	2–40	4–14	13–705
Stenonema (*pulchellum* gp.) spp.[49]	<0.01–0.89	5.5–8.4	4–213	1–185	3–11	7–233
Stenonema (*tripunctatum* gp.) spp.[6]	0.01–0.77	7.2–8.4	47–175	3–845	8–11	60–800
Stenonema (*vicarum* gp.) spp.[2]	2.89	7.3–7.9	39–88	1–6	8–9	66–95
Baetidae						
Baetis spp.[39]	<0.01–0.89	5.6–8.5	5–213	1–2750	4–14	16–1000
Callibaetis	<0.01–0.45	5.6–8.3	5–220	7–2500	4–14	13–900
Centroptilum sp.[2]	<0.01–0.02	6.0–8.8	2–74	27–34	9–10	70–117
Cloeon sp.[1]	0.86	6.6	20	3	7	21
Neocloeon alamance? Trav.[7]	0.01–2.89	6.7–7.4	32–88	6–2750	7–11	66–1000
Pseudocloeon sp.[10]	<0.01–0.90	6.6–8.4	20–97	1–33	6–12	20–216
Heterocloeon sp.[6]	<0.01–0.72	6.2–7.9	47–128	10–55	8–14	87–193
Leptophlebiidae						
Choroterpes sp.[2]	<0.01–0.01	8.8–8.8	73–74	27–27	9–9	114–117
Habrophlebiodes americana (Banks)[1]	2.89	7.3	39	6	8	66
Leptophlebia austrinus (Trav.)[2]	0.11–0.13	6.7–6.8	18–20	2–3	9–9	12–12
Leptophlebia poss. *intermedius* (Trav.)[1]	0.85	6.6	20	5	7	21
Leptophlebia (*Blasturus*) sp.[3]	0.11–0.85	6.6–6.8	18–20	2–5	7–10	12–21
Paraleptophlebia praepedita (Eaton)[1]	0.72	7.2	47	14	11	87
Paraleptophlebia volitans? McD.[1]	0.86	6.6	20	3	7	21
Paraleptophlebia guttata Mc D.[1]	0.02	7.9	205	10	9	705
Paraleptophlebia sp.[3]	0.51–0.76	5.5–5.6	4–6	6–8	3–5	11–16
Traverella presidiana (Trav.)[6]	<0.01–0.02	8.0–8.1	174–190	59–88	7–7	173–222
Ephemerellidae						
Ephemerella argo Burks[1]	0.13	6.8	18	2	10	9
Ephemerella bicolor Clem.[1]	1.8	7.2	61	23	10	322
Ephemerella deficiens Morg.[4]	<0.01–0.05	6.8–8.0	20–97	2–11	6–12	15–124
Ephemerella hirsuta Bern.	0.86	6.6	20	3	7	21
Ephemerella lita? Burks[1]	2.89	7.3	39	6	8	66
Ephemerella simplex McD.[1]	0.1	6.9	22	2	7	15
Ephemerella temporalis McD.[8]	<0.01–2.89	6.8–8.4	5.97	1–14	4–11	6–216
Ephemerella trilineata Bern.[5]	0.82–0.90	6.6–6.7	1–20	3–7	7–8	21–22
Ephemerella tuberculata Morg.[1]	<0.01	8.0	97	10	11	124
Tricorythidae						
Leptohyphes robacki Allen[3]	<0.01	7.0–8.0	27–100	1–20	7–11	10–159
Leptohyphes dolani Allen[7]	<0.01–0.24	6.8–7.0	15–34	1–4	7–9	8–15
Tricorythodes spp.[29]	<0.01–0.77	7.1–8.5	26—220	7–845	5–14	18–800
Caenidae						
Branchycercus nitidus (Trav.)[2]	0.10–0.82	6.4–6.7	9–20	1–3	7–8	6–21
Brachycerus lacustris? Ndm.[1]	2.89	7.3	39	6	8	66
Brachycerus sp.[1]	<0.01	8.0	97	11	6–12	124
Caenis spp.[27]	<0.01–0.67	5.4–8.5	3–220	5–72	2–14	6–705

[a]All results are in parts per million (ppm) except pH. Superscript numbers following species names indicate the number of records comprising the following ranges.

RANGES OF CHEMICAL ANALYSES OF WATERS IN WHICH SPECIES OF THE ORDER EPHEMEROPTERA WERE COLLECTED[a]

Ca	Mg	NH_3	NO_3	NO_2	PO_4	SO_4	BOD	Turbidity
3	1	<0.01	—	<0.01	.05	3.3	—	99
2–64	1–17	<0.01–0.97	0.36–2.30	<0.01–0.01	<0.01–0.32	<1.0–135.0	0.6–6.0	8–>72,000
3	1	0.04	—	<0.01	0.03	2.6	0.9	50
2–37	< 1–17	0.01–1.10	0.29	<0.01–0.07	<0.01–0.33	<1.0–72.8	0.5–6.0	10–>72,000
36–70	15–27	<0.01–1.09	0.08–2.30	<0.01–0.01	0.02–0.56	31.2–251.0	0.8–7.5	6–>72,000
4–220	1–39	0.01–1.09	0.03–1.18	<0.01–0.01	<0.01–0.56	< 1.0–450.0	0.4–6.0	6–> 72,000
2–37	<1–43	<0.01–1.10	0.03–0.06	<0.01–0.04	<0.01–0.86	<1.0–72.8	0.5–6.0	3–548
27–398	4–36	0.03–5.00	0.06–0.90	<0.01–0.09	< 0.01–0.62	18.6–370.0	0.7–2.2	1–12
22–25	3–8	0.02–0.09	0.50	0.01	0.03–0.05	7.5–25.0	0.8–1.1	20–34
3–398	1–200	<0.01–5.00	0.03–0.90	0.01–0.17	<0.01–0.62	<1.0–570.0	0.3–15.4	3–>72,000
3–180	6–180	<0.01–0.15	0.03–1.18	<0.01–0.02	<0.01–0.30	<1.0–480,0	1.1–4.4	8–66
37	6	<0.01	—	<0.01	<0.01–0.12	25.1–45.2	1.7	18–39
6	1	0.97	—	<0.01	<0.01	3.6	—	25
22–92	3–200	0.09–1.09	0.08–1.18	<0.01	<0.01–0.30	25.0–570.0	0.8–1.5	6–20
6–64	1–15	<0.01–1.10	0.12–1.18	<0.01–0.01	<0.01–0.56	3.5–135.0	1.1–3.8	10–44
27–36	4–15	0.01–0.17	0.21–0.51	<0.01–0.01	0.03–0.35	20.5–43.2	1.6–3.8	12–>72,000
36–37	6–6	<0.01	—	<0.01–0.01	0.12–0.12	24.6–25.1	1.7–1.8	37–39
22	3	0.09	0.50	—	0.05	25.0	0.8	20
2–3	1–1	0.02–0.05	—	<0.01–0.01	—	2.2–2.5	0.8–1.1	120–140
6	1	0.85	—	<0.01	<0.01	3.5	—	15
2–6	1–1	0.02–0.85	—	<0.01–0.01	<0.01	2.2–3.5	0.8–1.1	15–140
27	4	0.17	0.21	—	0.03	26.5	1.6	12
6	1	0.97	—	<0.01	<0.01	3.6	—	25
220	39	0.08	—	<0.01	<0.01	450.0	0.4	7
3–4	1–2	<0.10–0.12	—	0.01–0.01	0.01–0.07	<1.0	2.1–2.5	39–54
51–58	11–20	0.02–0.05	—	<0.01–0.02	<.01–.02	12.1–20.5	—	59–106
2	1	0.02	—	<0.01	—	2.5	0.8	120
92	22	0.21	0.12	—	0.04	313.0	1.5	27
3–35	1–9	<0.01–0.02	—	<0.01–0.01	<0.01–0.05	2.5–32.7	0.5–4.1	25–98
6	1	0.97	—	<0.01	<0.01	3.5	—	35
22	3	—	0.50	—	0.05	25.0	0.8	20
4	2	0.02	—	<0.01	<0.01	2.7	0.6	34
3–64	1–14	0.01–0.27	0.21–2.30	<0.01–0.01	<0.01–0.56	2–135.0	0.6–2.0	12–89
6–6	1–1	0.60–1.10	—	<0.01	<0.01	3.5–3.6	—	10–15
35	9	0.01	—	0.01	0.04	32.7	3.8	34
2–37	1–16	<0.01–0.08	0.08–0.30	<0.01–0.03	<0.01–0.10	21.0–42.4	0.3–6.0	33–259
2–4	<1–2	<0.01–0.16	—	<0.01–0.02	<0.01–0.05	2.3–3.3	0.8–2.0	25–259
4–398	2–43	<0.01–5.00	0.36–0.90	<0.01–0.04	<0.01–0.72	1.3–450.0	0.4–7.5	10–>72,000
2–6	1	0.01–1.10	—	<0.01	<.01	2.5–3.5	0.6	5–35
22	3	0.09	0.50	—	0.05	25.0	0.8	20
35	9	0.01	—	0.01	0.04	32.7	3.8	34
2–220	<1–43	<0.01–0.34	0.03–1.18	<0.01–0.04	<0.01–0.87	<1.0–450.0	0.4–7.5	3–>72,000

(continued)

TABLE VI (*continued*)

Ephemeroptera	Fe	pH	Alkalinity	Cl	DO	Total hardness
Neoephemeridae						
Neoephemera youngi Bern.[5]	0.05–0.13	6.4–6.8	9–20	1–2	9–10	6–15
Potamanthidae						
Potamanthus sp.[18]	<0.01–0.51	7.6–8.4	73–124	1–28	6–14	95–162
Behningiidae						
Dolania americana Edm.[3]	0.03–0.10	6.3–6.4	9–13	1–2	8–9	4–7
Ephemeridae						
Ephemera simulans Walk.[1]	<0.01	8.0	88	10	6–9	119
Hexagenia atrocaudata McD.[7]	<0.01	7.6–8.0	88–116	1–20	5–14	95–149
Hexagenia limbata (Serville)[4]	<0.01	6.0–7.9	2–7	33–72	5–10	70–233
Hexagenia munda eleganns? Trav.[4]	0.28–0.90	6.6–7.1	20–20	3–5	7–9	20–23
Pentagenia vittigera (Walsh)[4]	<0.01	7.8–7.9	5–130	56–72	5–8	193–233
Polymitarcidae						
Ephoron leukon Wlmsn.[4]	<0.01	7.9–8.0	87–97	10	6–12	116–124
Campsurus sp.[1]	<0.01	7.8	220	25	5	208
Tortopus incertus (Trav.)[5]	0.01–0.10	6.8–7.0	15–28	1–4	7–9	8–15
Tortopus primus (McD)[1]	<0.01	8.0	184	71	7	212
Baetiscidae						
Baetisca lacustris? Mcd.[2]	0.10–0.10	6.4–6.4	9–10	1	9	6

D. Respiration

The nymphs of the stoneflies take in oxygen either cutaneously or by means of tracheal gills. The Chloroperlidae lack external gills. In those forms with gills, they may be present on the mentum, submentum, neck, thoracic segments (base of coxae), abdominal segments one through three, or around the anus.

E. Chemical Parameters

Table XIV shows that while the Plecoptera are not too sensitive to high pH, they are sensitive to most other parameters. None was found at pH less than 4.5, alkalinity greater than 210 ppm, dissolved oxygen less than 4 ppm, chloride greater than 2000 ppm plus magnesium greater than 150 ppm. At chloride greater than 1000 ppm, only a *Paragnetina* species was present, and it may have washed in. At total hardness greater than 300 ppm, *Perlesta placida* and *Phasganophora capitata* were present. *Phasganophora capitata* was also present while sulfate was greater than 400 ppm.

TABLE VI (*continued*)

Ca	Mg	NH_3	NO_3	NO_2	PO_4	SO_4	BOD	Turbidity
—	—	0.01–0.05	—	<0.01–0.01	<0.01	2.2–2.5	0.6–1.1	5–120
25–49	6–17	<0.01–0.97	0.44	<0.01–0.05	0.02–0.48	7.5–72.8	1.0–6.0	11–125
2–2	1–1	0.01–0.63	—	<0.01	<0.01–0.06	2.3–2.5	0.5–0.6	5–36
34	8	<0.01	—	0.01	0.05	31.2	4.1	98
25–36	8–15	<0.01–0.02	—	<0.01–0.01	0.03–0.15	7.5–43.2	1.1–4.1	25–98
23–35	26–43	<0.01	—	<0.01–0.04	<0.01–0.34	34.9–45.2	2.0–2.0	18–103
—	—	0.05–1.10	—	<0.01	<0.01	3.2–3.6	—	10–35
23–35	26–43	<0.01	—	0.03–0.30	0.19–0.34	34.9–44.0	2.0–2.8	71–>72,000
34–35	8–9	<0.01–0.02	—	0.01	0.04–0.05	31.2–32.7	3.4–4.1	32–98
30	33	<0.01	—	<0.01	0.04	29.7	2.0	66
3	1–2	<0.01–0.06	—	<0.01–0.03	0.02–0.05	2.6–4.4	0.5–1.5	25–548
64	19	0.02	—	0.02	0.02	20.0	—	67
2	1	0.01–0.16	—	<0.01	<0.01	2.4–2.5	0.55–0.83	5–6

V. Hemiptera

A. Life History

With few exceptions, adult Hemiptera overwinter as adults. The eggs are laid, generally in rows, on or in aquatic vegetation, on rocks, on objects at the water's edge, or on floating objects. The incubation period of the eggs may vary from one week to a month. There are, with few exceptions, five nymphal instars, and nymphal development may take from 25 to over 50 days. The nymphs resemble the adults and occupy the same habitats. The general life cycle occupies one year, although in most aquatic Hemiptera there are overlapping broods. Flight is common in the Corixidae, Notonectidae, and Belostomatidae; rare in the Naucoridae, Gerridae, and Veliidae; absent in the Nepidae.

B. Habitats

The aquatic Hemiptera are essentially slow water, pond forms. In rivers they are found along the margins in shallow water (Corixidae); on or among aquatic vegetation along stream margins or in backwaters (Notonectidae,

TABLE VII

	Fe	pH	Alkalinity	Cl	DO	Total hardness
Pteronarcidae						
Pteronarcys spp.[7]	0.03–0.76	5.5–7.0	4–28	1–6	5–8	4–16
Pteronarcys biloba Newm.[1]	0.51	8.0	124	9	9	130
Pteronarcys dorsata Say[7]	<0.01–0.76	5.5–8.8	4–92	1–28	5–14	11–149
Nemouridae						
Nemoura (A.) wui Claas.[1]	2.90	7.3	39	6	8	66
Nemoura sp.[4]	0.11–0.16	6.7–6.9	18–20	2–3	10–11	9–12
Capniidae						
Nemocapnia carolina Banks[5]	0.10–0.16	6.4–6.9	10–20	1–3	9–11	7–12
Taeniopterygidae						
Taeniopteryx lita Frison[4]	0.11–0.16	6.7–6.9	8–20	2–3	10–11	9–12
Taeniopteryx nivalis (Fitch)[2]	0.02–0.08	6.0–6.8	2–20	33–34	10	70–82
Perlodidae						
Isogenus? sp.[1]	2.90	7.3	39	6	8	66
Perlidae						
Neoperla clymene (Newm.)[11]	<0.01–0.60	6.3–8.2	9–213	1–72	5–13	4–233
Perlinella drymo (Newm.)[6]	0.04–0.13	6.3–6.9	9–20	1–2	8–10	6–15
Perlesta placida (Hagen)[21]	<0.01–1.80	5.5–8.4	4–122	1–845	5–12	6–800
Phasganophora capitata (Pict.)[17]	<0.01–0.76	5.5–8.8	4–205	1–30	5–12	6–705
Acroneuria abnormis (Newm.)[17]	<0.01–0.76	5.5–8.0	4–97	1–11	5–14	6–147
Acroneuria evoluta (Klap.)[3]	<0.01–0.02	8.7–8.8	73–74	27–28	9	114–117
Acroneuria internata (Walk.)[2]	<0.01–0.01	8.8	73–74	27	9	114–117
Acroneuria mela Frison[4]	<0.01–0.27	7.6–8.4	102–116	19–20	6–15	144–159
Acroneuria nr. mela Frison[6]	0.08–0.13	6.4–6.9	9–22	1–2	7–10	6–15
Acroneuria ruralis (Hagen)[5]	<0.01–2.90	7.1–8.8	36–74	6–28	6–9	26–117
Paragnetina immarginata (Say)[1]	<0.01	8.8	73	27	9	114
Paragnetina media (Walk.)[2]	0.05–0.54	5.6–7.0	5–21	1–7	4–7	10–16
Paragnetina kansensis (Banks)[22]	<0.01–0.28	6.4–7.8	9–128	1–56	7–11	4–193
Paragnetina sp.[1]	0.35	7.6	86	5500	10	5000
Isoperlidae						
Isoperla sp.[9]	0.05–0.16	6.4–6.9	9–22	1–3	7–11	6–15

[a]All results are in parts per million (ppm) except pH. Superscript numbers following species names indicate the number of records comprising the following ranges.

Naucoridae, Belostomatidae, and Nepidae); on the water surface in slow water (Gerridae and Veliidae); and on the surface in faster water (some Veliidae). Some Naucoridae are found under rocks in fast water.

C. Food

The majority of the Hemiptera families considered here are predaceous, feeding on chironomids, mosquito larvae (most families), mayfly nymphs, larger aquatic arthropods, tadpoles, and fish (Belostomatidae). The Cori-

RANGES OF CHEMICAL ANALYSES OF WATERS IN WHICH 24 SPECIES OF THE ORDER PLECOPTERA WERE COLLECTED[a]

Ca	Mg	NH_3	NO_3	NO_2	PO_4	SO_4	BOD	Turbidity
2–4	<1–1	0.02–0.64	—	<0.01–0.01	<0.02–0.07	<1.0–2.8	0.6–2.5	8–548
49	17	0.97	—	0.02	0.48	14.2	2.1	11
2–37	1–15	0.04–0.14	—	<0.01–0.01	<0.02–0.27	<1.0–65.3	0.9–7.5	50–140
22	3	0.09	0.50	—	0.05	25.0	0.8	20
2–3	1	0.02–0.05	—	<0.01–0.01	—	2.2–2.5	0.8–1.1	110–140
2–3	1	0.02–0.16	—	<0.01–0.01	—	2.2–2.5	0.8–2.5	6–140
2–3	1	0.02–0.05	—	<0.01–0.01	—	2.2–2.5	0.8–1.1	110–140
—	—	<0.01	—	<0.01–0.01	<0.01–0.53	45.1–45.2	—	18–44
22	3	0.09	0.50	—	0.05	25.0	0.8	20
2–56	1–43	<0.01–2.50	0.44–0.47	<0.01–0.04	<0.01–0.72	2.3–32.7	0.6–4.1	5–103
2–4	<1	0.01–0.16	—	<0.01–0.01	<0.01–0.02	2.3–2.5	0.6–0.6	5–140
2–398	<1–21	<0.01–5.00	0.47–2.30	<0.01–0.02	<0.01–0.72	<1.0–135.0	0.6–3.8	5–98
2–220	1–39	<0.01–1.09	0.06–0.50	<0.01–0.01	<0.01–0.27	<1.0–450.0	0.4–6.0	3–125
2–36	1–14	<0.01–0.05	—	<0.01–0.01	<0.01–0.15	<1.0–41.6	0.6–4.1	5–110
36–37	6	<0.01	0.07–0.08	<0.01–0.01	0.12–0.27	24.6–25.1	1.0–1.8	37–50
36–37	6	<0.01	0.07–0.08	<0.01–0.01	0.12	24.6–25.1	1.6–1.8	37–39
35–36	14–17	<0.01–0.02	—	0.01	0.06–0.13	41.6–56.3	2.8–5.5	18–100
2–4	1–2	.01–.05	—	<0.01	<0.01	2.2–3.1	0.6–0.8	5–120
6–37	3–6	<0.01–0.06	0.07–0.50	<0.01	<0.01–0.12	12.6–25.1	0.8–1.7	20–39
36	6	<0.01	0.07	<0.01	0.11	24.6	1.8	37
3	1–2	0.05–0.12	—	<0.01–0.01	0.03–0.07	<1.0	1.1–2.1	26–39
2–7	<1–2	0.01–0.64	0.36–0.41	<0.01–0.02	<0.01–0.35	2.2–43.9	0.5–2.0	6–>72,000
2000	32	13.40	1.10	0.37	0.46	129.4	1.8	3
2–4	1–2	.01–.16	—	<0.01	<0.01	2.4–3.1	0.8–1.0	5–120

xidae feed on bottom ooze, algae, and diatoms, but do take in rotifers, chironomids, and mosquito larvae (and appear to require some of the latter).

The corixids lack the typical piercing-sucking hemipteran mouth parts; their stylets pierce and rasp and the mouth opening is broad, permitting the ingestion of small whole cells and organisms.

D. RESPIRATION

The aquatic Hemiptera are atmospheric air breathers. A few nymphal naucorids can engage in cutaneous respiration, but the majority are dependent on atmospheric oxygen. The Nepidae have a pair of caudal appendages which are held together to form a "snorkel" tube which they stick out of the water in order to breath. The Belostomatidae break the surface film with their caudal straplike appendages, and air is taken in between the hemelytra and the dorsal abdominal surface. The mature Naucoridae use the same mechanism. The Notonectidae also break the surface with the tip of the abdomen, and the ventral air troughs are covered by hydrofuge hairs. Corixidae break the surface with the head and pronotum, and the air spreads over the extensive plastron surface on the body. Corixids can remain under water longer by using oxygen which diffuses into the plastron air bubble.

E. Chemical Parameters

As would be expected, the atmospheric oxygen-breathing Hemiptera are more tolerant of environmental extremes than most insects. They are not, however, as tolerant as either the Coleoptera or Diptera. Most of the tolerant species (except with regard to chloride and sulfate) are not Gerridae and Veliidae. One would expect these completely air-breathing insects to be almost independent of water chemistry, but this does not appear to be the case.

The number of Gerridae + Veliidae in relation to all Hemiptera in each category are as follows: pH 8.5, one of three; pH 4.5, one of three; alkalinity greater than 210 ppm, one of four; chloride greater than 1000 ppm, five of five; chloride greater than 2000 ppm plus magnesium greater than 150 ppm, two of two; iron greater than 5.00 ppm, one of two; dissolved oxygen less than 4 ppm, two of eight; total hardness greater than 300 ppm, four of nine; sulfate greater than 400 ppm, six of seven; BOD greater than 5.9 ppm, four of eight.

Belostoma fluminea is one of the more tolerant Hemiptera, appearing on five of the ten extreme tolerance lists (see pp. 363–370). The Hydrometridae and Gelastocoridae are, in my opinion, too peripherally aquatic to be of any significance, and are not considered in this work.

In summary, although the Hemiptera seem to be more responsive to chemical extremes than one would theoretically expect, I could not consider them to be a really significant group in the evaluation of damaged situations.

VI. Neuroptera

A. Life History

The eggs, laid in crevices on an object (branch, rock, etc.) overhanging the water, hatch in about 8 to 14 days and the larvae drop into the water. In the water the larvae move about until they find a sponge colony. Presumably those that fail to find a sponge, die. The larvae enter the osteoles of the sponge and feed on the sponge tissues. There are three larval instars, the first of which lasts about one week. The third instar leaves the water to pupate, spinning a silken cocoon in the ground or on some sheltered spot. The pupal stage lasts five to six days.

Two or three complete cycles may be completed in a year, with the winter being passed in the larval or prepupal stage.

B. Habitats

The larvae of the Sisyridae, the only strictly aquatic Neuroptera, are parasitic on freshwater sponges, primarily *Spongilla*, *Eunapius*, and *Ephydatia*. The larvae move through the interior passages of the sponge.

C. Food

The larvae feed by drawing nourishment from the sponge cells by means of styletlike mouth parts.

D. Respiration

Respiration of the second and third instar in both *Sisyra* and *Climacia* is by ventral, jointed, tracheal gills on the abdomen. Respiration in the first instar appears to be cutaneous.

E. Chemical Parameters

The larvae of the Sisyridae do not seem to be especially tolerant of the extemes of water chemistry. *Climacea areolaris* was found at alkalinity greater than 210 ppm, total hardness greater than 300 ppm, and sulfate greater than 400 ppm. This does not imply that the sponge hosts are similarly sensitive, as not all sponge colonies are parastitized. (See also discussions of the sponge hosts in Chapter 2.)

VII. Megaloptera

A. Life History

The eggs of the Megaloptera are laid on objects out of, but hanging over the water. They hatch in nine to ten days for *Sialis*, and five to six days for *Chauliodes*. The larvae drop into the water and start feeding immediately. When the larvae are mature they leave the water and dig into the ground or rotten logs to form the pupal cell. The pupal period is about two weeks. The length of the life cycle is not clear; for *Sialis* it is probably one year; for *Corydalis* and *Chauliodes*, it may be as long as three years.

TABLE VIII

	Fe	pH	Alkalinity	Cl	DO	Total hardness
Corixidae						
Callicorixa audeni Hung.[1]	0.74	4.4	0	7	9	250
Hesperocorixa spp.[9]	0.32–16.10	3.3–8.7	0–95	6–47	8–10	66–322
Hesperocorixa lucida (Abbott)[1]	0.02	8.3	107	20	5	176
Hesperocorixa nitida (Fieb.)[1]	0.10	6.4	9	1	9	6
Palmacorixa buenoi Abbott[1]	—	6.0	2	34	10	70
Sigara modesta Abbott[8]	<0.01–0.23	6.0–8.4	2–205	10–34	5–13	70–705
Trichocorixa calva (Say)[19]	<0.01–0.60	7.6–8.4	19–180	1–56	6–14	8–570
Trichocorixa kanza Sailer[2]	0.05–0.82	6.7–7.1	20–34	3–16	6–8	21–25
Trichocorixa macroceps (Kirk.)[1]	—	6.8	20	33	10	82
Notonectidae						
Notonecta indica (L.)[1]	0.08	6.8	20	2	8	13
Notonecta irrorata Uhl.[2]	0.51–0.54	5.6–5.6	5–6	7–8	3–4	15–16
Notonecta lunata Hung.[3]	0.37–2.20	5.2–7.4	2–88	11–34	10–10	70–210
Notonecta uhleri Kirk.[2]	0.03–0.66	6.3–6.4	10–13	1	8–8	4–7
Pleidae						
Plea striola Fieb.[2]	0.02–0.06	7.3–8.3	107–146	20–62	2–5	167–176
Nepidae						
Nepa apiculata Uhler[1]	0.30	8.3	113	21	9	162
Ranatra australis Hung.[3]	<0.01–0.82	6.7–7.8	20–220	3–62	2–8	21–208
Ranatra buenoi Hung.[2]	0.10–0.24	7.0	28–34	2–3	7	8–9
Ranatra fusca P. de B.[1]	0.08	6.8	20	33	10	82
Ranatra kirkaldyi Bueno[7]	0.03–0.54	5.6–6.8	5–28	1–33	3–10	4–82
Ranatra nigra Herr.-Sch.[8]	<0.01–0.90	6.7–8.3	20–122	4–21	6–14	20–162
Naucoridae						
Pelocoris femoratus P. de B.[6]	0.06–0.51	5.6–7.9	5–213	7–965	2–10	13–370
Belostomatidae						
Belostoma fluminea Say[23]	<0.01–2.20	5.6–8.8	5–213	7–395	2–14	13–600
Belostoma lutarium Stal.[2]	0.04–0.10	6.9–6.9	19–22	2–2	7–8	15
Lethocerus americanus (Leidy)[1]	0.72	7.2	47	14	11	87
Gerridae						
Gerris alacris Hussey[1]	0.10	6.9	22	2	7	15
Gerris canaliculalus Say[4]	0.10–0.76	5.5–7.0	4–28	2–8	4–7	8–16
Gerris comatus Dk. & Harr.[6]	<0.01–0.61	8.0–8.5	97–180	11–22	5–12	124–600
Gerris conformis Uhl.[3]	0.03–0.45	5.6–7.8	5–108	1–8	4–8	4–120
Gerris dissortis Dk. & Harr.[8]	<0.01	8.4	93	22	10	171
Gerris marginatus Say[9]	0.02–16.10	4.4–8.4	0–205	1–5000	8–11	6–5000
Gerris remigis (Say)[3]	< 0.01–2.89	7.3–8.0	39–205	6–11	6–12	66–705
Limogonus nr. *hesione* Kirk.[2]	<0.01	8.8	73–74	27–27	9	114–117
Metrobates hesperius Uhl.[26]	<0.01–0.60	6.3–8.3	20–130	1–185	6–14	7–193
Rheumatobates hungerfordi Wiley[2]	0.05	7.0–7.8	21–108	1–8	7	11–190
Rheumatobates rileyi Berg.[7]	0.05–1.01	5.9–8.4	10–122	1–21	6–10	26–162
Rheumatobates tenuipes Meinert[10]	<0.01–0.28	6.9–8.0	21–116	1–2500	6–14	9–900
Trepobates sp.[4]	<0.01–0.51	5.5–7.9	6–220	1–25	3–9	15–208
Trepobates knighti Dk. & Harr.[3]	0.32–1.80	6.7–7.4	36–88	14–30	10	197–322
Trepobates inermis Esaki[24]	<0.01–0.77	5.4–8.3	22–124	1–5500	7–12	8–5000
Trepobates pictus (Herr.-Sch.)[4]	0.04–1.91	6.7–8.0	36–175	11–395	9–10	116–600
Veliidae						
Microvelia americana Uhl.[3]	0.02–0.35	6.4–7.6	13–205	1–5500	8–10	8–5000
Rhagovelia choreutes Hussey[2]	—	7.9	203–206	69–72	5–6	227–233
Rhagovelia obesa (Uhl.)[27]	0.02–2.89	5.6–8.3	2–205	1–2630	3–10	4–2100

[a]All results are in parts per million (ppm) except pH. Superscript numbers following species names indicate the number of records comprising the following ranges.

Ranges of Chemical Analyses of Waters in Which 43 Species of the Order Hemiptera Were Collected[a]

Ca	Mg	NH_3	NO_3	NO_2	PO_4	SO_4	BOD	Turbidity
76	16	0.35	0.20	—	0.04	322.0	0.2	2
22–92	3–22	< 0.01–1.09	0.08–0.50	0.01	0.02–0.12	25.0–315.0	0.4–1.7	3–39
51	12	0.15	0.03	0.01	0.10	16.1	2.0	16
2	1	0.01	—	<0.01	<0.01	2.5	0.6	5
—	—	<0.01	—	<0.01	<0.01	45.2	—	18
36–220	11–39	<0.01–0.15	0.03	<0.01–0.01	<0.01–0.13	16.1–450.0	0.7–5.7	6–96
2–170	<1–34	<0.01–2.50	0.36–0.47	<0.01–0.02	<0.01–0.72	2.6–350.0	0.9–6.0	25–>72,000
6–6	1–2	0.06–1.10	—	<0.01	<0.01	3.5–13.0	0.3	18–35
—	—	<0.01	—	0.01	0.53	45.1	—	18
3	1	0.05	—	<0.01	<0.01	3.1	0.8	52
3–4	2–2	0.12–0.12	—	0.01	0.01–0.07	<1.0	2.1–2.4	39–41
55–62	13–15	<0.01–0.21	0.19–1.18	<0.01	<0.01–0.30	45.2–220.0	0.4–1.8	18–33
2	<1	0.05–0.64	—	<0.01	0.02–0.07	2.3	0.6	8–36
46–51	12–13	0.02–0.15	0.03–0.05	0.01–0.04	0.10–0.87	16.1–25.3	2.0–4.8	16–128
37	17	0.01	—	<0.01	0.05	72.8	6.0	125
6–46	1–33	<0.01–1.10	0.05	<0.01–0.04	<0.01–0.04	3.5–29.7	2.0–4.8	35–128
2–3	<1	0.06	—	<0.01–0.01	0.02	—	1.1–1.5	204–548
—	—	<0.01	—	0.01	0.53	45.1	—	44
2–4	<1–2	<0.01–0.64	—	<0.01–0.01	0.01–0.53	<1.0–45.1	0.6–2.4	8–548
6–56	1–14	<0.01–2.50	—	<0.01–0.02	<0.01–0.72	3.5–72.8	2.8–6.0	15–125
3–55	1–73	<0.01–0.15	0.05–1.18	<0.01–0.04	<0.01–0.87	<1.0–240.0	1.8–4.8	9–103
3–180	1–43	<0.01–2.50	0.03–1.18	<0.01–0.37	0.01–0.72	<1.0–370.0	0.4–6.5	4–98
4	1–2	0.02–0.17	—	<0.01	<0.01–0.02	2.3–2.7	0.6–0.9	34–96
27	4	0.17	0.21	—	0.03	26.5	1.6	12
4	2	0.02	—	<0.01	<0.01	2.7	0.6	34
2–4	<1–2	0.06–0.12	—	<0.01–0.01	<0.02–0.07	<1.0	1.5–2.5	39–548
35–180	9–36	0.01–0.21	0.03–2.30	<0.01–0.01	<0.01–0.56	16.1–370.0	1.1–3.8	16–40
2–32	<1–7	0.10–0.64	—	<0.01–0.02	0.02–0.07	<1.0–13.5	0.6–4.4	11–49
49	12	0.09	0.06	0.01	<0.01	22.1	0.7	3
2–2000	1–39	0.01–13.40	0.12–2.30	<0.01–0.37	<0.01–0.50	2.5–450.0	0.6–2.0	3–20
22–220	3–39	0.01–0.09	0.50	<0.01–0.01	<0.01–0.05	25.0–450.0	0.8–3.8	7–34
36–37	6–6	<0.01	0.07–0.08	<0.01–0.01	0.11–0.12	24.6–25.1	1.7–1.8	37–39
2–56	1–17	<0.01–2.00	0.36–0.50	<0.01–0.02	<0.01–0.72	2.5–72.8	0.3–6.0	8–>72,000
3–32	1–7	0.05	—	<0.01–0.02	0.02–0.03	13.5	1.1–4.4	11–261
6–56	3–17	0.01–2.50	0.16	<0.01–0.02	<0.01–0.72	7.5–161.0	0.5–6.0	5–125
2–60	<1–180	<0.01–0.17	—	<0.01–0.01	<0.01–0.15	2.3–480.0	0.8–3.3	8–261
4–32	2–33	<0.01–0.12	—	<0.01–0.02	0.01–0.04	<1.0–29.7	2.0–4.4	11–66
55–92	15–27	0.15–1.09	0.08–1.18	—	0.03–0.30	115.0–313.0	0.8–1.8	6–27
2–2000	<1–200	<0.01–13.40	0.20–2.30	<0.01–0.37	<0.01–0.72	13.0–570.0	0.9–2.8	3–548
33–180	6–36	0.10–6.13	0.08	<0.01–0.37	0.01–0.05	26.0–370.0	0.8–15.4	4–37
2–2000	<1–39	0.08–13.40	1.10	<0.01–0.37	<0.01–0.46	129.4–450.0	0.4–1.8	3–36
35–47	26–35	—	—	0.03	0.28–0.34	37.8–44.0	—	71–77
2–865	<1–39	0.01–5.80	0.08–2.30	<0.01–0.05	<0.01–0.72	<1.0–450.0	0.4–6.0	3–261

TABLE IX

	Fe	pH	Alkalinity	Cl	DO	Total hardness
Neuroptera–3 species						
Sisyridae						
Climacea areolaris Hagen[11]	<0.01–0.33	6.5–7.8	17–243	2–22	6–9	12–416
Sisyra vicaria Walk[1]	0.01	7.7	202	18	5	245
Sisyra poss. *apicalis* Banks.[1]	0.86	6.6	20	3	7	21
Megaloptera–4 species						
Corydalidae						
Corydalis cornutus (L.)[34]	<0.01–0.76	5.5–8.8	4–206	1–72	5–14	4–233
Chauliodes sp.[13]	<0.01–0.82	5.6–8.4	2–116	1–34	3–14	7–159
Nigronia sp.[10]	0.01–1.80	4.4–8.8	0—97	1–30	6–11	26–322
Sialidae						
Sialis sp.[24]	0.03–16.10	3.3–8.8	0–180	2–185	5–14	11–600

[a] All results are in parts per million (ppm) except pH. Superscript numbers following species names indicate the the number of records comprising the following ranges.

B. Habitats

I have consistently found *Sialis* in soft mud bottom. It has been reported under stones, but I have never found any species in such a situation.

Larvae of the Corydalidae prefer coarse or rubble bottoms. They are most commonly collected on the undersides of rocks in moderate-to-fast water. *Nigronia* is commonly found in very quiet water.

C. Food

All the Megaloptera are active predators, feeding mostly on smaller insects.

D. Respiration

Respiration is cutaneous, by means of lateral gill processes on the abdomen. In *Corydalis* there is an additional gill tuft at the base of each of these projections.

E. Chemical Parameters

Of the four genera treated here, only *Sialis* can be considered to be tolerant of damaged situations. It is found on six of the ten lists of extreme chemical ranges. It is very common in acid-mine drainages, tolerating iron of 16.10

RANGES OF CHEMICAL ANALYSES OF WATERS IN WHICH SPECIES OF THE ORDER NEUROPTERA AND MEGALOPTERA WERE COLLECTED[a]

Ca	Mg	NH_3	NO_3	NO_2	PO_4	SO_4	BOD	Turbidity
3–83	1–50	0.02–0.49	0.02–0.51	<0.01–0.04	<0.01–0.41	2.0–227.8	0.6–2.8	5–157
64	21	0.33	0.07	0.02	0.01	19.5	—	140
6	1	0.97	—	<0.01	<0.01	3.6	—	25
2–55	<1–35	<0.01–0.64	0.07–1.18	<0.01–0.03	<0.01–0.35	<1.0–115.0	0.3–6.0	5–>72,000
2–36	<1–17	<0.01–1.10	—	<0.01–0.01	<0.01–0.15	<1.0–56.3	0.6–5.5	6–100
6–92	3–22	<0.01–1.09	0.07–2.30	<0.01	<0.01–0.56	7.5–322.0	0.2–1.8	2–37
3–180	1–36	0.02–1.09	0.08–0.50	<0.01–0.01	<0.01–0.12	<1.0–370.0	0.2–6.0	1–125

ppm and pH of 3.3. I have commonly found it in organic ooze. The others are more sporadic in occurrence at the extremes and can be considered more sensitive.

VIII. Coleoptera

A. LIFE HISTORY

The life cycles of the Haliplidae, Dytiscidae, Gyrinidae, and Hydrophilidae are very similar. All have one-year life cycles with three larval instars. The eggs are laid on or in aquatic vegetation, sometimes in mud (Dytiscidae) and in egg cases attached to aquatic vegetation (Hydrophilidae). The eggs hatch in from five days (some haliplids) to three weeks (some gyrinids). There are not many data on the duration of the larval instars, but for one haliplid, Instar I lasted six days; Instar II, eight to ten days; and Instar III, seven to ten days—the entire larval life being three to four weeks.

Pupation takes place in damp soil or sand along stream margins, and the pupal stage may last two to six weeks. Overwintering is in the adult stage, although a very few cases of larval overwintering have been reported in the Haliplidae, Gyrinidae, and Hydrophilidae.

Very little is known regarding the life cycle of the Dryopidae, Psephenidae, and Elmidae. The elmids seem to have at least five instars. They pupate under stones or in wood in the water, and the pupal stage lasts about two weeks.

They overwinter in all stages, and possibly have a two-year life cycle. The Psephenidae lay their eggs on rocks in the water. The pupae are found on stones above the water line.

In the Donaciinae the eggs are laid on the undersides of the host plant, either by the adult cutting a hole in the leaf or crawling under the edge of the leaf. Pupation takes place at the larval feeding sites underwater.

B. Habitats

The Haliplidae and most Dytiscidae and Hydrophilidae are inhabitants of ponds or pondlike situations in streams. They generally live among marginal vegetation, algae, and debris. The genera *Bidessus*, *Deronectes*, and *Agabus* (Dytiscidae) can be found in fast streams as well as slow streams and ponds. The adults of the Haliplidae, Dytiscidae, and Hydrophilidae are all capable of swimming through the water—the Dytiscidae being the best adapted of the three. The Haliplidae and Hydrophilidae are not so well adapted and can crawl about on the substrate. The Gyrinidae adults swim on the surface film and only submerge to avoid attack, lay eggs, or to hibernate in the winter.

The larvae of these four families are also found in the same habitats, although more closely associated with the marginal vegetation. The larvae crawl around on the vegetation.

The adults and larvae of the Dryopidae and Elmidae cannot swim and are generally found crawling on rocks and wood in well-aerated, flowing water. They are, at best, slow moving. The latter is true of the adult Psephenidae, which are terrestrial and may enter the water to lay their eggs. The flattened *Psephenus* larvae cling to stones in moderate-to-fast flowing water. The larvae of the Donaciinae are found on the leaves and stems of *Nuphar* and *Nymphaea*. The adults are not truly aquatic, and are found on the dorsal surface of the waterlily leaves.

C. Food

The adults and larvae of the Haliplidae are omnivorous, feeding on plant and animal material. The Dytiscidae are predaceous in both stages. The larvae are especially voraceous predators. The Gyrinidae adults are omnivorous, but the larvae are predaceous. In the Hydrophilidae the same picture applies. Some hydrophilid adults will eat animal tissue; the larvae, except *Berosus*, are predators.

The adults of the Dryopidae and Elmidae are vegetative feeders, feeding on plant tissue, roots (Dryopidae larvae), algae, and moss.

The adults and larvae of the Donaciinae feed on a variety of aquatic plants, although some forms have very specific preferences. The plants fed on are *Nuphar*, *Nymphaea*, *Myriophyllum*, *Sagittaria*, etc.

D. Respiration

The adults of the Haliplidae, Dytiscidae, Gyrinidae, and Hydrophilidae all carry their air under the elytra. The first three renew their air by touching the water surface with the tip of the abdomen and taking in air at the groove at the posterior end of the lateral margin of the elytra. In the Hydrophilidae the antennae are used to form a respiratory air funnel with the prothorax. This funnel of air is connected to the subelytral air chambers.

The larvae of the Haliplidae (except *Peltodytes*) and most Dytiscidae have to come to the surface to renew their air supply. They have caudal spiracles. Matheson (1912) reported that *Peltodytes* larvae may be able to take in air through its tracheated spines. The larvae of *Coptotomus* have lateral gills, and some larval Hydroporinae can obtain their air from the water, although they have no gills. The larvae of the Hydrophilidae also have to surface, although they have air reservoirs in the tracheal system and can stay under water longer. *Berosus* larvae have lateral gills. The larvae of the Gyrinidae have lateral tracheal gills and are truly aquatic.

The adults of the Dryopidae and Elmidae depend on oxygen diffused from the water. The body is covered with hydrofuge hairs, forming a respiratory plastron. It is rare that this has to be renewed at the surface; generally it can be renewed underwater, if necessary. The mode of respiration of the Dryopidae larvae is not clear. Elmidae larvae have caudal respiratory gills; *Psephenus* larvae, lateral abdominal gills.

The larvae of the Donaciinae pierce the stems of aquatic plants to obtain their oxygen.

E. Chemical Parameters

In spite of the fact that a few larvae are aquatic (especially Gyrinidae) I do not consider the Haliplidae, Dytiscidae, Gyrinidae, and Hydrophilidae to be of great significance in water quality studies. They are entirely too mobile as adults, being able to enter and leave a body of water at will, and my experience has indicated that they lack significance in this type of work. For this reason, in Table X they are treated only at the generic level. Adding many pages of tables to treat them at the specific level would not be justifiable. This is not true of the other families treated. The Dryopidae, Elmidae, and larval Psephenidae are truly aquatic and of significance (see Sinclair, 1964).

At pH greater than 8.5, six elmids and eight other beetle genera were found. Below pH 4.5 there were no elmids or related families, while nine genera of other beetles were present. At alkalinity greater than 210 ppm, only two species of *Stenelmis* were present. Only *Helichus* and five other non-dryopoid genera were present with chlorides greater than 1000 ppm. In brackish water with iron greater than 5.00 ppm no dryopoids or related

TABLE X

	Fe	pH	Alkalinity	Cl	DO	Total hardness
Haliplidae						
Haliplus spp.[7]	0.01–0.76	5.5–8.5	4–205	2–22	3–12	11–705
Peltodytes spp.[21]	<0.01–0.76	5.5–8.8	4–220	1–72	2–14	11–705
Dytiscidae						
Acilius spp.[1]	16.10	3.0	0	9	8	165
Agabus spp.[6]	1.00–16.10	3.0–7.3	0–39	6–47	8–10	66–250
Agaporus spp.[1]	8.75	3.3	0	47	8	162
Bidessus spp.[6]	<0.01–2.89	4.4–8.5	0.206	7–69	5–12	66–705
Canthydrus spp.[3]	<0.01	7.7	106–108	45–48	6–6	152–164
Coelambus spp.[2]	<0.01–0.54	5.6–8.5	5–205	1–185	3–12	7–705
Copelatus spp.[3]	0.11–0.76	5.5–6.9	4–21	2–6	5–10	11–14
Coptotomus spp.[10]	<0.01–0.86	6.0–8.8	2–206	1–69	5–10	8–227
Cybister spp.[1]	0.10	6.4	10	1	9	7
Desmopachria spp.[4]	<0.01–0.90	6.7–7.7	21–108	2–45	6–8	14–162
Graphoderes spp.[1]	0.72	7.2	47	14	11	87
Hydaticus spp.[1]	0.02	7.0	19	2	8	16
Hydroporus spp.[18]	<0.01–16.10	3.0–8.0	0–205	1–47	3–12	4–705
Hydrovatus spp.[1]	0.02	8.3	107	30	5	176
Ilybius spp.[5]	<0.01–8.75	3.3–8.0	0–88	1–48	6–10	13–210
Laccophilus spp.[28]	<0.01–16.10	3.0–8.8	0–206	6–5500	3–14	11–5000
Thermonectes spp.[2]	0.35–0.37	7.4–7.6	86–88	14–5500	10	197–5000
Noteridae						
Hydrocanthus spp.[2]	<0.01	7.9–7.9	203–206	69–72	5–6	227–233
Gyrinidae						
Dineutes spp.[32]	<0.01–0.77	5.5–8.8	4–213	1–845	2–14	4–800
Gyretes spp.[2]	<0.01–0.82	6.7–7.8	20–203	3–72	6–8	21–233
Gyrinus spp.[30]	<0.01–2.20	5.5–8.5	2–124	1–5500	3–14	8–5000
Hydrophilidae						
Anacaena spp.[1]	2.89	7.3	39	6	8	66
Berosus spp.[22]	<0.01–0.77	7.0–8.8	21–213	2–5500	5–14	8–5000
Cercyon spp.[3]	0.04–0.13	6.8–8.0	18–175	2–11	8–10	9–600
Cymbiodyta spp.[2]	0.03	6.4–7.8	13–26	2–7	8	4–18
Enochrus spp.[16]	<0.01–2.89	4.4–8.8	0–180	1–56	5–11	9–600
Helophorus spp.[7]	0.08–16.10	3.0–7.2	0–61	2–23	8–11	13–322
Hydrobius spp.[12]	<0.01–16.10	3.0–8.4	0–206	1–72	5–11	7–600
Hydrochus spp.[1]	<0.01	7.9	206	69	5	237
Laccobius spp.[6]	0.02–2.89	7.3–8.4	39–205	6–11	8–11	66–705
Paracymus spp.[7]	0.10–2.89	6.7–7.3	18–88	2–30	7–11	8–322
Phaenonotum spp.[3]	0.08–0.77	6.8–8.2	20–64	2–845	8–8	13–800
Tropisternus spp.[30]	<0.01–2.89	5.5–8.8	2–206	1–5500	2–14	4–5000
Eubriidae						
Ectopria nervosa (Melsh.)[2]	0.03–0.06	6.3–6.4	10–13	1	8	4–7
Dryopidae						
Helichus fastigiatus (Say)[4]	0.03–0.06	5.5–7.4	4–88	6–30	5–10	11–322
Helichus lithophilus (Germ.)[17]	<0.01–0.60	6.9–8.4	19–213	2–2630	5–14	11–2100
Psephenidae						
Psephenus herricki DeKay[11]	<0.01–0.30	7.6–8.4	87–116	10–22	5–14	116–171
Psephenus lecontei Lec.[1]	0.04	8.0	175	11	9	600

[a]All results are in parts per million (ppm) except pH. Superscript numbers following species names names indicate the number of records comprising the following ranges.

RANGES OF CHEMICAL ANALYSES OF WATERS IN WHICH 35 GENERA AND 33 SPECIES OF THE ORDER COLEOPTERA WERE COLLECTED[a]

Ca	Mg	NH_3	NO_3	NO_2	PO_4	SO_4	BOD	Turbidity
3–220	1–39	0.02–0.15	0.03	<0.01–0.01	<0.01–0.10	<1.0–450.0	0.4–2.8	4–54
3–220	1–39	<0.01–0.17	0.02–0.08	<0.01–0.04	<0.01–0.87	<1.0–450.0	0.4–4.8	7–128
46	12	0.17	0.26	—	0.05	315.0	0.6	3
22–62	3–19	0.09–0.32	0.16–0.50	—	0.02–0.05	25.0–315.0	0.4–1.6	1–33
43	13	0.32	0.38	—	0.02	198.0	1.0	13
22–220	3–39	<0.01–0.21	0.03–0.20	<0.01–0.03	<0.01–0.28	44.0–450.0	0.2–2.8	2–27
40–42	12–15	0.02–0.02	0.51–0.52	0.01	0.18–0.19	55.6–57.2	0.8–1.3	120–245
2–220	<1–39	0.02—1.10	0.03–0.29	<0.01–0.01	<0.01–0.33	<1.0–450.0	0.4–2.8	4–>72,000
2–3	1	0.01–0.02	—	<0.01	<0.01–0.02	<1.0–2.7	0.8–2.5	44–110
2–47	<1–26	<0.01–0.97	0.07	<0.01–0.03	<0.01–0.28	3.6–45.2	0.8–2.0	18–540
2	1	0.16	—	<0.01	—	2.4	0.8	6
3–42	1–13	0.02–0.88	0.51	<0.01–0.01	<0.01–0.19	2.7–57.2	0.8	15–245
27	4	0.17	0.21	—	0.03	26.5	1.6	12
4	1	0.02	—	<0.01	0.03	2.3	1.4	120
2–220	<1–39	<0.01–0.64	0.12–0.38	<0.01–0.01	<0.01–0.07	<1.0–450.0	0.2–4.1	2–98
51	12	0.15	0.03	0.01	0.10	16.1	2.0	16
3–62	1–13	0.01–0.32	0.19–0.38	<0.01–0.01	0.02–0.05	2.6–220.0	0.4–4.1	13–98
3–2000	1–36	<0.01–13.40	0.07–2.30	<0.01–0.37	<0.01–0.56	<1.0–370.0	0.8–2.3	3–>72,000
55–2000	15–32	0.15–13.40	1.18	0.37	0.30–0.46	115.0–129.4	1.8–1.8	3–19
35–47	26–35	<0.01	—	0.03	0.28–0.34	37.8–44.0	—	71–77
2–398	<1–21	<0.01–5.00	0.04–2.30	<0.01–0.04	0.01–0.87	<1.0–135.0	0.6–6.0	5–>72,000
6–35	1–35	<0.01–1.10	—	<0.01–0.03	<0.01–0.34	3.5–37.8	—	35–77
3–2000	<1–200	0.01–13.40	0.03–1.10	<0.01–0.37	<0.01–0.62	0.6–570.0	0.3–4.1	3–548
22	3	0.09	0.50	—	0.05	25.0	0.8	20
2–2000	<1–43	<0.01–13.40	0.07–1.10	<0.01–0.37	<0.01–0.62	0.6–450.0	0.4–6.0	3–548
2–180	1–36	0.02–0.10	—	<0.01	<0.01–0.01	2.5–370.0	0.8–1.1	4–120
2–4	<1–2	0.09–0.64	—	0.01	0.07	1.3	0.6	36
2–180	1–36	<0.01–0.35	0.03–2.30	<0.01–0.09	<0.01–0.56	16.1–370.0	0.2–6.0	2–>72,000
3–92	1–22	0.05–1.10	0.12–0.50	<0.01	<0.01–0.05	3.1–315.0	0.6–1.6	3–52
2–180	<1–36	<0.01–1.10	0.20–2.30	<0.01–0.03	<0.01–0.56	2.3–370.0	0.2–2.0	2–548
47	26	<0.01	—	0.03	0.28	44.0	—	71
22–220	3–39	0.08–0.97	0.44–2.30	<0.01–0.02	<0.01–0.56	14.2–450.0	0.4–2.1	4–29
2–92	<1–27	0.05–1.09	0.08–1.18	0.01	0.02–0.30	2.5–313.0	0.8–1.8	6–548
3–398	1–21	0.02–5.00	0.90	<0.01–0.01	<0.01–0.62	2.7–46.1	0.8–1.8	10–52
3–2000	<1–36	<0.01–13.40	0.03–1.18	<0.01–0.37	0.04–0.87	<1.0–370.0	0.4–5.5	3–245
2	<1	0.05–0.64	—	<0.01	0.02–0.07	2.3	0.6	8–36
3–92	1–27	<0.10–1.09	0.08–1.18	<0.01	<0.02–0.30	<1.0–313.0	0.8–2.5	6–54
2–865	1–43	<0.01–5.80	0.20–0.50	<0.01–0.05	<0.01–0.72	2.3–72.8	0.5–6.0	3–>72,000
34–49	8–17	<0.01–0.14	0.03–0.06	<0.01–0.01	<0.01–0.15	22.1–72.8	0.7–6.0	2–125
180	36	0.10	—	<0.01	0.01	370.0	1.1	4

(*continued*)

TABLE X (*continued*)

	Fe	pH	Alkalinity	Cl	DO	Total hardness
Elmidae						
Ancyronyx variegatus (Germ.)[14]	0.05–0.76	5.5–7.1	4–36	1–185	5–9	6–87
Optioservus fastiditus (Lec.)[4]	0.01–0.77	8.0–8.8	64–124	9–845	8–9	114–800
Optioservus ovalis (Lec.)[3]	0.60–0.77	7.2–8.2	47–122	14–845	9–11	87–800
Promoresia sp.[1]	0.72	7.2	47	14	11	87
Machronychus glabratus Say[26]	<0.01–0.90	5.5–8.3	4–130	1–185	3–14	6–193
Dubiraphia quadrinotata Say[8]	0.02–2.89	5.5–8.2	4–205	2–23	5–10	8–705
Dubiraphia vittata (Melsh.)[7]	<0.01–2.89	7.2–8.3	36–113	6–30	8–12	66–322
Stenelmis antennalis Sand.[5]	0.08–0.28	6.4–7.1	9–34	1–3	7–10	6–24
Stenelmis beameri Sand.[5]	<0.01–0.02	7.9–8.7	74–213	10–67	5–14	116–233
Stenelmis bicarinata Lec.[14]	<0.01–0.70	6.8–8.8	21–124	2–845	6–14	14–800
Stenelmis concinna Sand.[1]	<0.01	8.8	74	27	9	117
Stenelmis crenata (Say)[24]	0.01–0.76	5.5–8.8	4–213	6–72	5–14	11–705
Stenelmis decorata Sand.[4]	<0.01–0.30	7.0–8.3	28–113	2–21	6–10	8–162
Stenelmis douglasensis Sand.[2]	0.23–0.27	8.3–8.4	102	18–19	10–11	154–159
Stenelmis exilis Sand.[2]	0.23–0.27	8.4–8.4	102–103	18–19	11–13	159–165
Stenelmis fuscata Blatch.[2]	0.82–0.86	6.6–6.7	20	3	7–8	21
Stenelmis grossa Sand.[2]	0.28–0.51	5.6–7.1	6–28	3–8	3–9	15–23
Stenelmis hungerfordi Sand.[1]	0.05	7.1	34	16	6	25
Stenelmis knobeli Sand.[3]	0.01–0.02	8.7–8.8	73–74	27–28	9	114–117
Stenelmis mera Sand.[1]	0.02	8.7	74	28	9	116
Stenelmis quadrimaculata Horn[1]	<0.01	8.4	96	21	10	170
Stenelmis sexlineata Sand.[1]	<0.01	7.8	130	56	8	193
Stenelmis sinuata Lec.[1]	0.32	6.7	36	30	10	287
Stenelmis tarsalis Sand.[1]	0.02	8.7	74	28	9	116
Stenelmis vittipennis Zimm.[7]	< 0.01	7.6–8.0	87–116	10–20	6–14	116–149
Chrysomelidae						
Donacia palmata Oliv.[2]	0.85–0.86	6.6	20	3–5	7	21
Donacia sp.[1]	2.89	7.3	39	6	8	66
Curculionidae						
Hyperodes sp.[1]	0.01	8.8	73	27	9	114

forms were found, although the other beetle genera were represented by one and five genera, respectively. Only *Machronychus glabratus* and one species of *Stenelmis* could tolerate dissolved oxygen less than 4 ppm, while eight other genera of "nonaquatic" beetles were found. The dryopoids are quite abundant in hard water (greater than 300 ppm) as are other beetles. Where sulfates were greater than 400 ppm, only two elmid species were found, along with eight other beetle genera. The dryopoids seem to be more tolerant of high organic loading than one would expect: one Psephenidae, four Elmidae, and one Dryopidae could tolerate BOD greater than 5.9 ppm, although none were found where BOD was greater than 10.00 ppm.

TABLE X (*continued*)

Ca	Mg	NH_3	NO_3	NO_2	PO_4	SO_4	BOD	Turbidity
2–11	<1–14	0.01–0.17	—	< 0.01	<0.01–0.02	< 1.0–5.5	0.5	5–120
36–398	6–21	<0.01–5.00	0.07–0.90	0.01–0.02	0.12–0.62	14.2–46.1	1.0–1.8	3–50
27–398	4–21	0.17–5.00	0.21–0.090	0.01–0.02	0.03–0.72	13.3–46.1	1.6–2.8	10–15
27	4	0.17	0.21	—	0.03	26.5	1.6	12
2–37	<1–17	<0.01–0.97	0.35–0.50	<0.01–0.01	<0.01–0.34	<1.0–72.8	0.3–6.0	8–>72,000
2–220	<1–39	0.06–2.50	0.12–0.47	<0.01–0.02	<0.01–0.72	<1.0–450.0	0.8–2.8	15–548
22–92	3–27	0.01–1.09	0.08–0.50	<0.01–0.01	0.02–0.13	25.0–313.0	0.8–6.0	6–125
2–7	<1–1	0.01–0.06	—	<0.01	<0.01–0.02	2.2–3.2	0.6–0.8	5–204
23–37	6–43	<0.01–0.02	0.08	0.01–0.04	0.05–0.27	25.0–34.8	1.0–4.1	23–103
3–398	1–21	<0.01–5.00	0.03–0.90	<0.01–0.02	<0.01–0.72	2.8–72.8	0.5–6.0	6–125
37	6	<0.01	0.08	0.01	0.12	25.1	1.7	39
3–220	1–43	<0.01–2.50	0.07–0.47	<0.01–0.04	<0.01–0.72	<1.0–450.0	0.4–5.5	4–>72,000
2–37	<1–17	<0.01–0.06	—	<0.01	0.02–0.05	31.2–72.8	1.5–6.0	98–548
36–37	15–17	0.02	—	0.01	0.02–0.06	56.3	5.3–5.5	100–110
36–39	17–17	0.01–0.02	—	0.01	0.06	55.5–56.3	5.5–5.7	96–100
6	1	0.97–1.10	—	<0.01	<0.01	3.5–3.6	—	25–35
4–7	1–2	0.05–0.12	—	<0.01–0.01	<0.01–0.01	<1.0–3.2	2.4	15–41
6	2	0.06	—	<0.01	<0.01	13.0	0.3	18
36–37	6	<0.01	—	<0.01–0.01	0.11–0.27	24.6–25.1	1.0–2.0	37–50
37	6	<0.01	0.08	0.01	0.27	25.0	1.0	50
49	11	0.14	0.03	0.01	<0.01	28.8	0.7	6
—	—	0.03	0.36	0.02	0.30	43.9	2.8	>72,000
70	27	1.09	0.08	—	0.03	251.0	0.8	6
37	6	<0.01	0.08	0.01	0.27	25.0	1.0	50
34–36	8–15	<0.01–0.02	—	0.01	0.04–0.15	31.2–43.2	2.8–4.1	18–98
6	1	0.85–0.97	—	<0.01	<0.01	3.5–3.6	—	15–25
22	3	0.09	0.50	—	0.05	25.0	0.8	20
36	6	<0.01	0.07	<0.01	0.11	24.6	1.8	37

IX. Lepidoptera

A. Life History

Not a great deal is known of the life history details of the North American aquatic Pyralidae, e.g., egg duration, length of instars, and pupal stage.

In those species living on rocks, the female enters the water and deposits the eggs on the rocks. The larvae and pupae live on the rocks. The pupal cases have a waterproof inner lining with an escape slit cut by the larvae for the adult.

The species that live on plants lay the eggs on or under the aquatic vegetation. In the case of *Nymphala maculalis*, the eggs are laid on the underside of water-lily leaves. The egg holes of *Donacia* are often used. The eggs hatch in 11 days, and the larvae live on the underside of leaves and on stems of the lilies. They build cases of excised leaf parts, and pupate on the lower leaf surface.

Another, *Nymphula serralinealis*, has a larval life of four weeks and pupal duration of five to ten days. It may have two broods per year. The adults of the leaf forms may emerge under water.

B. Habitats

Most of the species of *Nymphula*, *Parapoynx*, and *Synclita* are found on and among vegetation such as alligator weed and *Echinodorus*, generally in slowly flowing water. Many construct flattened tubular cases of two half-tubes of leaf (most *Parapoynx* I have collected) or oblong flattened cases of excised plant parts (*Nymphula* and *Synclita*). *Parargyractis* lives on rocks in fast water and constructs a weblike case under which it lives.

C. Food

The larvae of *Parargyractis* feed on algae and diatoms, while those of *Nymphula*, *Parapoynx*, and *Synclita* appear to be typical Lepidoptera larvae, feeding on aquatic vegetation such as alligator weed.

D. Respiration

Most of the aquatic Pyralidae obtain their oxygen cutaneously (especially early instars) or by means of gills (*Parargyractis*). As in the case-making caddisfly larvae, the Lepidoptera larvae increase water flow past the body

TABLE XI

	Fe	pH	Alkalinity	Cl	DO	Total hardness
Parargyractis spp.[14]	< 0.01–0.88	5.4–8.2	2–24	2–2630	4–13	42–2100
Parapoynx spp.[17]	< 0.01–1.00	5.6–8.0	5–124	3–2630	3–13	11–2100
nr. *Parapoynx* spp.[5]	<0.01	7.7–8.1	87–102	10–20	6–12	116–145
Nymphula sp.[4]	<0.01–0.02	7.8–8.5	93–211	15–22	5–12	164–188
Poss. *Synclita* sp.[2]	0.45–0.86	5.6–6.6	5–20	3–7	4–7	13–21

[a]All results are in parts per million (ppm) except pH. Superscript numbers following species names indicate the number of records comprising the following ranges.

by body motion. In some plant miners, air is obtained from the plant intercellular spaces, and there is some evidence of plastron respiration in one species of *Nymphula*.

E. Nomenclature

The generic names used by Lange (1956) and Welch (1959) do not agree. The following list gives the corresponding names.

Lange, 1956	Welch, 1959 (see Edmondson *et al.*, 1959 under General key references)
Parargyractis	*Catalysta* (*Elophila*)
Parapoynx	*Nymphula* (*Paraponyx*)
Nymphula	*Nymphula* (*Hydrocampa*)

F. Chemical Parameters

The records of the aquatic Lepidoptera are sparse, but do provide some data. *Parapoynx* has been found where chloride exceeded 1000 ppm and where dissolved oxygen was less than 4 ppm; *Parargyractis* where chloride exceeded 1000 ppm and BOD was greater than 5.9 ppm but not over 10.0. *Nymphula* was found where alkalinity exceeded 210 ppm, and a species of a genus near *Parapoynx* where BOD was greater than 5.9 ppm but below 10.0 ppm. These are mostly single records, and in general the aquatic pyralids seem to occupy a middle ground with respect to chemical extremes.

Ranges of Chemical Analyses of Waters in Which 5 Species of the Order Lepidoptera Were Collected[a]

Ca	Mg	NH_3	No_3	No_2	PO_4	SO_4	BOD	Turbidity
13–865	1–22	<0.01–5.80	0.10–1.18	<0.01–0.05	<0.01–0.72	1.8–115.0	0.9–6.0	3–137
3–865	1–27	<0.01–5.80	0.08–1.10	<0.01–0.05	<0.01–0.72	<1.0–251.0	0.5–5.9	3–63
30–40	8–17	<0.01–0.04	0.30–0.43	0.01–0.03	<0.01–0.17	31.2–65.0	2.8–6.0	32–63
47–62	8–12	0.04–0.15	0.53	<0.01–0.01	<0.01–0.03	18.5–28.8	0.7–2.8	3–54
3–6	1	0.97–1.00	—	<0.01	<0.01–0.04	<1.0–3.5	2.2	25–49

X. Trichoptera

A. Life History

Female caddisflies of the noncase-making group enter the water and deposit their eggs in strips on a hard substrate. The female case-makers extrude the eggs and form them into a mass at the tip of the abdomen, then attach them to submerged stones, logs, or vegetation. Most caddisfly eggs hatch in 10 to 24 days. Some species will overwinter in the egg stage. The larvae, as far as can be determined, go through five to six instars before the prepupal, or resting, stage. There are few data on the length of the larval stages—either length of each instar or overall duration. Many species appear to overwinter in the larval stage when the little activity or feeding takes place. Pupation takes place under water.

The noncase-makers construct a stone case or a cocoon lined with silk. In the case-makers the case is anchored to a support, the top is closed off, and then the pupal stage lasts about two to three weeks. At the end of the pupal stage the pupa leaves the case, swims to the surface, and usually crawls out of the water onto a firm substrate. The adult then emerges from the pupal skin.

Most caddisflies complete their life cycle in one year. Emergence may take place from May to October. In many species there are overlapping broods.

B. Habitats

Ross (1944) has, for Illinois, given an excellent account of stream and pond types and their inhabitants. These extensive data need not be repeated here. The list below will give some idea of the locations within a body of water where caddisfly larvae live. Naturally these categories are not mutually exclusive and a great deal of overlap is to be expected.

Fast Water and riffles:

a. On or under stones—*Rhyacophila*, *Cheumatopsyche*, *Chimarra*, *Hydropsyche*, *Brachycentrus*, *Hesperophylax*, and *Helicopsyche*
b. Attached to upper surface on stones—*Leucotrichia*
c. On wood substrate—*Cheumatopsyche*, *Hydropsyche*, *Chimarra*, *Macronemum*, *Athripsodes*, and *Cyrnellus*

Moderate flow:

a. Trailing root masses—*Neureclipsis*, *Leptocella*, and *Oecetis*
b. Rocks or twigs—*Pycnopsyche*, *Agapetus*, *Platycentropus*, *Neophylax*, *Psilotreta*, *Oecetis*, *Lepidostoma*, *Glossosoma*, and *Cyrnellus*
c. In tubes in sandy bottom—*Phylocentropus*
d. In sandy bottom (case makers)—*Molanna* and *Athripsodes*

Pondlike conditions:

a. On submerged vegetation or sticks—*Ptilostomis*, *Triaenodes*, *Mystacides*, *Limnephilus*, *Athripsodes*, *Hydroptila*, and *Oxyethira*
b. On sandy beach area—*Mystacides* and *Molanna*
c. Rocks in larger lakes—*Athripsodes* and *Hydropsyche*

C. Food

Except for a few obligate predators such as *Rhyacophila* and *Oecetis*, most caddisfly larvae are omniverous. The net builders, especially those in faster water, eat a predominance of planktonic forms but will also eat other small organisms which may be found in stream drift. Some of the more phytophagous forms, if crowded, may turn cannibalistic. The above mentioned predators feed primarily on chironomid larvae.

D. Respiration

Respiration in the caddisfly larvae is either cutaneous or by means of gills. The genera *Rhyacophila* and *Chimarra*, the family Hydroptilidae, and some Psychomyiidae, for example, lack external gills. Most Hydropsychidae and the case-makers possess some abdominal gills. The net-builders generally live in moderate to fast flowing water, while the case-makers renew the oxygen in the water around the gills by undulations of the abdomen, creating a current through the case.

E. Retreat Types

Except for the free-living Rhyacophilidae, most caddisfly larvae construct a netlike retreat or case. The following list is modified from Ross (1944).

Net:
- Fingerlike—Many Philopotamidae
- Trumpetlike—Most Psychomyiidae
- Box or fanlike, with caudal retreat—Hydropsychidae

Tube of fine sand: *Phylocentropus*

Saddle case: *Glossosomatidae*

Purse case: Hydroptilidae

Portable cases (stones, wood, leaves, stem fragments, silk, hollowed out twigs): Phryganeidae, Limnephilidae, Molannidae, Odontoceridae, Leptoceridae, Calamoceratidae, Lepidostomatidae, Brachycentridae, Sericostomatidae, and Helicopsychidae.

F. Chemical Parameters

As can be seen from extreme tolerance lists 1–10, the tolerant caddisfly larvae are mostly noncase-makers existing where pH is greater than 8.5,

TABLE XII

	Fe	pH	Alkalinity	Cl	DO	Total hardness
Glossosomatidae						
Agapetus prob. *illini* Ross[1]	2.89	7.3	39	6	8	66
Philopotamidae						
Chimarra feria Ross[1]	0.28	7.1	28	3	9	23
Chimarra obscura (Walk.)[6]	<0.01–0.51	7.6–8.3	97–175	9–21	6–14	124–600
Chimarra socia Hagen[11]	0.01–0.77	6.3–8.7	9–124	1–845	8–10	4–800
Chimarra aterrima Hagen[1]	0.02–0.50	6.8–6.8	17–19	4–6	7–9	15–19
Psychomyiidae						
Cyrnellus fraternus (Banks)[4]	0.03–0.14	7.3–7.7	40–58	1–19	5–10	54–75
Lype prob. *diversa* (Banks)[22]	0.72	7.2	47	14	11	87
Neureclipsis prob. *crepuscularis* (Walk)[22]	0.01–0.77	5.5–8.5	4–116	1–845	6–14	8–800
Nyctiophylax prob. *vestitus* (Hagen)[2]	1.80–2.89	7.2–7.3	39–61	6–23	8–10	66–322
Nyctiophylax sp. A.? Flint[18]	<0.01–0.90	6.6–7.9	20–220	2–2500	5–14	21–1000
Phylocentropus placidus (Banks)[6]	0.06–0.82	5.6–7.0	5–27	1–8	3–9	7–21
Polycentropus prob. *crassicornis* (Walk.)[4]	0.32–2.89	5.2–7.3	10–47	6–30	8–10	66–287
Polycentropus remotus? (Banks)[8]	<0.01–0.90	5.5–8.4	4–113	4–21	3–14	11–170
Hydroptilidae						
Hydroptila sp.[3]	<0.01–0.02	8.3–8.5	93–107	20–22	5–12	164–176
Leucotrichia sp.[1]	0.37	7.4	88	14	10	197
Oxyethira sp.[3]	<0.01–0.26	7.1–7.8	32–220	26–965	5–8	87–370
Phryganeidae						
Ptilostomis sp.[6]	0.32–8.75	3.3–7.4	2–95	10–47	8–10	70–287
Limnephilidae						
Limnephilus spp.[3]	0.02–0.10	6.4–8.5	9–97	1–22	9–12	6–164
Neophylax nacatus Denn.[2]	0.72–2.89	7.2–7.3	39–47	6–14	8–11	66–87
Platycentropus nr. *radiatus* (Say)[1]	2.89	7.3	39	6	8	66
Pycnopsyche spp.[11]	<0.01–0.10	6.0–8.8	2–205	1–34	8–14	4–705
Pycnopsyche prob. *lepida* (Hagen)[3]	0.72–2.89	7.2–7.3	39–61	6–23	8–11	66–322
Pycnopsyche guttifer? (Walk.)[4]	0.18–2.89	7.2–7.3	36–61	6–30	8–11	66–322
Pycnopsyche scabripennis? (Ramb.)[2]	0.32–0.72	6.7–7.2	36–47	14–30	10–11	87–287
Molannidae						
Molanna sp.[3]	<0.01–0.72	6.4–8.4	13–96	1–21	8–11	4–170
Hydropsychidae						
Cheumatopsyche spp.[35]	0.02–2.90	6.0–8.5	2–180	1–845	6–14	6–800
Hydropsyche betteni Ross[11]	0.23–1.80	5.9–8.5	10–113	11–2630	8–11	154–2100
Hydropsyche (*bifida* gp.) spp.[18]	<0.01–1.80	7.5–8.8	61–205	8–2630	8–11	114–2100
Hydropsyche cuanis Ross[3]	<0.01–0.02	8.7–8.8	73–74	27–28	9	114–117
Hydropsyche frisoni Ross[11]	<0.01–0.24	6.9–7.8	19–130	1–56	6–8	8–193
Hydropsyche hageni Banks[2]	0.27–0.30	8.3–8.4	102–113	19–21	9–11	159–162
Hydropsyche orris Ross[9]	<0.01–0.13	6.8–7.9	20–213	2–72	5–10	12–233
Hydropsyche phalerata Hagen[15]	<0.01–0.30	7.6–8.8	73–116	10–28	9–14	114–162
Hydropsyche recurvata Banks[1]	0.30	8.3	113	21	9	162
Hydropsyche simulans Ross[12]	0.01–0.16	6.3–8.8	18–74	2–30	8–11	9–117
Macronemum carolina (Banks)[16]	0.01–0.30	6.4–8.5	10–113	1–20	7–14	4–165
Macronemum transversum (Walk)[1]	<0.01	8.0	116	19	6–14	147
Macronemum zebratum (Hagen)[2]	<0.01–0.30	7.0–8.8	27–116	1–28	6–14	8–162
Potamyia flava (Hagen)[1]	0.03	6.4	13	1	8	4

[a]All results are in parts per million (ppm) except pH. Superscript numbers following species names indicate the number of records comprising the following ranges.

Ranges of Chemical Analyses of Waters in Which 56 Species of the Order Trichoptera Were Collected[a]

Ca	Mg	NH_3	NO_3	NO_2	PO_4	SO_4	BOD	Turbidity
22	3	0.09	0.50	—	0.05	25.0	0.8	20
7	1	0.05	—	<0.01	<0.01	3.2	—	15
35–180	9–36	<0.01–0.97	—	<0.01–0.02	<0.01–0.48	14.2–370.0	1.1–6.0	4–125
2–398	<1–21	<0.01–5.00	0.08–0.90	<0.01–0.01	<0.01–0.62	2.3–25.0	0.6–2.1	5–39
3	2–3	<0.01	0.14–0.19	<0.01	0.05–0.13	2.6–7.7	0.9–1.2	59–161
16–23	4–4	0.04–0.35	0.06–2.90	0.01–0.02	0.01–0.38	4.8–26.8	0.6–2.3	8–126
27	4	0.17	0.21	—	0.03	26.5	1.6	12
2–398	<1–21	<0.10–5.00	0.03–0.90	<0.01–0.02	0.02–0.62	<1.0–72.8	0.7–6.0	7–259
22–92	3–22	0.09–0.21	0.12–0.50	—	0.04–0.05	25.0–313.0	0.8–1.5	20–27
6–68	1–200	<0.01–1.10	—	<0.01–0.09	<0.01–0.16	3.5–570.0	0.3–3.5	7–66
2–6	<1–2	0.01–1.10	—	<0.01–0.02	<0.01–0.02	<1.0–3.5	0.6–2.4	5–259
22–70	3–27	0.09–1.09	0.08–0.51	—	0.02–0.05	25.0–161.0	0.5–1.6	5–20
3–49	1–14	<0.10–0.88	0.03–0.06	<.01–0.02	<0.01–0.05	<1.0–43.2	0.7–4.1	6–98
47–51	11–12	0.09–0.15	0.03–0.06	0.03–0.06	<0.01–0.01	16.1–24.0	0.7–2.8	3–19
55	15	0.15	1.20	—	0.30	115.0	1.8	19
11–30	14–73	<0.01–0.10	—	<0.01	<0.01	24.0–29.7	2.0	8–66
27–70	4–27	<0.01–1.09	0.08–1.18	<0.01	<0.01–0.30	26.5–251.0	0.4–1.8	5–19
2–47	1–11	0.01–0.16	0.03	<0.01	<0.01–0.03	2.4–24.0	0.6–2.8	5–19
22–27	3–4	0.09–0.17	0.21–0.50	—	0.03–0.05	25.0–26.5	0.8–1.6	12–20
22	3	0.09	0.05	—	0.05	25.0	0.8	20
2–220	<1–39	<0.01–0.64	0.07	<0.01–0.01	<0.01–0.52	2.4–450.0	0.3–3.3	5–44
22–92	3–22	0.09–0.21	0.12–0.50	—	0.03–0.05	25.0–313.0	0.8–1.6	12–27
22–92	3–27	0.09–1.09	0.08–0.50	—	0.03–0.05	25.0–313.0	0.8–1.5	6–27
27–70	4–27	0.17–1.09	0.08–0.14	—	0.03	26.5–251.0	0.8–1.6	6–12
2–49	<1–11	0.14–0.64	0.03–0.21	<0.01–0.01	<0.01–0.07	26.5–28.8	0.7–1.6	6–36
2–398	<1–36	<0.01–5.00	0.12–2.30	<0.01–0.09	<0.01–0.56	2.3–370.0	0.3–7.5	4–>72,000
36–865	12–27	0.01–5.80	0.12–2.30	<0.01–0.05	0.02–0.62	56.3–313.0	0.8–6.0	3–125
49–865	6–39	<0.01–5.80	0.03–2.30	<0.01–0.05	<0.01–0.72	13.3–450.0	0.7–5.3	3–110
36–37	6	<0.01–<0.01	0.07–0.08	<0.01–0.01	0.11–0.27	24.6–25.1	1.0–1.8	37–50
2–6	<1–3	0.03–0.17	0.29–0.50	<0.01–0.02	<0.01–0.35	2.3–43.9	0.2–2.8	18–>72,000
36–37	17–17	0.01–0.62	—	<0.01–0.01	0.05–0.06	56.3–72.8	5.5–6.0	100–125
3–35	1–43	<0.01–0.15	—	<0.01–0.09	<0.01–0.34	2.2–44.0	0.3–2.4	7–120
34–37	16–17	<0.01–0.02	0.07–0.08	<0.01–0.02	0.02–0.27	24.6–72.8	0.9–6.0	37–125
37	17	0.01	—	<0.01	0.05	72.8	6.0	125
2–37	1–6	<0.01–1.10	0.07–0.08	<0.01–0.01	<0.01–0.27	0.6–25.1	0.5–1.8	39–140
2–39	<1–17	0.02–0.64	—	<0.01–0.01	<0.01–0.07	2.3–72.8	0.5–6.0	8–261
36	14	<0.01	—	0.01	0.15	41.6	3.3	23
2–37	<1–17	<0.01–0.15	0.07–0.08	<0.01–0.02	<0.01–0.27	12.6–72.8	0.3–6.0	18–548
2	<1	0.64	—	<0.01	0.07	—	0.6	36

(continued)

TABLE XII (*continued*)

	Fe	pH	Alkalinity	Cl	DO	Total hardness
Odontoceridae						
Psilotreta sp.[2]	0.72	7.2–7.3	39–47	6–14	8–11	66–87
Leptoceridae						
Athripsodes sp.[7]	<0.01–2.89	6.0–8.8	2–97	6–34	6–12	66–124
Athriposodes nr. *alagmus* Ross[3]	0.82–0.86	6.6–6.7	20–20	3–5	7–8	21
Athripsodes nr. *transversus* (Hagen)[6]	0.05–0.24	6.7–7.0	18–34	2	7–10	8–15
Leptocella candida Hagen[8]	<0.01–0.82	6.4–8.2	13–213	1–2630	5–8	4–2100
Leptocella diarina Ross[4]	<0.01	7.7–7.8	124–130	40–56	8	178–193
Mystacides prob. *sepulchralis* (Walk.)[2]	<0.01–2.89	7.3–8.4	39–96	6–21	8–11	66–170
Oecetis avara (Banks)[2]	0.01	8.4	93–96	21–22	10	170–171
Oecetis nr. *cinerascens* (Hagen)[4]	<0.01–0.08	6.8–8.4	20–107	2–21	5–10	13–176
Oecetis eddlestoni? Ross[5]	<0.01–0.06	7.3–8.0	87–146	10–62	2–14	116–167
Oecetis inconspicua (Walk)[3]	0.86–2.89	6.6–7.3	20–39	3–6	7–8	20–66
Triaenodes sp. B Ross[3]	0.32–2.89	6.7–7.3	2–39	6–34	8–10	66–287
Triaenodes injusta (Hagen)[12]	<0.01–0.97	6.6–8.8	20–124	3–845	7–12	21–800
Lepidostomatidae						
Lepidostoma sp.[3]	0.10–2.89	6.4–7.3	9–39	1–6	8–10	6–66
Brachycentridae						
Brachycentrus numerosus (Say)[6]	0.03–0.10	6.3–6.9	9–20	1–2	8–9	4–15
Micrasema sp.[3]	0.03–0.24	6.4–7.0	10–34	1–2	7–8	4–8
Helicopsychidae						
Helicopsyche prob. *borealis* (Hagen)[4]	<0.01–0.03	7.9–8.4	96–205	10–21	8–10	170–705

chloride exceeds 1000 ppm, SO_4 exceeds 400 ppm, and BOD is greater than 5.9 ppm. The net-builders outnumber the case-makers by at least 3:1. At high BOD's (greater than 5.9 ppm), only net-builders were found and were abundant in numbers of species and populations. None, however, was found where BOD exceeded 10.0 ppm. At high chloride, four caddisflies were present (three net-builders), but under brackish situations (chloride treater than 2000 ppm; magnesium greater than 150 ppm) only *Nyctiophylax* (sp. A. of Flint) was found. At low dissolved oxygen, only one Psychomyiidae and one Leptoceridae were present. Where pH was less than 4.5 and iron was greater than 5.0 ppm the phryganeid *Ptilostomis* was abundant and was the only caddisfly present. High hardness does not seem to be too significant as far as the caddisflies are concerned. Eleven species were found (five net-builders, one hydroptilid, and five case-makers), although it is noteworthy that no *Hydropsyche* species were present. The net-builders as a whole seem to be tolerant of organic loading, but not of toxic pollutants.

TABLE XII (*continued*)

Ca	Mg	NH_3	NO_3	NO_2	PO_4	SO_4	BOD	Turbidity
22–27	3–4	0.09–0.17	0.21–0.50	—	0.03–0.05	25.0–26.5	0.8–1.6	12–20
22–37	3–9	<0.01–0.09	0.08–0.50	<0.01	<0.01–0.27	25.0–45.2	0.8–4.1	18–98
6	1	0.85–1.10	—	<0.01	<0.01	3.5	—	15–35
2–4	<1–2	0.02–0.06	—	<0.01–0.01	0.02–0.02	2.5–3.1	0.6–1.5	34–548
2–865	<1–43	<0.01–5.80	0.29–0.50	<0.01–0.05	<0.01–0.90	3.5–58.8	0.6–2.1	3–>72,000
—	—	0.02–0.03	0.29–0.50	<0.01–0.01	0.33–0.35	39.2–43.9	2.3–2.8	>72,000
22–49	3–11	0.09–0.14	0.03–0.50	0.01	<0.01–0.05	25.0–28.8	0.7–0.8	6–20
49	11–12	0.09–0.14	0.03–0.06	0.01	<0.01	22.1–25.5	0.7	3–6
3–51	1–12	0.05–0.15	0.03	<0.01–0.01	<0.01–0.10	3.1–28.8	0.7–2.0	6–52
34–46	8–14	<0.01–0.02	0.05	0.01–0.04	0.05–0.87	25.3–43.2	3.3–4.8	23–128
6–22	1–3	0.09–0.97	0.05	<0.01	<0.01–<0.05	3.5–25.0	0.8	15–35
22–70	3–27	<0.01–1.09	0.08–0.50	<0.01	<0.01–0.05	25.0–251.0	0.8	6–20
6–398	1–21	<0.01–5.00	0.07–0.90	<0.01–0.02	<0.01–0.72	3.5–46.1	1.8–4.1	10–98
2–22	1–3	0.01–0.09	0.50	<0.01–0.01	<0.01–0.05	2.5–25.0	0.6–1.1	5–140
2–4	<1–1	0.01–0.64	—	<0.01	<0.01–0.06	2.3–2.5	0.6–0.9	5–96
2–3	<1–1	0.06–0.64	—	<0.01	0.02–0.07	2.4	0.6–0.8	6–204
49–220	11–39	0.08–0.14	0.03	<0.01–0.01	<0.01	28.8–450.0	0.4–1.1	6–25

XI. Diptera

A. Life History

The details of aquatic Diptera life histories are very poorly known, and they are such a variable group that it is difficult to generalize among them.

There are few detailed histories of Tipulidae. In *Eriocera* the eggs are dropped in the water and hatch on the bottom. The larval life is spent under rocks, and the species overwinters in the larval stage. The pupa burrows in loose gravel and emerges in about seven days, or longer if the weather is cool. The larval stage in general lasts some months for double brooded species, and almost a year for single brooded species. The pupal stage may last six to eight days.

The Simuliidae lay their eggs on floating vegetation or rocks under water, usually in fast flowing streams. The eggs hatch in a few days, and the larvae attach to rocks or plants in riffles and start to feed. Some species may overwinter in the egg stage or in the larval stage. There are six larval moults of varying length, and there may be one to several generations per year. Pupation takes place on the rocks, and the larvae spin a very characteristic cocoon. The pupal stage may last from four to seven days. The adult emerges in a gas bubble and flies off.

In the Chironomidae few detailed histories are known, but Sadler (1934) reared *Chironomus tentans* and his data illustrate a typical life cycle. The eggs are laid in large masses in the water—about 2300 eggs per mass. They hatch in 3 to 17 days, depending on the temperature. There are four larval instars: the first lasts 5 to 9 nine days; the second, 6 to 8 days; the third, 6 to 10 days; and the fourth, 4 to 21 days. The pupa is formed in the loose silken larval tube and emerges after about three days. Sadler was able to rear four generations per year under artificial conditions, but I have seen no evidence that this occurs under natural conditions. Most Chironomidae appear to overwinter in the larval stage.

The results of Miller (1941) indicate that species such as *Micropsectra dubia*, most *Chironomus* species, *Polypedilum halterale*, *P. scalaenum*, *Procladius bellus*, and *P. culiciformis* have one generation per year. Many Orthocladiinae, *Tanytarsus confuses*, *T. viridiventris*, *Lauterborniella agrayloides,* and *Corynoneura celeripes* have two generations per year. Some forms, such as *Endochironomus nigricans* and *Chironomus staegeri*, appear to have a two-year cycle.

The eggs of the Tabanidae are laid on a plant or wood substrate near water or a damp area. The life cycle appears to be one year, but is variable. Pupation takes place in drier areas.

The 50 to 200 cigar-shaped eggs of *Bezzia* and *Probezzia* are laid on algal mats and moss. The larvae start feeding on chironomid larvae. The larval period is about 40 days, and up to two months elapse from egg to adult. The larvae pupate in algae or logs and debris at the water surface. The pupal period is three to four days.

B. Habitats

The aquatic Diptera larvae, mostly Chironomidae, are found in almost all aquatic habitats. The tipulid larvae are present in moss, leaf packs, mud, and debris along the margins of streams and lakes and on rocks in fast water (*Antocha*). The simuliid larvae are found on firm substrates, rocks, wood, or leaves in fast flowing water. The stratiomyid larvae and *Chrysops* larvae are generally found in mud and debris along stream margins, often in shallow water. *Tabanus* is less aquatic and may occur in stream banks and damp soil away from water. *Atherix* is found in rotting wood and on rocks and moss in streams. The Ceratopogonidae larvae most commonly found in streams and lakes in limnological work belong to the *Palpomyia* group of genera. These are generally present in bottom mud and debris.

The chironomid larvae which constitute the majority of aquatic Diptera species are found in almost every imaginable habitat—in mud at all levels from the deepest lakes to shallow streams; on plants; mining in leaves and stems of plants; on all sides of rocks in all degrees of flow; on or wood; in sponges and bryozoans; parasitic on other insects and snails; in algal mats; and in sand and debris between rocks in fast water. Some build cases resebling those of the hydroptilid caddisflies. The chironomids are undoubtedly the most ubiquitous of all aquatic insects.

The larvae of the syriphid, *Tubifera tenex*, lives in bottom mud, usually of very high organic content.

Hemerodromia, of the Empididae, lives under rocks in small streams.

C. Food

The larvae of most of the Tipulidae are feeders on moss, decaying leaves, wood, algae, diatoms, and plant roots. Some larvae, especially in the Hexotomini, are predators, feeding on other Diptera larvae, worms, and dragonfly nymphs.

The larvae of the Simuliidae are omnivorous filter feeders, using their fans to strain plankton and other drift from the water. The Culicidae are also filter-feeders, feeding on diatoms, protozoa, and other small organisms. Some can forage over the bottom for their food. *Chaoborus* (the phantom midge) is predaceous. Most of the common aquatic Ceratopogonidae, *Palpomyia*, *Bezzia*, *Probezzia*, and *Culicoides*, are predaceous on each other, Chironomidae, and small crustacea.

In the Chironomidae there is a complete range of feeding habits. Most Tanypodinae are predaceous on each other, other midge larvae, Crustacea, and small worms. Some *Stenochironomus* feed on wood. Many species in several genera are leaf miners in *Potomogeton* and water lilies (both leaves and stems). Algae and diatoms are common food for many Orthocladiinae and Chironominae.

The majority of the larvae of aforementioned subfamilies are what I would call "vacuum cleaner" feeders. Everything of reasonable size—sand, algae, fungal, spores, protozoa, etc.—is taken into the digestive tract, although which are essential to their nutrition is not clear. Some specialized forms feed on sponge, *Nostoc*, and *Pectinatella*, or draw their nutrition from mayfly nymphs.

Most of the *Tabanus* larvae appear to be predaceous, while *Chrysops* are probably feeders on vegetative material. Stratiomyid larvae appear to feed on decaying vegetable matter.

D. Respiration

Except for *Antocha*, which is truly aquatic and breathes by means of tracheal gills, most tipulid larvae have caudal spiracles and must come to the surface occasionally. Some have hairs around the spiracles, which trap

TABLE XIII

	Fe	pH	Alkalinity	Cl	DO	Total hardness
Tipulidae						
Antocha saxicola O.S.[3]	0.61	7.8–8.4	88–108	11	9–11	95–216
Tipula abdominalis Say[3]	0.74–2.89	4.4–7.3	0–61	6–23	8–10	66–322
Psychodidae						
Psychoda alternata Say[1]	0.45	5.6	5	7	4	13
Chaoboridae						
Chaoborus punctipennis (Say)[6]	<0.01–0.89	5.6–7.8	4–130	5–965	5–8	11–370
Chironomidae						
Tanypodinae						
Tanypus punctipennis (Meig.)[2]	0.37–0.61	7.4–8.4	88–97	11–14	10–11	197–216
Tanypus stellatus Coq.[4]	<0.01	7.6–7.9	111–213	20–72	5–11	149–233
Macropelopia sp.[1]	0.10	6.4	9	1	9	6
Procladius (*Ps.*) *bellus* (Loew)[8]	<0.01–0.89	6.7–7.8	20–220	7–2750	2–10	22–1000
Procladius (*Pr.*) sp. (? *freemani* Subl.)[7]	0.23–2.89	4.4–8.3	0–102	7–18	4–10	13–250
Clinotanypus pinguis (Loew)[9]	<0.01–0.89	5.5–7.9	4–220	3–2500	3–14	11–900
Coelotanypus spp. (Coq.)[10]	<0.01–0.90	6.6–8.0	20–116	3–2750	6–14	20–1000
Ablabesmyia aspera ? Roback[1]	0.01	8.8	73	27	9	114
Ablabesmyia mallochi (Walley)[4]	<0.01	7.0–8.8	21–113	8–30	9–14	21–144
Ablabesmyia monilis ? (L.)[7]	<0.01–0.76	5.5–7.9	4–216	6–72	3–6	11–233
Conchapelopia spp.[25]	<0.01–2.89	6.8–8.8	10–205	1–56	6–14	7–705
Conchapelopia cornuticaudata (Walley)[2]	1.80–2.89	7.2–7.3	39–61	6–23	8–10	66–322
Labrundinia pilosella (Loew)[2]	0.10–0.54	5.6–6.4	5–9	2–7	4–9	6–16
Zavrelimyia sp.[2]	0.72	7.2–7.4	24–47	7–14	9–11	18–87
Diamesinae						
Diamesa sp.[1]	2.89	7.3	39	6	8	66
Potthastia longimanus ? (Kieff.)[1]	0.10	6.4	10	1	9	7
Prodiamesa nr. *bathyphila* (Kieff.)[1]	—	7.8	108	8	7	120
Orthocladiinae						
Brillia par (Coq.)[2]	0.02–0.10	6.4–7.9	9–205	1–10	9	6–705
Brillia par var. A Joh.[2]	0.03–0.06	6.3–6.4	10–13	1–1	8	4–7
Cardiocladius obscurus ? Joh.[9]	<0.01–0.74	4.4–8.8	0–124	6–28	6–12	114–250
Corynoneura (*C.*) *celeripes* ? Winn.[1]	0.10	6.4	9	1	9	6
Corynoneura (*C.*) *scutellata* ? Winn.[1]	0.72	7.2	47	14	11	87
Corynoneura (*C.*) *taris* Roback[1]	<0.01	7.6	111	20	8	149
Corynoneura (*T.*) sp. b[2]	0.10–0.82	6.4–6.7	9–20	1–3	8–9	6–21
Corynoneura (*T.*) *xena* Roback[1]	2.89	7.3	39	6	8	66
Cricotopus bicinctus (Meig.)[13]	<0.01–2.89	6.3–8.8	2–97	7–2630	6–12	18–2100
Cricotopus nr. *exilis* Joh.[4]	0.35–0.61	7.6–8.4	86–97	11–5500	10–11	216–5000
Cricotopus (*sylvestris* gp.) sp.[2]	0.37–0.61	7.4–8.4	88–97	11–14	10–11	197–216
Eukiefferiella sp.[4]	<0.01–0.02	6.0–8.7	20–88	10–34	6–10	82–119
Hydrobaenus nivoriundus Joh.[2]	0.10–0.90	6.4–6.7	9–20	1–4	7–9	6–20
Metriocnemus lundbecki Joh.[1]	2.89	7.3	39	6	8	66
Trichocladius sp. 3 Roback[1]	0.51	5.6	6	8	3	15
Psectrocladius sp. 3 Roback[7]	<0.01–0.30	7.9–8.8	74–116	10–27	6–14	117–162
Psectrocladius nr. *pilosus* Roback[1]	0.01	8.8	73	27	9	114
Psectrocladius nr. *simulans* (Joh.)[1]	0.08	7.2–7.8	26–43	2–7	7–8	18–59

[a] All results are in parts per million (ppm) except pH. Superscript numbers following species names indicate the number of records comprising the following ranges.

RANGES OF CHEMICAL ANALYSES OF WATERS IN WHICH 84 SPECIES OF THE ORDER DIPTERA WERE COLLECTED[a]

Ca	Mg	NH_3	NO_3	NO_2	PO_4	SO_4	BOD	Turbidity
25–64	7–14	0.02–0.21	2.30	0.01–0.02	0.02–0.56	7.5–135.0	1.1–4.4	11–34
22–92	3–22	0.09–0.35	0.12–0.50	—	0.04–0.05	25.0–322.0	0.2–1.5	2–27
3	1	0.10	—	0.01	<0.04	<1.0	2.2	49
3–28	1–73	0.10–0.85	0.36	<0.01	<0.01–0.30	<1.0–43.9	2.5–2.8	9–>72,000
55–64	14–15	0.15–0.21	1.20–1.30	—	0.30–0.56	115.0–135.0	1.8–2.0	19–29
23–47	15–43	<0.01–0.02	—	0.01–0.04	0.13–0.34	42.7–44.0	2.8	18–103
2	1	0.01	—	<0.01	<0.01	2.5	0.6	5
5–70	1–200	<0.01–1.10	0.05–1.18	<0.01–0.04	<0.01–0.87	0.6–570.0	0.8–4.8	6–128
3–76	1–16	0.10–0.35	0.19–1.18	0.01	0.02–0.30	<1.0–322.0	0.2–5.3	2–110
3–60	1–180	<0.01–0.97	—	<0.01–0.01	<0.01–0.04	<1.0–480.0	2.0–3.5	8–66
6–68	1–200	<0.01–1.10	—	<0.01–0.01	<0.01–0.15	3.0–520.0	—	9–23
36	6	<0.01	0.07	<0.01	0.11	24.6	1.8	37
5–37	2–14	<0.01–1.10	0.08	0.01–0.02	0.02–0.11	0.6–43.2	1.7–4.4	11–39
3–47	1–43	<0.01–0.12	—	<0.01–0.04	<0.01–0.34	<1.0–44.0	0.5–2.5	22–103
2–220	<1–39	<0.01–1.09	0.07–0.50	<0.01–0.01	0.02–0.53	2.3–450.0	0.4–6.0	6–>72,000
22–92	3–22	0.09–0.21	0.12–0.50	—	0.04–0.05	25.0–313.0	0.8–1.5	20–27
2–3	1–2	0.01–0.12	—	<0.01–0.01	<0.01–0.07	<1.0–2.5	0.6–2.1	5–39
4–27	2–4	0.17–1.00	0.21	—	0.03	0.7–26.5	1.6	12
22	3	0.09	0.50	—	0.05	25.0	0.8	20
2	1	0.16	—	<0.01	—	2.4	0.8	6
32	7	0.05	—	0.02	0.02	13.5	4.0	11
3–220	1–39	0.01–0.08	—	<0.01	<0.01	2.5–40.0	0.4–0.6	5–7
2–2	<1	0.05–0.64	—	<0.01	0.02–0.06	2.3	0.6	8–36
35–76	6–17	<0.01–2.50	0.07–2.30	<0.01–0.02	0.04–0.72	13.3–322.0	0.8–3.8	11–50
2	1	0.01	—	<0.01	<0.01	2.5	0.6	5
27	4	0.17	0.21	—	0.03	26.5	1.6	12
36	15	0.02	—	0.01	0.13	42.7	2.8	18
2–6	1	0.01–1.10	—	<0.01	<0.01	2.5–3.5	0.6	5–35
22	3	0.09	0.50	—	0.05	25.0	0.8	20
4–865	2–22	<0.01–5.80	0.07–0.90	<0.01–0.05	0.04–0.72	<1.0–251.0	0.8–3.8	3–84
64–2200	14–32	0.21–13.40	1.10–2.30	0.37	0.46–0.56	129.4–135.0	1.8–2.0	3–29
55–64	14–15	0.15–0.21	1.18–2.30	—	0.30–0.56	115.0–135.0	1.8–2.0	19–29
34–37	6–8	<0.01	0.08	0.01–0.01	0.05–0.53	25.0–45.1	1.0–4.1	44–98
2–6	1	0.01–0.88	—	<0.01	<0.01	2.5–3.5	0.6	5–15
22	3	0.09	0.50	—	0.05	25.0	0.8	20
4	2	0.12	—	0.01	0.01	<1.0	2.4	41
34–37	6–17	<0.01–0.02	0.08	<0.01–0.01	0.05–0.15	25.1–72.8	1.7–6.0	23–125
36	6	<0.01	0.07	<0.01	0.11	24.6	1.8	37
4	2	0.03–0.90	—	0.01	0.01	1.3–20.7	1.7	7

(*continued*)

TABLE XIII (*continued*)

	Fe	pH	Alkalinity	Cl	DO	Total hardness
Psectrocladius nr. *elatus* Roback[2]	1.01–2.20	5.2–5.9	2–10	10–11	10	177–210
Rheorthocladius sp.[11]	<0.01–0.76	5.5–7.9	4–213	1–72	5–10	11–233
Chironomus attenuatus (Walk.)[21]	<0.01–16.10	3.0–8.4	0–220	6–62	2–14	18–250
Chironomus riparius (Meig.)[5]	0.03–0.04	7.4–8.0	82–180	8–395	1–9	116–600
Cryptochironomus nr. *argus* Roback[4]	<0.01–0.23	7.1–8.4	34–213	16-72	5–13	25–233
Cryptochironomus nr. *fulyus* (Joh.)[11]	<0.01–0.72	5.6–8.8	6–116	8–2630	3–14	15–2100
Dicrotendipes fumidus (Joh.)[4]	<0.01	6.8–7.8	20–220	3–33	5–10	21–208
Dicrotendipes modestus (Say)[12]	<0.01–0.77	7.1–8.2	32–220	8–5500	5–14	26–5000
Dicrotendipes nervosus (Staeger)[7]	0.01–0.24	6.3–8.3	17–56	2–2020	3–11	18–386
Endochironomus nigricans (Joh.)[16]	<0.01–0.89	6.4–8.0	10–213	2–2500	5–9	7–900
Glyptotendipes sp.[16]	0.01–0.85	6.6–8.5	20–180	3–2500	6–14	21–900
Harnischia nr. *abortiva* (Mall.)[6]	<0.01–0.23	7.0–8.4	21–146	2–62	2–14	21–167
Harnischia pseudotener (Goetg.)[2]	0.76	5.5–7.9	4–88	1–6	5–9	11–95
Microtendipes caducus Townes[4]	<0.01–0.72	7.6–8.4	47–124	9–21	6–11	87–170
Microtendipes pedellus DeGeer[5]	<0.01–0.51	5.6–8.4	5–205	1–22	3–10	4–705
Paratendipes albimanus Meig.)[1]	2.89	7.3	39	6	8	66
Polypedilum fallax Joh.[8]	0.03–0.51	4.6–8.0	9–180	1–30	7–9	6–570
Polypedilum nr. *fallax* Joh.[2]	1.44–1.80	4.7–7.2	61–65	7–23	9–10	250–322
Polypedilum halterale (Coq.)[6]	<0.01–0.10	6.4–8.0	9–126	1–55	6–14	6–188
Polypedilum illinoense (Mall.)[37]	<0.01–16.10	3.0–8.8	0–220	1–2750	6–14	6–2100
Polypedilum nr. *scalaenum* (Schr.)[16]	<0.01–16.10	3.0–8.2	0–206	3–69	5–14	11–227
Polypedilum parascalaenum ? Beck[2]	<0.01–0.01	8.7–8.8	73–74	27–28	9–9	114–116
Sergentia flavipes (Meig.)[2]	0.37	7.4–7.8	88–108	8–14	7–10	120–197
Stenochironomus sp.[6]	0.06–0.16	6.3–6.8	9–21	1–3	8–11	6–14
Stictochironomus sp.[3]	0.03–2.89	7.2–8.0	39–180	6–14	8–11	66–570
Stictochironomus devinctus (Say)[3]	0.10–0.24	6.4–7.0	10–34	1–2	7–10	7–12
Tribelos jucundus (Walk.)[6]	0.01–0.60	5.6–8.8	5–122	1–27	3–9	6–135
Xenochironomus festivus (Say)[1]	0.05	7.7	47	3	8	60
Xenochironomus scopula ? Townes[1]	<0.01	8.4	93	22	10	171
Xenochironomus xenolabis (Kieff.)[3]	<0.01–0.08	7.2–7.9	43–213	2–69	5–7	59–233
Atanytarsus sp. 2 Roback[1]	0.51	8.0	124	9	9	130
Rheotanytarsus exiguus ? (Joh.)[18]	<0.01–0.51	5.5–8.8	2–128	1–55	3–14	4–193
Calopsectra nr. *guerla* Roback[12]	<0.01–2.89	5.5–8.8	4–220	6–72	3–12	11–322
Paratanytarsus sp. 1 nr. *dissimilis* (Joh.)[1]	0.10	6.4	9	1	9	6
Paratanytarsus sp. 2 nr. *dissimilis* (Joh.)[2]	<0.01	7.9–7.9	203–213	67–72	5–6	208–233
Zavrelia sp.[1]	0.10	6.4	9	1	9	6
Ceratopogonidae						
Atrichopogon sp.[1]	0.10	6.9	22	2	7	15
Palpomyia gp. spp.[22]	<0.01–.72	5.6–8.0	6–213	8–2750	2–14	15–1000
Stratiomyidae						
Odontomyia sp.[2]	0.03–0.04	8.0	175–180	11	8–9	570–600
Stratiomys sp.[2]	0.37–1.80	7.2–7.4	61–88	14–23	10	197–322
Rhagionidae						
Atherix variegata (Walk.)[4]	0.03–0.10	6.3–6.4	9–13	1–1	8–9	4–7
Atherix nr. *variegata* (Walk.)[4]	0.51–0.77	7.9–8.2	64–124	1–845	9	95–800
Tabanidae						
Chrysops spp.[4]	0.72–2.89	5.5–7.3	2–47	6–34	5–11	11–87
Tabanus spp.[7]						
Syrphidae	0.02–0.86	6.6–8.2	20–205	3–185	7–9	21–705
Tubifera tenax (L.)[3]	0.04	7.4–8.0	82–175	8–34	<1–9	116–600

TABLE XIII (*continued*)

Ca	Mg	NH_3	NO_3	NO_2	PO_4	SO_4	BOD	Turbidity
51–62	12–13	0.21–0.22	0.16–0.19	—	0.02–0.03	161.0–220.0	0.4–0.5	5–33
3–35	1–43	<0.01–0.43	—	<0.01–0.04	<0.01–0.53	<1.0–51.9	0.3–2.5	3–77
4–46	2–33	< 0.01–1.00	0.05–2.30	<0.01–0.04	0.02–0.87	0.7–315.0	0.8–6.0	3–128
32–180	6–36	0.05–6.13	—	<0.01–0.37	<0.01–0.05	13.5–370.0	1.1–15.44	4–11
6–39	2–43	<0.01–0.06	—	<0.01–0.04	<0.01–0.33	13.0–55.5	0.3–5.7	18–103
4–865	2–22	<0.01–5.80	0.07–0.51	0.01–0.05	0.01–0.50	<1.0–58.8	0.8–4.4	3–245
5–30	2–33	<0.01–1.10	—	<0.01–0.01	0.02–0.53	0.6–45.1	1.1–2.0	10–66
6–2000	2–35	<0.01–13.40	0.20–1.10	<0.01–0.37	<0.01–0.72	12.6–129.4	0.3–3.3	3–77
4–35	2–73	0.03–0.80	0.04–1.85	<0.01–0.68	<0.01–0.18	4.8–157.4	0.8–12.9	18–165
2–180	1–180	<0.01–0.60	0.50	<0.01–0.04	<0.01–0.34	2.4–480.0	0.3–2.3	4–>72,000
6–180	1–180	0.03–0.85	0.03	<0.01–0.09	<0.01–0.10	3.2–480.0	0.4–7.5	4–125
5–46	2–17	<0.01–1.10	0.05	0.01–0.04	0.01–0.87	0.6–55.5	1.7–5.7	7–128
3–25	1–8	<0.01–0.02	—	<0.01–0.01	<0.02–0.03	<1.0–7.5	1.1–2.5	34–54
27–49	4–17	0.02–0.97	0.03–0.44	0.01–0.02	<0.01–0.48	14.2–42.7	0.7–2.8	6–18
2–220	< 1–39	0.08–0.64	0.06	<0.01–0.01	<0.01–0.07	<1.0–450.0	0.6–2.4	3–49
22	3	0.09	0.50	—	0.05	25.0	0.8	20
4–170	1–34	0.01–1.10	0.44	<0.01–0.02	<0.01–0.48	0.6–350.0	0.6–2.1	5–25
69–92	19–22	0.21–0.24	0.12–0.20	—	0.04–0.05	280.0–313.0	0.8–1.5	1–27
2–36	1–14	<0.01–0.16	0.51	<0.01–0.01	0.05–0.34	2.4–43.2	0.6–4.1	5–>72,000
2–865	<1–200	<0.01–5.80	0.07–0.90	<0.01–0.05	0.01–0.62	<1.0–570.0	0.5–6.0	3–> 72,000
3–56	1–26	<0.01–2.50	0.26–2.30	<0.01–0.06	<0.01–0.72	<1.0–315.0	0.6–3.3	3–> 72,000
36–37	6–6	<0.01	0.07–0.08	<0.01–0.01	0.11–0.27	24.6–25.1	1.0–1.8	37–50
32–55	7–15	0.05–0.15	1.18	0.02	0.02–0.30	13.5–130.0	1.8–4.4	11–19
2–3	<1–1	0.01–0.16	—	<0.01–0.01	<0.01–0.02	2.3–2.7	0.6–1.0	5–120
22–170	3–34	0.09–0.17	0.21–0.50	<0.01–0.01	<0.01–0.05	<25.0–350.0	0.8–1.6	12–25
2–3	<1–1	0.05–0.16	—	<0.01–0.01	0.02	2.4–2.5	0.6–1.1	6–140
2–56	1–10	<0.01–2.50	0.07–0.47	<0.01–0.02	0.01–0.72	<1.0–24.6	0.8–2.8	6–41
—	—	0.03	—	0.09	0.02	18.6	2.2	8
49	12	0.09	0.06	0.01	<0.01	22.1	0.7	3
23–47	26–43	<0.01–0.03	—	0.01–0.04	0.01–0.28	20.7–44.0	1.7	7–103
49	17	0.97	0.44	0.02	0.48	14.2	2.1	11
2–37	<1–15	<0.01–0.97	0.08–4.1	<0.01–0.06	<0.01–0.35	<1.0–45.2	0.3–4.4	5–>72,000
3–92	1–35	<0.01–1.09	0.07–0.50	<0.01–0.02	0.01–0.34	<1.0–313.0	0.8–3.4	6–77
2	1	0.01	—	<0.01	<0.01	2.5	0.6	5
30–35	33–35	<0.01	—	<0.01–0.03	0.04–0.34	29.7–37.8	2.0	66–77
2	1	0.01	—	<0.01	<0.01	2.5	0.6	5
4	2	0.02	—	<0.01	<0.01	2.7	0.6	34
4–220	1–200	<0.01–1.10	0.05–1.18	<0.01–0.34	<0.01–0.87	<1.0–570.0	0.4–4.8	7–128
170–180	34–36	0.10–0.12	—	<0.01	<0.01–0.01	350.0–370.0	1.1	4–25
55–92	15–22	0.15–0.21	0.12–1.18	—	0.04–0.30	115.0–313.0	1.5–1.8	19–27
2–2	<1–1	0.01–0.64	—	<0.01	<0.01–0.06	2.3–2.5	0.6–0.8	5–36
25–398	8–21	0.02–5.00	0.44–0.90	0.01–0.02	0.03–0.72	7.5–46.1	1.1–2.8	15–34
3–27	1–4	<0.01–0.17	0.21–0.50	<0.01	<0.01–0.05	<1.0–45.2	0.8–2.5	12–54
6–220	1–39	0.06–2.50	0.47	<0.01–0.02	<0.01–0.72	3.6–450.0	0.4–2.8	4–25
32–180	7–36	0.05–6.13	—	<0.01–0.17	0.01–0.04	13.5–370.0	1.1–15.4	4–37

a plastron of air which permits them to remain under water for extended periods. Stratiomyidae and *Atherix* larvae use a similar mode of respiration. The *Tabanus* and *Chrysops* larvae are not truly aquatic, and must extend their caudal spiracles out of the water.

The larvae of the Chironomidae, Simuliidae, Ceratopogonidae (*Palpomyia* group), and *Chaoborus* are truly aquatic and breath by cutaneous repiration or by means of blood gills. The larvae of the Culicidae must come to the surface to breath, although the larvae of *Mansonia* can extract air from aquatic plants.

The rat-tailed maggots (*Tubifera*) have an extensile breathing tube to reach the water surface.

Some chironomid larvae have a form of hemoglobin in their blood which permits them to function efficiently at lower oxygen levels and to recover more rapidly from low oxygen conditions.

E. Chemical Parameters

As can be seen from Table XIV, the Diptera, represented chiefly by the Chrionomidae, have representatives tolerant of many chemical extremes. At pH greater than 8.5, all 13 Diptera are Chironomidae, with all subfamilies represented. Below pH 4.5, *Tipula abdominalis* is the only non-chironomid. No orthocladiinae are represented here. At alkalinity greater than 210 ppm, except for a *Palpomyia* group species, only Chironomidae (again a few Orthcladiinae) are represented. The same is true where chloride exceeds 1000 ppm. In brackish situations, six of the seven species present are Chironomidae. *E. nigricans* and *P. illinoense* were particularly common. Where iron exceeded 5.00 ppm, only three Chironomidae were present. *Chironomus attenuatus* was especially abundant in stations with high iron and low pH. Where dissolved oxygen was less than 4 ppm, 13 of the 15 Diptera were Chironomidae. The *Palpomyia* group species was again present, together with *T. tenax*, long known to occur in low dissolved oxygen-high BOD situations. With high hardness, *Tipula abdominalis*, *Chaoborus punctipennis*, and two Stratiomyidae were present, along with nine Chironomidae. Where sulfate was greater than 400 ppm, *Palpomyia* was again present, along with *Tabanus* species and eight Chironomidae. No Orthocladiinae were represented. At BOD greater than 5.9 ppm or 10.0 ppm, *Tubifera tenax* was the only non-chironomid, *C. riparius* and *D. nervosus* were present where the BOD was greater than 10 ppm. Both *T. tenax* and *C. riparius* have long been incriminated in this situation.

It is unfortunate that, at present, the larvae of *Palpomyia*, *Bezzia*, and *Probezzia* are not readily separable. These genera are very common, expecially in places where chemical extremes exist.

Extreme Tolerance Lists

The following lists were extracted from the Tables V through XIII in order to enable the reader to see at a glance those species or genera which can tolerate the extremes of some of the chemical parameters given.

LIST 1

SPECIES FOUND AT pH LESS THAN 4.5

Hemiptera
Hesperocorixa sp.
Callicorixa audeni Hung.
Gerris marginatus Say
Megaloptera
Nigronia sp.
Sialis sp.
Coleoptera
Agabus spp.
Agaporus spp.
Bidessus spp.
Ilybius spp.
Laccophilus spp.
Hydroporus spp.
Enochrus spp.
Helophorus spp.
Hydrobius spp.
Trichoptera
Ptilostomis sp.
Diptera
Tipula abdominalis (Say)
Procladius freemani? (Sublette)
Chironomus attenuatus (Walk.)
Polypedilum illinoense (Mall.)
Polypedilum nr. *scalaenum* (Schrank)

LIST 2

SPECIES FOUND AT pH GREATER THAN 8.5

Odonata
Enallagma traviatum Selys
Ephemeroptera
Isonychia sp.
Centroptilum sp.
Choroterpes sp.
Plecoptera
Pteronarcys dorsata Say
Phasganophora capitata (Pict.)
Acroneuria evoluta (Klap.)
Acroneuria internata (Walk.)
Acroneuria ruralis (Hagen)
Paragnetina immarginata (Say)
Hemiptera
Hesperocorixa sp.
Belostoma fluminea (Say)
Limogonus hesione Kirk.
Megaloptera
Corydalis cornutus (L.)
Nigronia sp.
Sialis sp.
Stenelmis bicarinata Lec.
Stenelmis crenata (Say)
Stenelmis knobeli Sand.
Stenelmis mera Sand.
Senelmis tarsalis Sand.
Hyperodes sp.
Trichoptera
Chimarra socia Hagen
Hydropsyche (*bifida* gr.) sp.
Hydropsyche cuanis Ross
Hydropsyche phalerata Hagen
Hydropsyche simulans Ross
Macronemum zebratum (Hagen)
Pycnopsyche sp.
Triaenodes injusta (Hagen)
Diptera
Ablabesmyia mallochi (Walley)
Conchapelopia spp.
Cardiocladius obscurus (Joh.)
Cricotopus bicinctus (Meig.)
Eukiefferiella sp.

LIST 2 (*continued*)

Coleoptera	*Psectrocladius* sp. 3 Roback
Peltodytes spp.	*Psectrocladius* nr. *pilosus* Roback
Coptotomus spp.	*Cryptochironomus* nr. *fulvus* Joh.
Laccophilus spp.	*Polypedilum illinoense* (Mall.)
Dineutes spp.	*Polypedilum parascalaenum* Beck
Berosus spp.	*Tribelos jucundus* (Walk.)
Enochrus spp.	*Rheotanytarsus exiguus* (Joh.)
Tropisternus spp.	*Calopsectra* nr. *guerla* Roback
Optioservus fastiditus (Lec.)	

LIST 3

SPECIES FOUND AT ALKALINITY GREATER THAN 210 PPM

Odonata
- *Argia apicalis* (Say)
- *Argia moesta* (Hagen)
- *Argia sedula* (Hagen)
- *Argia translata* Hagen
- *Enallagma basidens* Calvert
- *Enallagma exsulans* (Hagen)
- *Enallagma signatum* Hagen
- *Enallagma traviatum* Selys
- *Enallagma vesperum* Calvert
- *Ischnura posita* (Hagen)
- *Progomphus obscurus* (Ramb.)
- *Dromogomphus spinosus* Selys
- *Boyeria vinosa* (Say)
- *Didymops transversa* (Say)
- *Epicordulia princeps* (Hagen)

Ephemeroptera
- *Stenonema* (*pulchellum* gp.) spp.
- *Baetis* spp.
- *Callibaetis* spp.
- *Tricorythodes* spp.
- *Caenis* spp.
- *Campsurus* sp.

Hemiptera
- *Ranatra australis* Hung.
- *Pelocoris femoratus* P. de B.
- *Belostoma fluminea* (Say)
- *Trepobates* sp.

Neuroptera
- *Climacea areolaris* Hagen

Coleoptera
- *Peltodytes* spp.
- *Dineutes* spp.
- *Berosus* spp.
- *Stenelmis beameri* Sand.
- *Stenelmis crenata* (Say)

Lepidoptera
- *Nymphula* sp.

Trichoptera
- *Nyctiophylax* sp. A Flint
- *Oxyethira* sp.
- *Leptocella candida* Hagen

Diptera
- *Tanypus stellatus* Coq.
- *Procladius bellus* (Loew)
- *Clinotanypus pinguis* (Loew)
- *Ablabesmyia monilis*? (L.)
- *Rheorthocladius* sp.
- *Chironomus attenuatus* (Walk.)
- *Cryptochironomus* nr. *argus* Roback
- *Dicrotendipes fumidus* (Joh.)
- *Dicrotendipes modestus* (Say)
- *Endochironomus nigricans* (Joh.)
- *Polypedilum illinoense* (Mall.)
- *Xenochironomus xenolabis* (Kieff.)
- *Calopsectra* nr. *guerla* Roback
- *Paratanytarsus* nr. *dissimilis* (Joh.)
- *Palpomyia* group spp.

LIST 4

SPECIES FOUND AT CHLORIDE GREATER THAN 1000 PPM

Odonata
- *Hetaerina americana* (Fabr.)
- *Argia moesta* (Hagen)
- *Argia sedula* (Hagen)
- *Argia violacea* (Hagen)
- *Enallagma civile* (Hagen)
- *Enallagma signatum* (Hagen)
- *Enallagma traviatum* Selys
- *Ischnura posita* (Hagen)
- *Erpetogomphus designatus* Hagen
- *Dromogomphus spinosus* Selys
- *Gomphus vastus* Walsh
- *Gomphus lividus* Selys
- *Basiaeschna janata* (Say)
- *Boyeria vinosa* (Say)
- *Erythemis simplicicollis* (Say)

Ephemeroptera
- *Baetis* spp.
- *Callibaetis* spp.
- *Neocloeon alamance*? Trav.

Plecoptera
- *Paragnetina* sp.

Hemiptera
- *Gerris marginatus* Say
- *Rheumatobates tenuipes* Meinert
- *Trepobates inermis* Esaki
- *Microvelia americana* Uhl.
- *Rhagovelia obesa* (Uhl.)

Coleoptera
- *Laccophilus* spp.
- *Thermonectes* spp.
- *Gyrinus* spp.
- *Berosus* spp.
- *Tropisternus* spp.
- *Helichus lithophilus* (Germ.)

Lepidoptera
- *Parargyractis* sp.
- *Parapoynx* sp.

Trichoptera
- *Nyctiophylax* sp. A Flint
- *Hydropsyche betteni* Ross
- *Hydropsyche* (*bifida* gp.) sp.
- *Leptocella candida* Hagen

Diptera
- *Procladius bellus* (Loew)
- *Coelotanypus concinnus* (Coq.)
- *Clinotanypus pinguis* (Loew)
- *Cricotopus bicinctus* (Meig.)
- *Cricotopus* nr. *exilis* Joh.
- *Cryptochironomus* nr. *fulvus* Joh.
- *Dicrotendipes modestus* (Say)
- *Dicrotendipes nervosus* (Staeger)
- *Endochironomus nigricans* (Joh.)
- *Glyptotendipes* sp.
- *Polypedilum illinoense* (Mall.)
- *Palpomyia* gp. spp.

LIST 5

SPECIES FOUND AT CHLORIDE GREATER THAN 2000 PPM PLUS MAGNESIUM GREATER THAN 150 PPM

Odonata
- *Ischnura posita* (Hagen)
- *Erythemis simplicicollis* (Say)

Ephemeroptera
- *Baetis* spp.
- *Callibaetis* spp.
- *Neocloeon alamance*? Trav.

Hemiptera
- *Rheumatobates tenuipes* Meinert
- *Trepobates inermis* Esaki

Coleoptera
- *Gyrinus* spp.

Trichoptera
- *Nyctiophylax* sp. A Flint

Diptera
- *Procladius bellus* (Loew)
- *Coelotanypus concinnus* (Coq.)
- *Clinotanypus pinguis* (Loew)
- *Endochironomus nigricans* (Joh.)
- *Glyptotendipes* sp.
- *Polypedilum illinoense* (Mall.)
- *Palpomyia* gp. spp.

LIST 6

SPECIES FOUND AT IRON GREATER THAN 5.00 PPM

Hemiptera
- *Hesperocorixa* sp.
- *Gerris marginatus* Say

Megaloptera
- *Sialis* sp.

Coleoptera
- *Agabus* sp.
- *Ilybius* spp.
- *Laccophilus* spp.
- *Helophorus* spp.
- *Hydrobius* spp.

Trichoptera
- *Ptilostomis* sp.

Diptera
- *Chironomus attenuatus* (Walk.)
- *Polypedilum illinoense* (Mall.)
- *Polypedilum* nr. *scalaenum* (Schrank)

LIST 7

SPECIES FOUND AT DISSOLVED OXYGEN LESS THAN 4 PPM

Odonata
- *Ischnura posita* (Hagen)
- *Pachydiplax longipennis* (Burm.)

Ephemeroptera
- *Paraleptophlebia* sp.
- *Caenis* sp.

Hemiptera
- *Notonecta irrorata* Uhl.
- *Plea striola* Fieb.
- *Ranatra australis* Hung.
- *Ranatra kirkaldyi* Bueno
- *Pelocoris femoratus* P. de B.
- *Belostoma fluminea* Say
- *Trepobates* sp.
- *Rhagovelia obesa* Uhl.

Megaloptera
- *Chauliodes* sp.

Coleoptera
- *Haliplus* spp.
- *Peltodytes* spp.
- *Coelambus* spp.
- *Laccophilus* spp.
- *Hydroporus* spp.
- *Dineutes* spp.
- *Gyrinus* spp.
- *Tropisternus* spp.
- *Machronychus glabratus* Say
- *Stenelmis grossa* Sand.

Lepidoptera
- *Parapoynx* sp.

Trichoptera
- *Polycentropus remotus*? (Banks)
- *Oecetis eddlestoni*? Ross

Diptera
- *Procladius bellus* (Loew)
- *Clinotanypus pinguis* (Loew)
- *Ablabesmyia monilis* (L.)
- *Trichocladius* sp. 3 Roback
- *Chironomus attenuatus* (Walk.)
- *Chironomus riparius* (Meig.)
- *Cryptochironomus* nr. *fulvus* (Joh.)
- *Dicrotendipes nervosus* (Staeger)
- *Harnischia* nr. *abortiva* (Mall.)
- *Microtendipes pedellus* DeGeer
- *Tribelos jucundus* (Walk.)
- *Rheotanytarsus exiguus* (Joh.)
- *Calopsectra* nr. *guerla* Roback
- *Palpomyia* gp. spp.
- *Tubifera tenax* (L.)

LIST 8

SPECIES FOUND AT TOTAL HARDNESS GREATER THAN 300 PPM AND CHLORIDE LESS THAN 1000 PPM

Odonata
- *Argia tibialis* (Ramb.)
- *Enallagma exsulans* (Hagen)
- *Ischnura verticalis* (Say)
- *Progomphus obscurus* (Ramb.)
- *Lanthus albistylus* (Hagen)
- *Anax junius* (Drury)
- *Nasiaeshna pentacantha* (Ramb.)
- *Libellula luctuosa* Burm.
- *Plathemis lydia* (Drury)

Ephemeroptera
- *Stenonema interpunctatum* gp.
- *Tricorythodes* spp.
- *Caenis* spp.

Plecoptera
- *Perlesta placida* (Hagen)
- *Phasganophora capitata* (Pict.)

Hemiptera
- *Hesperocorixa* spp.
- *Sigara modesta* Abbott
- *Trichocorixa calva* (Say)
- *Pelocoris femoratus* P. de B.
- *Belostoma fluminea* Say
- *Gerris comatus* Dk. & Harr.
- *Gerris remigis* (Say)
- *Trepobates knighti* Dk. & Harr.
- *Trepobates pictus* (Herr.-Sch.)

Neuroptera
- *Climacea areolaris* Hagen

Megaloptera
- *Sialis* sp.

Coleoptera
- *Haliplus* spp.
- *Peltodytes* spp.
- *Bidessus* spp.
- *Coelambus* spp.
- *Hydroporus* spp.
- *Dineutes* spp.
- *Cercyon* spp.
- *Enochrus* spp.
- *Helophorus* spp.
- *Hydrobius* spp.
- *Laccobius* spp.
- *Paracymus* spp.
- *Phaenonotum* spp.
- *Helichus fastigiatus* (Say)
- *Optioservus fastidiatus* (Lec.)
- *Optioservus ovalis* (Lec.)
- *Dubiraphia quadrinotata* (Say)
- *Dubiraphia vittata* (Melsh.)
- *Stenelmis bicarinata* Lec.
- *Stenelmis crenata* (Say)
- *Psephenus lecontei* Lec.

Trichoptera
- *Chimarra socia* (Hagen)
- *Chimarra obscura* (Walk.)
- *Nyctiophylax* prob. *vestitus* (Hagen)
- *Neurecipsis* prob. *crepuscularis* (Walk.)
- *Cheumatopsyche* spp.
- *Oxyethira* sp.
- *Pycnopsyche* spp.
- *Pycnopsyche* prob. *lepida* (Hagen)
- *Pycnopsyche guttifer* (Walk.)
- *Triaenodes injusta* (Hagen)
- *Helicopsyche* prob. *borealis* (Hagen)

Diptera
- *Tipula abdominalis* (Say)
- *Chaoborus punctipennis* (Say)
- *Conchapelopia* spp.
- *Conchapelopia cornuticaudata* (Walley)
- *Brillia par* (Coq.)
- *Chironomus riparius* (Meig.)
- *Microtendipes pedellus* DeGeer
- *Polypedilum fallax* Joh.
- *Polypedilum* near *fallax* Joh.
- *Stictochironomus* sp.
- *Calopsectra* nr. *guerla* Roback
- *Odontomyia* sp.
- *Stratiomys* sp.

LIST 9

SPECIES FOUND AT SULFATE GREATER THAN 400 PPM

Odonata
Ischnura posita (Hagen)
Gomphus lividus Selys
Plathemis lydia (Drury)
Erythemis simplicicollis (Say)

Ephemeroptera
Stenonema (*interpunctatum* gp.) spp.
Baetis spp.
Callibaetis spp.
Neocloeon alamance? Trav.
Paraleptophlebia guttata (McD.)
Tricorythodes spp.
Caenis spp.

Plecoptera
Phasganophora capitata (Pict.)

Hemiptera
Sigara modesta (Abbott)
Gerris marginatus Say
Gerris remigis (Say)
Rheumatobates tenuipes Meinert
Trepobates inermis Esaki
Microvelia americana Uhl.
Rhagovelia obesa (Uhl.)

Neuroptera
Climacea areolaris Hagen

Megaloptera
Sialis sp.

Coleoptera
Haliplus spp.
Peltodytes spp.
Bidessus spp.
Coelambus spp.
Hydroporus spp.
Gyrinus spp.
Berosus spp.
Laccobius spp.
Dubiraphia quadrinotata (Say)
Stenelmis crenata (Say)

Trichoptera
Nyctiophylax sp.A, Flint
Hydropsyche (*bifida* gp.) spp.
Hydropsyche prob. *borealis* (Hagen)
Pycnopsyche spp.

Diptera
Procladius bellus (Loew)
Coelotanypus concinnus (Coq.)
Clinotanypus pinguis (Loew)
Conchapelopia spp.
Endochironomus nigricans (Joh.)
Glyptotendipes sp.
Microtendipes pedellus DeGeer
Polypedilum illinoense (Mall.)
Palpomyia gp. spp.
Tabanus spp.

LIST 10

SPECIES FOUND AT BOD GREATER THAN 5.9 PPM

Odonata
- *Calopteryx maculata* (Beauv.)
- *Hetaerina americana* (Fabr.)
- *Argia apicalis* (Say)
- *Argia moesta* (Hagen)
- *Argia tibialis* (Rambur)
- *Ischnura verticalis* (Say)[a]
- *Dromogomphus spinosus* Selys
- *Gomphus hybridus* Willmsn.
- *Gomphus vastus* Walsh
- *Anax junius* Drury
- *Didymops transversa* (Say)
- *Neurocordulia virginiensis* Davis
- *Epicordulia princeps* (Hagen)
- *Libellula luctuosa* Burm.

Ephemeroptera
- *Isonychia* spp.
- *Heptagenia* spp.
- *Stenonema* (*bipunctatum* gp.) spp.
- *Stenonema* (*interpunctatum* gp.) spp.
- *Stenonema* (*pulchellum* gp.) spp.
- *Baetis* spp.[a]
- *Leptohyphes robacki* Allen
- *Tricorythodes* spp.
- *Caenis* spp.
- *Potamanthus* sp.

Plecoptera
- *Pteronarcys dorsata* Say
- *Phasganophora capitata* (Pict.)

Hemiptera
- *Trichocorixa calva* (Say)
- *Ranatra nigra* Herr.-Sch.
- *Belostoma fluminea* Say
- *Nepa apiculata* Uhler
- *Metrobates hesperius* Uhler
- *Rheumatobates rileyi* Bergr.
- *Trepobates pictus* (Herr.-Sch.)[a]
- *Rhagovelia obesa* Uhler

Megaloptera
- *Corydalis cornutus* (L.)
- *Sialis* sp.

Coleoptera
- *Dineutes* spp.
- *Berosus* spp.
- *Enochrus* spp.
- *Helichus lithophilus* (Germ.)
- *Psephenus herricki* DeKay
- *Machronychus glabratus* Say
- *Dubiraphia vittata* (Melsh.)
- *Stenelmis bicqrinata* Lec.
- *Stenelmis decorata* Sand.

Lepidoptera
- *Parargyractis* sp.
- nr. *Parapoynx* sp.

Trichoptera
- *Chimarra obscura* (Walk.)
- *Neureclipsis* prob. *crepuscularis* (Walk.)
- *Cheumatopsyche* spp.
- *Hydropsyche betteni* Ross
- *Hydropsyche hageni* Banks
- *Hydropsyche phalerata* Hagen
- *Hydropsyche recurvata* Banks
- *Macronemum carolina* (Banks)
- *Macronemum zebratum* (Hagen)

Diptera
- *Conchapelopia* spp.
- *Psectrocladius* sp. 3 Roback
- *Chironomus attenuatus* (Walk.)
- *Chironomus riparius* (Meig.)[a]
- *Dicrotendipes nervosus* (Staeger)[a]
- *Glyptotendipes* sp.
- *Polypedilum illinoense* (Mall.)
- *Tubifera tenax* (L.)[a]

[a]BOD greater than 10 ppm.

TABLE XIV
SUMMARY OF EXTREME TOLERANCE LISTS BY INSECT ORDERS[a]

Order	pH <4.5	pH >8.5	Alkalinity >210	Cl >1000	Cl >1000 +Mg >150	Fe >5.00	DO <4	Total hardness >300 + Cl <1000	SO_4 >400	BOD >5.9	BOD >10.0
Odonata	0	1	15	15	2	0	2	9	4	14	1
Ephemeroptera	0	3	6	3	3	0	2	3	7	10	1
Plecoptera	0	6	0	1	0	0	0	2	1	2	0
Hemiptera	3	3	4	5	2	2	8	9	7	8	1
Neuroptera	0	0	1	0	0	0	0	1	1	0	0
Megaloptera	2	3	0	0	0	1	1	1	1	2	0
Coleoptera	9	14	5	6	1	5	10	21	10	9	0
Trichoptera	1	8	3	4	1	1	2	11	4	9	0
Lepidoptera	0	0	1	2	0	0	1	0	0	2	0
Diptera	5	13	15	12	7	3	15	13	10	8	3
	20	51	50	48	16	12	41	70	45	64	6

[a] Number of taxa in each order tolerating the limits set forth at the heads of the columns are given.

References

The following references are grouped under general keys (comprehensive works with keys to all or most of the aquatic insects); general ecology and "pollution" biology (works on sampling or the ecology and relationship to "pollution" of aquatic insects, generally not restricted to a single order); and references on the systematics and ecology of the individual orders of aquatic insects. For additional discussion of the arrangement of these references, see Section I,C.

General Keys

Edmondson, W. T. *et al.* (1959). "Fresh-Water Biology" (O. Ward and O. Whipple), 2 nd. ed. *Wiley, New York.*

Parrish, F. K. (ed.) (1968). Keys to water quality indicative organisms (Southeastern United States). Ephemeroptera (Berner) 10 pp., Plecoptera (Hanson) 6 pp.; Trichoptera (Wallace) 19 pp.; Chironomidae (Beck) 22 pp. Fed. Water Pollut. Contr. Administration, U.S. Dept. of the Interior.

Pennak, R. W. (1953). "Fresh-Water Invertebrates of the United States." Ronald Press, New York.

Peterson, A. (1961). "Larvae of Insects," Vol. 2. Edwards Brothers, Ann Arbor, Michigan.

Usinger, R. L. *et al.* (1956). "Aquatic Insects of California." Univ. of California Press, Berkeley and Los Angeles.

General Ecology—Pollution Biology

Allanson, B. R., and Kerrich, J. E. (1961). A statistical method for estimating the number of animals found in field samples drawn from polluted rivers. *Verh. Int. Vereining. Theroet. Angew. Limnol.* **14**, 491–494.

Berg, K. *et al.* (1948). Biological studies on the river Susaa. *Folia Limnol. Scand.* **4**,

Brinkhurst, R. O. (1965). Observations on the recovery of a British river from gross organic pollution. *Hydrobiologia* **25**, 9–51.

*Cairns, J. C. Jr. and Dickson, K. L. (1973). Biological methods for the assessment of water quality. *Amer. Soc. for Testing and Materials, Phila, Pa. Spec. Tech. Publ.* **528** (VII) + 256 pp.

*Cairns, J. C. Jr., Lanza, G. R., and Parker, B. C. (1972). Pollution related structural and functional changes in aquatic communities with emphasis on freshwater Algae and Protozoa. *Proc. Acad. Natur. Sci. Philadelphia* **124**, 79–127.

Cummins, K. W. (1966). A review of stream ecology with special emphasis on organism-substrate relationships. *Spec. Publ. Pymatuning Lab. Field Biol.* **4**, 2–51.

Doudoroff, P., and Warren, C. E. (1957). Biological indices of water pollution with special reference to fish populations. *In* "Biological Problems in Water Pollution" (Transactions, 1956 Seminar), pp. 144–162. U.S. Dept. of Health Education and Welfare, Washington, D.C.

Gaufin, A. R. (1957). The use and value of aquatic insects as indicators of organic enrichment. "Biological Problems in Water Pollution," pp. 139–149. U.S. Public Health Serv., Washington, D.C.

Gaufin, A. R. (1958). The effects of pollution on a midwestern stream. *Ohio J. Sci.* **58(4)**, 197–208.

Gaufin, A. R., and Tarzwell, C. M. (1952). Aquatic invertebrates as indicators of stream pollution. *Publ. Health Rep.* **67**, 57–67.

*Hynes, H. B. N. (1960). "The Biology of polluted Water." Liverpool Univ. Press, Liverpool.

*Hynes, H. B. N. (1970). "The Ecology of Running Waters." Univ. of Toronto Press, Toronto.

Ingram, W. M., Mackenthun, K. M., and Bartsch, A. G. (1966). Biological field investigative data for water pollution surveys. U.S. Dept. of the Interior, Fed. Water Pollut. Contr. administration.

*Macan, T. T. (1963). "Freshwater Ecology." Wiley, New York.

Mellanby, K. (1967). "Pesticides and Pollution." The New Naturalist–Collins, London.

Patrick, R. (1949). A proposed biological measure of stream conditions based on a survey of the Conestoga Basin, Lancaster Co., Pa. *Proc. Acad. Natur. Sci. Philadelphia* **101**, 277–341.

Patrick, R., Cairns, J., and Roback, S. S. (1967). An ecosystematic study of the fauna and flora of the Savannah River. *Proc. Acad. Natur. Sci. Philadelphia* **118**, 109–407.

Purdy, W. C. (1926). The biology of polluted water. *J. Amer. Waterworks Ass.* **16**, 45–54.

Reid, G. K. (1961). "Ecology of Inland Waters and Estuaries." Reinhold, New York.

Richardson, R. E. (1928). The bottom fauna of the middle Illinois River 1913–1925. Its distribution, abundance, valuation and index value in the study of stream pollution. *Bull. Ill. Natur. Hist. Surv.* **17**, 387–475.

Roback, S. S., and Richardson, J. W. (1969). The effects of acid mine drainage on aquatic insects. *Proc. Acad. Natur. Sci. Philadelphia* **121**, 81–107.

Wilbur, C. G. (1969). "The Biological Aspects of Water Pollution. Thomas, Springfield, Illinois.

Wurtz, C. B. (1960). Quantitative sampling. *Nautilus* **73**, 131–135.

Odonata

Byers, C. F. (1930). A contribution to the knowledge of Florida Odonata. *Univ. Fl. Publ. Biol. Sci. Ser.* **1**, 1–237.

Corbett, P. S. (1962). "A Biology of Dragonflies." Quadrangle Books, Chicago, Illinois.

Kennedy, C. H. (1917). Notes on the life history and ecology of the dragonflies (Odonata) of central California and Nevada. *Proc. U.S. Nat. Mus.* **52**, No. 2192, 483–635.

Needham, J. G., and Heywood, H. B. (1929). "A Handbook of the Dragonflies of North America." Thomas, Springfield, Illinois.

Needham, J. G., and Westfall, M. W. Jr. (1955). "A Manual of the Dragonflies of North American (Anisoptera)." Univ. California Press, Berkeley and Los Angeles.

Roback, S. S., and Westfall, M. J. Jr. (1967). New records of Odonata nymphs from the United States and Canada with water quality data. *Trans. Amer. Entomol. Soc.* **93**, 101–124.

Snodgrass, R. E. (1954). The dragonfly larva. *Smithsonian Misc. Collect.* **123(2)**, publ. 4175, 3–38.

Walker, E. M. (1953). "The Odonata of Canada and Alaska," Vol. 1, General, The Zygoptera. Univ. of Toronto Press, Toronto.

Walker, E. M. (1958). "The Odonata of Canada and Alaska," Vol. 2, The Anisoptera. Univ. of Toronto Press, Toronto.

Wright, M., and Peterson, A. (1944). A key to the genera of dragonfly nymphs of the United States and Canada. *Ohio J. Sci.* **39**, 151–166.

Ephemeroptera

Allen, P. K., and Edmunds, G. F. Jr. Revision of the Genus Ephemerella.

(1959). I. The Subgenus *Timpanoga. Can. Entomol.* **91**, 51–58.

(1961). II. The Subgenus *Caudatella. Ann. Entomol. Soc. Amer.* **54**, 603612.

(1961). III. The Subgenus *Attenuatella. J. Kansas Entomol. Soc.* **34**, 161–173.

(1962). IV. The Subgenus *Dannella. J. Kansas Entomol. Soc.* **35**, 333–338.

(1962). V. The Subgenus *Drunella* in North America. *Misc. Publ. Entomol. Soc. Amer.* **3(5)**, 147–179.

(1963). VI. The Subgenus *Serratella. Ann. Entomol. Soc. Amer.* **56**, 583–600.

(1963). VII. The Subgenus *Eurylophella. Can. Entomol.* **95**, 597–623.

(1965). VIII. The Subgenus *Ephemerella* in North America. *Misc. Publ. Entomol. Soc. Amer.* **4(6)**, 244–282.

Berner, L. (1950). The mayflies of Florida. *Univ. Fl. Stud. Biol. Sci. Ser.* **IV(2)**, XII + 267 pp.

Berner, L. (1959). A tabular summary of the biology of North American mayfly nymphs (Ephemeroptera). *Bull. Fl. State Mus. (Biol. Sci.)* **4(1)**, 58 pp.

Burks, B. D. (1953). The mayflies or Ephemeroptera of Illinois. *Bull. Ill. Natur. Hist. Surv.* **26(1)**, 1–216.

Edmunds, G. F., Jr. (1958). North American mayflies of the family Oligoneuriidae. *Ann. Entomol. Soc. Amer.* **54(4)**, 375–382.

Edmunds, G. F., Jr., Allen, R. K., and Peters, W. L. (1963). An annotated key to the nymphs of the families of mayflies (Ephemeroptera). *Univ. Utah Biol. Ser.* **13(1)**, 49 pp.

Edmunds, G. F., Jr., and Traver, J. R. (1959). The classification of the Ephemeroptera I. Ephemeroidea. Behningiidae. *Ann. Entomol. Soc. Amer.* **52**, 43–51.

Leonard, J. W. (1962). Environmental requirements of Ephemeroptera. Biological Problems in Water Pollution. Third seminar, Publ. Health Serv. Publ. 999-WP-25:110–117.

Leonard, J. W., and Leonard, F. A. (1962). Mayflies of Michigan Trout Streams. Cranbrook Institute of Science, Bloomfield Hills, Michigan.

Needham, J. G., Traver, J. R., and Yin-Chi Hsu (1935). "The Biology of Mayflies with a Systematic Account of North American Species." Comstock Publ. Ithaca, New York.

Thew, T. B. (1960). Revision of the genera of the family Caenidae. *Trans. Amer. Entomol. Soc.* **86**, 197–205.

Traver, J. R. (1932). Mayflies of North Carolina. *J. Elisha Mitchell Sci. Soc.* **47**, 85–161, 163–236.

Plecoptera

Claassen, P. W. (1931). Plecoptera nymphs of America (north of Mexico). *Thomas Say Foundation, Entomol. Soc. Amer.* **3**, 1–199.

*Claassen, P. W. (1940). A catalogue of the Plecoptera of the world. *Cornell Univ. Agr. Exp. Sta. Mem.* **232**, 1–235.

Frison, T. H. (1929). Fall and winter stoneflies or Plecoptera of Illinois. *Bull. Ill. Natur. Hist. Surv.* **18**, 340–409.

Frison, T. H. (1935). The stoneflies or Plecoptera of Illinois. *Bull. Ill. Natur. Hist. Surv.* **20**, 281–471.

Frison, T. H. (1942). Studies of North American Plecoptera with special reference to the fauna of Illinois. *Bull. Ill. Natur. Hist. Surv.* **22**, 235–355.

Gaufin, A. R. (1962). Environmental requirements of Plecoptera. Biological Problems in Water Pollution, Third seminar, Publ. Health Serv. Publ., 999-WP-25:105–109.

Gaufin, A. R. Nebeker, A. V., and Sessions, J. (1966). The stoneflies (Plecoptera) of Utah. *Univ. Utah Biol. Ser.* **XIV**, 1–89.

Gaufin, A. R., Ricker, W. E., Miner, H., Milam, P., and Hays, R. A. (1972). The stoneflies (Plecoptera) of Montana. *Trans. Amer. Entomol. Soc.* **98**, 91–121.

Ricker, W. E. (1952). Systematic studies in Plecoptera. *Ind. Univ. Publ. Sci. Ser.* **18**, 1–200.

Ross, H. H., and Ricker, W. E. (1971). The classification, evolution and dispersal of the winter stonefly genus *Allocapnia*. *Ill. Biol. Monogr. Urbana, Ill.* **45**, 240 pp.

Hemiptera–Heteroptera

Anderson, L. D. (1932). A monograph of the genus *Metrobates* (Hemiptera: Gerridae). *Kansas Univ. Sci. Bull.* **20**, 297–311.

Bacon, J. E. (1956). A taxonomic study of the genus *Rhagovelia* (Hemiptera, Veliidae) of the Western Hemisphere. *Kansas Univ. Sci. Bull.* **38**, 695–913.

Blatchley, W. S. (1926). "Heteroptera or True-bugs of Eastern North America." Nature Publ. Indianapolis, Indiana.

Brooks, A. R., and Kelton, L. A. (1967). Aquatic and semiaquatic Heteroptera of Alberta, Saskatchewan, and Manitoba (Hemiptera). *Mem. Entomol. Soc. Can.* **51**, 1–92.

Cummings, C. (1933). The giant water bugs (Belostomatidae, Hemiptera). *Kansas Univ. Sci. Bull.* **21**, 197–219.

Gould, G. E. (1931). The *Rhagovelia* of the Western Hemisphere with notes on world distribution (Hemiptera, Veliidae). *Kansas Univ. Sci. Bull.* **20**, 5–61.

Hungerford, H. B. (1919). The biology and ecology of aquatic and semiaquatic Hemiptera. *Kansas Univ. Sci. Bull.* **11**, 3–341.

Hungerford, H. B. (1922). The Nepidae of North America. *Kansas Univ. Sci. Bull.* **14**, 425–469.

Hungerford, H. B. (1933). The Genus *Notonecta* of the world. *Kansas Univ. Sci. Bull.* **21**, 5–195.

Hungerford, H. B. (1948). The Corixidae of the Western Hemisphere (Hemiptera). *Kansas Univ. Bull.* **32**, 1–827.

Hungerford, H. B. (1960). Key to subfamilies, genera and subgenera of the Gerridae of the world. *Kansas Univ. Sci. Bull.* **41**, 3–23.

Schroeder, H. O. (1931). The genus *Rheumatobates* and notes on the male genitalia of some Gerridae (Hemiptera, Gerridae). *Kansas Univ. Sci. Bull.* **20**, 63–99.

Sprague, I. B. (1967). Nymphs of the Genus *Gerris* (Heteroptera: Gerridae) in New England. *Ann. Entomol. Soc. Amer.* **60**, 1038–1044.

Truxal, F. E. (1953). A revision of the genus *Buenoa* (Hemiptera: Notonectidae). *Kansas Univ. Sci. Bull.* **35**, 1351–1523.

Neuroptera

Anthony, M. H. (1902). The metamorphosis of *Sisyra*. *Amer. Natur.* **36**, 615–631.

Brown, H. P. (1952). The Life History of *Climacea areolaris* (Hagen) a neuropterous 'parasite' of Fresh-water sponges. *Amer. Midl. Natur.* **47**, 130–160.

Carpenter, F. M. (1940). A revision of the Nearctic Hemerobiidae, Berothidae, Sisyridae, Polystoechotidae and Dilaridae (Neuroptera). *Proc. Amer. Acad. Arts Sci.* **74**, 193–280.

*Parfin, S. I., and Gurney, A. B. (1956). The spongilla-flies, with special reference to those of the Western Hemisphere (Sisyridae, Neuroptera). *Proc. U.S. Nat. Mus.* **105(3360)**, 421–529.

Porrier, M. A., and Argeneaux, Y. M. (1972). Studies on Southern Sisyridae (Spongilla-flies) with a key to the third instar larvae and additional sponge-host records. *Amer. Midl. Natur.* **88**, 455–458.

Megaloptera

Davis, K. C. (1903). Sialididae of North and South America. *Bull. N.Y. State Mus.* **7**, 442–486.

Ross, H. H. (1957). Nearctic alder flies of the genus *Sialis* (Megaloptera, Sialidae). *Bull. Ill. Natur. Hist. Surv.* **21(3)**, 57–78.

Townsend, L. H. (1935). Key to the larvae of certain families and genera of Nearctic Neuroptera. *Proc. Entomol. Soc. Washington* **37(2)**, 25–30.

Coleoptera

Arnett, R. H., Jr. (1963). "The Beetles of the United States." Catholic Univ. of Amer. Press., Washington, D.C.

Bertrand, H. P. I (1972). Larves et nymphes des Coléoptères aguatigues de globe. *Imp. F. Paillart*, Paris, 804 pp.

Brown, H. P. (1970). A key to the genera of the beetle family Dryopidae of the new world. *Entomol. News* **81**, 171–175.

Brown, H. P. (1972). Aquatic dryopoid beetles (Coleoptera) of the United States. Identification Manual 6. Water Pollu. Contr. Res. Ser., 18050 ELDO4/72, EPA.

Böving, A. G., and Craighead, F. C. (1931). An illustrated synopsis of the principal larval forms of the order Coleoptera. Entomol. Amer., Brooklyn Entomol. Soc., 21 (N. ser).

Dillon, E. S., and Dillon, L. S. (1961). "A Manual of Common Beetles of Eastern North America."—Row, Peterson, Evanston, Illinois.

Matheson, R. (1912). The Haliplidae of America North of Mexico. *J. N.Y. Entomol. Soc.* **20**, 156–193.

Roberts, C. H. (1895). The species of *Dineutes* of America North of Mexico. *Trans. Amer. Entomol. Soc.* 22, 279–288.

Roberts, C. H. (1913). Critical notes on the species of Haliplidae of America north of Mexico. *J. N.Y. Entomol. Soc.* **21**, 91–123.

Sanderson, M. W. (1938). A monographic revision of the North American species of *Stenelmis* (Dryopidae: Coleoptera). *Univ. Kansas Sci. Bull.* **25**, 635–717.

Sanderson, M. W. (1953–54). A revision of the Nearctic genera of Elmidae (Coleoptera). *J. Kansas Entomol. Soc.* **26(4)**, 148–163; **27(1)**, 1–13.

Sinclair, R. M. (1964). Water quality requirements of the family Elmidae (Coleoptera) with keys to the larvae and adults of the eastern genera. Tenn. Stream Pollut. Contr. B.

*Young, F. N. (1954). The water beetles of Florida. University of Florida Studies. *Biol. Sci. Ser.* V, No. 1, IX + 238 pp.

LEPIDOPTERA

Lange, W. H., Jr. (1956). A generic revision of the aquatic moths of North America. *Wasman J. Biol.* **14(1)**, 59–144.

Lloyd, J. T. (1914). Lepidopterous larvae from rapid streams. *J. Entomol. Soc.* **22**, 147–152.

Welch, P. S. (1916). Contribution to the biology of certain aquatic Lepidoptera. *Ann. Entomol. Soc. Amer.* **9**, 159–190.

TRICHOPTERA

Betten, C. (1934). The caddisflies or Trichoptera of New York State. *N.Y. State Mus. Bull.* **292**.

Flint, O. S. (1960). Taxonomy and biology of Nearctic limnephilid larvae (Trichoptera) with special reference to species in Eastern United States. *Entomol. Amer. Brooklyn Entomol. Soc.* **40**, 1–120.

Flint, O. S. (1962). Larvae of the caddisfly genus *Rhyacophila* in Eastern North America (Trichoptera; Rhyacophilidae). *Proc. U.S. Nat. Mus.* **113**, No. 3463, 465–493.

Flint, O. S. (1964). Notes on some Nearctic Psychomyiidae with special reference to their larvae, *Proc. U.S. Nat. Mus.* **115**, No. 3491, 467–481.

Hickin, N. E. (1967). "Caddis Larvae, Larvae of the British Trichoptera." Hutchinson, London.

*Lloyd, J. T. (1921). The biology of North American caddisfly larvae. *Bull. Lloyd Library No. 21, Entomol. Ser.* **1**, 1–124.

Nielsen, A. (1942). Über die Entwicklung und Biologie der Trichopteren. *Arch. Hydrobiol. Suppl. Band* **17**, 255–631.

Roback, S. S. (1962). Environmental requirements of Trichoptera—Biological problems in water pollution. Third seminar, Publ. Health Serv. Publ., 999-WP-25:118–126.

*Ross, H. H. (1944). The caddisflies, or Trichoptera of Illinois. *Bull. Ill. Natur. Hist. Surv.* **23(1)**, 1–326.

Vorhies, C. T. (1909). Studies of the Trichoptera of Wisconsin. *Trans. Wis. Acad. Sci. Arts Lett.* **16(1)**, No. 6, 647–739.

Diptera

Cook, E. F. (1956). The Nearctic Chaoborinae (Diptera: Culicidae). *Bull. Minn. Agri. Experiment Station*, 218:102 pp.

Curry, L. L. (1958). The larvae and pupae of the species of *Cryptochironomus* in Michigan. *Limnol. Oceanogr.* **3(4)**, 427–442.

Curry, L. L. (1962). A survey of environmental requirements for the midges. Biological Problems in Water Pollution. Third Seminar, Publ. Health Serv. Publ., 999-WP-25, 127–140.

Darby, R. E. (1962). Midges associated with California rice fields with special reference to their ecology. Hilgardia **32(1)**, 1–206.

Johannsen, O. A. (1934–37). Aquatic Diptera I-IV.—*Cornell Univ. Agr. Exp. Sta. Mem.* 164 (1934), 71 pp.; *Mem.* 177 (1935), 61 pp.; *Mem.* 205 (1937), 84 pp.; *Mem.* 210 (1937), 56 pp.

Malloch J. R. (1915). The Chironomidae or midges of Illinois, particular reference to the species occurring in the Illinois River. *Bull. Ill. State Lab. Natur. Hist.* **X**, 275–543.

Mason, W. R., Jr. (1968). An introduction to the identification of Chironomid larvae, pp. 1–89. *Div. of Pollut. Serv.*, F.W.P.C.A., Cincinnati, Ohio.

Miller, R. B. (1941). A contribution to the ecology of the Chironomidae of Costello Lake, Algonquin Park, Ontario. *Univ. Toronto Stud. Biol. Ser.* **49**, 1–63.

Roback, S. S. (1957). The immature tendipedids of the Philadelphia area (Diptera: Tendipedidae). *Monogr. Acad. Natur. Sci. Philadelphia* **9**, III + 140.

Roback, S. S. (1963). The genus *Xenochironomus* (Diptera: Tendipedidae). Kieffer, taxonomy and immature stages. *Trans. Amer. Entomol. Soc.* **88**, 235–245.

Roback, S. S. (1969). The immature stages of the genus *Tanypus* Meigen (Diptera: Chironomidae: Tanypodinae).—*Trans. Amer. Entomol. Soc.* **94**, 407–428.

Sadler, W. O. (1934). Biology of the midge *Chironomus tentans* Fabricius, and methods for its propagation. *Cornell Univ. Agr. Exp. Sta. Mem.* **173**, 1–25.

Saether, O. A. (1970). Nearctic and Palaearctic *Chaoborus* (Diptera: Chaoboridae). *Bull. Fish. Res. B. Can.* **174**, VIII + 57 pp.

Stone, A. (1964). Simuliidae and Thaumaleidae in Guide to the Insects of Connecticut. Part VI. The Diptera or true flies of Connecticut. *Conn. State Geol. Natur. Hist. Surv. Bull.* **97**, VIII + 126.

Stone, A., Sabrosky, C. W., Wirth, W. W., Foote, R. H., and Coulson, J. R. (1965). A Catalogue of the Diptera of America North of Mexico. Agr. Res. Serv., U.S. Dep. of Agr.

Teskey, H. J. (1969). Larvae and pupae of some eastern North American Tabanidae (Diptera). *Entomol. Soc. Can. Mem.* **63**, 1–147.

*Thienemann, A. (1954). *Chironomus*, Leben, Verbreitung und wirtschaftliche Bedeutung der Chironomiden. *Die Binnengewässer* **20**, XVI + 834 pp.

Thomsen, L. (1937). Aquatic Diptera. Part V. Ceratopogonidae. *Cornell Univ. Agr. Exp. Sta. Mem.* **210**, 57–80.

Subject Index

Because material relating to the chemical tolerances of over 400 insect species is presented in systematic order in nine tables and ten lists in Chapter 10, species references to that information are not included in this index. Appropriate table pages are included under the insect Orders. In addition, insect genera mentioned in text in Chapter 10 have not been indexed because of the strict systematic orientation of the discussions. However, references to insects in other chapters have been included in this index.

I

J

L

O

P

T

A 4
B 5
C 6
D 7
E 8
F 9
G 0
H 1
I 2
J 3